T0180922

Studies in Fuzziness and Soft Computing

Volume 366

Series editor

Janusz Kacprzyk, Polish Academy of Sciences, Warsaw, Poland
e-mail: kacprzyk@ibspan.waw.pl

The series "Studies in Fuzziness and Soft Computing" contains publications on various topics in the area of soft computing, which include fuzzy sets, rough sets, neural networks, evolutionary computation, probabilistic and evidential reasoning, multi-valued logic, and related fields. The publications within "Studies in Fuzziness and Soft Computing" are primarily monographs and edited volumes. They cover significant recent developments in the field, both of a foundational and applicable character. An important feature of the series is its short publication time and world-wide distribution. This permits a rapid and broad dissemination of research results.

More information about this series at http://www.springer.com/series/2941

Jana Krejčí

Pairwise Comparison Matrices and their Fuzzy Extension

Multi-criteria Decision Making with a New Fuzzy Approach

 Springer

Jana Krejčí
Department of Industrial Engineering
University of Trento
Trento
Italy

and

Faculty of Law, Business and Economics
University of Bayreuth
Bayreuth
Germany

ISSN 1434-9922   ISSN 1860-0808   (electronic)
Studies in Fuzziness and Soft Computing
ISBN 978-3-030-08519-3   ISBN 978-3-319-77715-3   (eBook)
https://doi.org/10.1007/978-3-319-77715-3

This Springer imprint is published by Springer Nature
The registered company is Springer International Publishing AG
The registered company address is: Gewerbestrasse 11, 6330 Cham, Switzerland

To my dad who has always lived for me and my sister, believed in us, and supported us in anything we chose to do.

Preface

Research on multi-criteria decision making (MCDM) methods based on pairwise comparison matrices (PCM) has been emerging rapidly since the introduction of the first well-known pairwise comparison (PC) methods in 1970s. Moreover, since the original PC methods were not designed to cope with large-dimensional PC problems and with uncertainty present in decision problems, the PC methods were also very soon extended to incomplete PCMs and fuzzy PCMs.

Hundreds of research papers on incomplete PC methods and fuzzy PC methods have been published; many of the methods base on the methods originally developed for PCMs. Moreover, many research papers criticizing and revealing mistakes in some fuzzy and incomplete PC methods have been published too. The research papers are usually very specialized and focused on one specific topic assuming a certain level of readers' knowledge (as expected from research papers). Naturally, also the notation used in the research papers varies from paper to paper. Thus, it is for most readers not simple to compare methods proposed in different research papers and identify relations among them.

Despite the huge number of the methods developed, no book providing a critical overview of the existing methods and presenting the research results in a broader and unified frame has been published so far. This book is the first step to close this gap. In particular, the book presents a detailed critical literature review of the fuzzy PC methods based on the fuzzy extension of the methods originally developed for PCMs. Differences and relations between various methods are emphasized, and the reviewed fuzzy PC methods and their drawbacks are illustrated by many numerical examples. In fact, it is shown in the book that the majority of the reviewed fuzzy PC methods violate the basic assumptions made on the original PC methods on which they are build. Thus, besides the literature review, the book also introduces constrained fuzzy arithmetic in fuzzy extension of PC methods in order to preserve the original assumptions. Based on constrained fuzzy arithmetic, new fuzzy PC methods and incomplete PC methods are introduced and critically compared with the reviewed methods.

The book is intended for researchers in MCDM as well as for graduate and Ph.D. students, in particular for those interested in AHP and other PC methods. The book is self-contained. Chapters 2 and 3 provide an overview of the concepts necessary for studying Chaps. 4 and 5. In particular, Chap. 2 provides a detailed critical overview of PC methods based on three well-known types of PCMs, while Chap. 3 provides an overview of the concepts from fuzzy set theory indispensable for the fuzzy extension of PC methods and a detailed introduction to constrained fuzzy arithmetic. Chapter 4 is then focused on fuzzy PC methods and Chap. 5 on incomplete PC methods. The readers with a solid background in PC methods and fuzzy set theory may wish to skip Chap. 2 and Sects. 3.1–3.3 in Chap. 3. Sections 3.4 and 3.5 are, however, highly recommended to all readers in order to fully understand the difference between standard and constrained fuzzy arithmetic, which plays a key role in Chap. 4. Actually, Sect. 3.5 on constrained fuzzy arithmetic is very interesting by itself, in particular then the examples provided in the section.

The book is suitable also for the readers interested only in studying standard PC methods, in particular then for graduate and Ph.D. students. These readers may read only Chap. 2 that provides a critical overview of the most-known PC methods developed for three different types of PCMs and shows relations between various methods. Further, the book is of value also for the researchers interested in studying constrained fuzzy arithmetic and applying it (not only) in MCDM. These readers may read only Example 1 on p. 9 and Sects. 3.4 and 3.5, or alternatively whole Chap. 3 in case they need to refresh the basics of fuzzy set theory.

Trento, Italy   Jana Krejčí
Bayreuth, Germany

Acknowledgments

This book was created as the extension of my Ph.D. thesis written within my 3-year double-degree Ph.D. studies at the Department of Industrial Engineering at the University of Trento and at the Faculty of Business, Economics and Law at the University of Bayreuth. It summarizes the research results achieved during my Ph.D. studies and published in a number of research papers acknowledged in the book.

I would like to acknowledge the significant contribution of my Ph.D. supervisors, professors Michele Fedrizzi and Johannes Siebert, to the research results published in this book as well as to the organization of the book. Further, I would like to acknowledge the research contribution of all co-authors of my research papers on which this book is based. Last but not least, I would like to thank professors Janusz Kacprzyk, José Luis García Lapresta, and Jörg Schlüchtermann for contributing to increasing the quality of the book, as they acted as evaluators of my Ph.D. thesis.

Contents

Abbreviations

AHP	Analytic hierarchy process
APCM	Additive pairwise comparison matrix
APCM-A	Additive pairwise comparison matrix with additive representation
APCM-M	Additive pairwise comparison matrix with multiplicative representation
DM	Decision-maker
EVM	Eigenvector method
FAPCM	Fuzzy additive pairwise comparison matrix
FAPCM-A	Fuzzy additive pairwise comparison matrix with additive representation
FAPCM-M	Fuzzy additive pairwise comparison matrix with multiplicative representation
FMPCM	Fuzzy multiplicative pairwise comparison matrix
FPCM	Fuzzy pairwise comparison matrix
GMM	Geometric mean method
LLSM	Logarithmic least squares method
MCDM	Multi-criteria decision making
MPCM	Multiplicative pairwise comparison matrix
PC	Pairwise comparison
PCM	Pairwise comparison matrix

Mathematical Symbols

$\emptyset$	Empty set
$\mathbb{N}$	Set of natural numbers
$\mathbb{R}$	Set of real numbers
$\mathbb{R}^+$	Set of positive real numbers greater than 0
$U \times V$	Cartesian product of two sets U and V
$\mathbb{R}^n$	n–ary cartesian power of set $\mathbb{R}$
$\mathcal{F}(\mathbb{R})$	Set of all fuzzy sets defined on $\mathbb{R}$
$\mathcal{F}_N(\mathbb{R})$	Set of all fuzzy numbers defined on $\mathbb{R}$
$\mathcal{F}_N(\mathbb{R}^+)$	Set of all positive fuzzy numbers
$\mathcal{F}_N(\mathbb{R})^n$	n–ary cartesian power of set $\mathcal{F}_N(\mathbb{R})$
$x \in \Omega$	Element belonging to set Ω
$|\Omega|$	Cardinality of set Ω
$\Omega \backslash Q$	Difference of sets Ω and Q
$\Omega \cap Q$	Intersection of sets Ω and Q
$\Omega \cup Q$	Union of sets Ω and Q
$(x_1, x_2, \ldots, x_n)$	n–tuple, i.e., an ordered list of n elements, $n \in \mathbb{N}$
$[a, b],\ \bar{c} = [c^L, c^U]$	Closed interval
$]a, b[$	Open interval
$\widetilde{c}$	Fuzzy set
$Supp\, \widetilde{c}$	Support of $\widetilde{c}$
$Core\, \widetilde{c}$	Core of $\widetilde{c}$
$\widetilde{c}_{(\alpha)}$	α–cut of $\widetilde{c}$
$\widetilde{c}_{(0)} = Cl(Supp\, \widetilde{c})$	Closure of the support of $\widetilde{c}$
$c \in \widetilde{c}$	Element belonging to the closure of the support of $\widetilde{c}$
$\widetilde{c} = (c^L, c^M, c^U)$	Triangular fuzzy number
$\widetilde{c} = (c^\alpha, c^\beta, c^\gamma, c^\delta)$	Trapezoidal fuzzy number
$A = \{a_{ij}\}_{i,j=1}^n$	Square matrix
$\widetilde{A} = \{\widetilde{a}_{ij}\}_{i,j=1}^n$	Square fuzzy matrix
$|A|$	Determinant of matrix A

A^T	Transpose of matrix A
$\widetilde{A}^T$	Transpose of fuzzy matrix A
$\lambda = EVM_\lambda(A)$	Maximal eigenvalue of matrix A
$\underline{w} = EVM_{\underline{w}}(A)$	Normalized maximal eigenvector of matrix A
$\underline{w} = (w_1, \ldots, w_n)^T$	Column vector
$\underline{w}^T = (w_1, \ldots, w_n)$	Row vector
$\underline{\widetilde{w}} = (\widetilde{w}_1, \ldots, \widetilde{w}_n)^T$	Column fuzzy vector
$\underline{\overline{w}} = (\overline{w}_1, \ldots, \overline{w}_n)^T$	Column interval vector
$\wedge$	Logical conjunction
$\vee$	Logical disjunction
$\ln$	Natural logarithm
$\log_9$	Logarithm of base 9
f^{-1}	Inverse of function f
$\arg f$	Argument of function f
$\lfloor x \rfloor$	Floor of $x \in \mathbb{R}$
$k!$	Factorial of number $k \in \mathbb{N}$

Summary

Pairwise comparison (PC) methods form a significant part of multi-criteria decision making (MCDM) methods. PC methods are based on structuring PCs of objects from a finite set of objects into a pairwise comparison matrix (PCM) and deriving priorities of objects that represent the relative importance of each object with respect to all other objects in the set. However, crisp PCMs are not able to capture uncertainty stemming from subjectivity of human thinking and from incompleteness of information about the problem that are often closely related to MCDM problems. That is why the fuzzy extension of PC methods has been of great interest.

In order to derive fuzzy priorities of objects from a fuzzy PCM (FPCM), the fuzzy extension based on standard fuzzy arithmetic is usually applied to the methods originally developed for crisp PCMs. However, such approach fails in properly handling uncertainty of preference information contained in the FPCM. Namely, reciprocity of the related PCs of objects in a FPCM and invariance of the given method under permutation of objects are violated when standard fuzzy arithmetic is applied to the fuzzy extension. This leads to distortion of the preference information contained in the FPCM and consequently to false results. This issue is a motivation to the first research question dealt with in this book: *"Based on a FPCM of objects, how should fuzzy priorities of these objects be determined so that they reflect properly all preference information available in the FPCM?"* This research question is answered by introducing an appropriate fuzzy extension of PC methods originally developed for crisp PCMs, i.e., such fuzzy extension that does not violate the reciprocity of the related PCs and invariance of PC methods under permutation of objects, and that does not lead to a redundant increase of uncertainty of the resulting fuzzy priorities of objects.

Fuzzy extension of three well-known types of PCMs—multiplicative PCMs, additive PCMs with additive representation, and additive PCMs with multiplicative representation—is examined in this book. In particular, construction of PCMs, verifying consistency, and deriving priorities of objects from PCMs are studied in detail for each type of these PCMs. First, well-known and in practice most often applied PC methods based on crisp PCMs are reviewed. Afterwards, fuzzy extensions of these methods proposed in the literature are reviewed in detail, and their drawbacks regarding the violation of reciprocity of the related PCs and of invariance of methods under permutation of objects are pointed out. It is shown that these drawbacks can be overcome by properly applying constrained fuzzy arithmetic to the computations instead of standard fuzzy arithmetic. In particular, we always have to look at a FPCM as a set of PCMs with different degrees of membership to the FPCM, i.e., we always have to consider only PCs that are mutually reciprocal. Constrained fuzzy arithmetic allows us to impose the reciprocity of the related PCs as a constraint on arithmetic operations with fuzzy numbers, and its appropriate application also guarantees invariance of the fuzzy PC methods under permutation of objects. Finally, new fuzzy extensions of the PC methods are proposed based on constrained fuzzy arithmetic, and it is proved that these methods do not violate the reciprocity of the related PCs and are invariant under permutation of objects. Because of these desirable properties, fuzzy priorities of objects obtained by the fuzzy PC methods proposed in this book reflect the preference information contained in FPCMs better in comparison to the fuzzy priorities obtained by the fuzzy PC methods based on standard fuzzy arithmetic.

Besides the inability to capture subjectivity, the PC methods are also not able to cope with situations where it is not possible or reasonable to obtain complete preference information from DMs. This problem occurs especially in the situations involving large-dimensional PCMs. When dealing with incomplete large-dimensional PCMs, a compromise between reducing the number of PCs required from the DM and obtaining reasonable priorities of objects is of paramount importance. This leads to the second research question: *"How can the amount of preference information required from the DM in a large-dimensional PCM be reduced while still obtaining comparable priorities of objects?"* This research question is answered by introducing an efficient two-phase PC method. Specifically, in the first phase, an interactive algorithm based on the weak-consistency condition is introduced for partially filling an incomplete PCM. This algorithm is designed in such a way that it minimizes the number of PCs required from the DM and provides a sufficient amount of preference information at the same time. The weak-consistency condition allows for providing ranges of possible intensities of preference for every missing PC in the incomplete PCM. Thus, at the end of the first phase, a PCM containing intervals for all PCs that were not provided by the DM is obtained. Afterward, in the second phase, the methods for obtaining fuzzy priorities of objects from FPCMs proposed in this book within the answer to the first research question are applied to derive interval priorities of objects from the incomplete PCM. The obtained interval priorities cover the priorities obtainable from all weakly consistent completions of the incomplete PCM and are very narrow. The performance of the method is illustrated by a real-life case study and by simulations that demonstrate the ability of the method to reduce the number of PCs required from the DM in PCMs of dimension 15 and greater by more than 60% on average while obtaining interval priorities comparable with the priorities obtainable from the hypothetical complete PCMs.

Chapter 1
Introduction

Abstract This chapter provides an introduction to multi-criteria decision making methods and specifies the focus of the book. It identifies critical issues related to multi-criteria decision making methods based on fuzzy pairwise comparison matrices and large-dimensional pairwise comparison matrices and derives the key research questions of the book. It provides a motivational example demonstrating the necessity of using constrained fuzzy arithmetic in the fuzzy extension of pairwise comparison methods. The example is explained in simple words and is easily understandable also to non-expert readers. This example is highly recommended to anyone interested in fuzzy pairwise comparison methods or in general in fuzzy arithmetic.

1.1 Multi-criteria Decision Making

Multi-criteria decision making (MCDM) is an extensive sub-discipline of operations research. Decision making is regarded as a process of evaluating decision alternatives (courses of action) based on preferences of a decision maker (DM). The DM is the subject in charge of making the decision; it can be, e.g., an individual, a group of people, a family, a company, a government, etc. The aim of MCDM, as its name suggests, is to deal with situations (problems) requiring a decision being made under the presence of multiple criteria. In some situations, considering only a single criterion may be sufficient to make a well-informed decision. Most real-life problems, however, are more complex, and considering only one criterion is too simplistic. In such cases multiple criteria have to be considered.

MCDM problems range from everyday decision problems with low impact (such as which dress to wear, which means of transport to use to commute to work, or on which week day to organize a business meeting) to important decision problems with substantial consequences and long-term impact (such as where to build a new power plant, whether to extend production capacity, which family house to buy, or which project to fund). We would probably not build an MCDM model for problems such as which dress to wear or which means of transport to use to commute to work. These problems are not complex enough; the DMs are able to consider all relevant criteria in their heads, and the consequences of such problems are not substantial.

© Springer International Publishing AG, part of Springer Nature 2018   1
J. Krejčí, *Pairwise Comparison Matrices and their Fuzzy Extension*, Studies
in Fuzziness and Soft Computing 366, https://doi.org/10.1007/978-3-319-77715-3_1

Contrarily, substantially more complex decision problems, such as where to build a new power plant, cannot be solved in ones head; an appropriate MCDM model needs to be used to support a well-informed decision.

In practice, DMs have difficulties to understand what they really want to achieve in the given problem and what options they have to achieve that (Bond et al. 2008; Siebert and Keeney 2015). DMs are not able, in most cases, to clearly formulate their preferences at the beginning of the decision-making process. Their preferences are formed during the decision-aiding process with the help of MCDM models and analysts (experts in the given approach of modeling acting as mediators between the DMs and the MCDM model). MCDM models help DMs to learn about their preferences and to construct a system of their preferences consistent with the assumptions of the MCDM model, thus guiding DMs in searching for the most preferred solution to the decision problem.

It is necessary to realize that no MCDM model can provide a "correct answer" to a decision problem. Unlike in single-criterion decision making, an objective optimal solution does not exist in the context of MCDM. Subjectivity is present in every MCDM problem, e.g., in the choice of the criteria relevant for the problem, in the relative importance assigned to each criterion, or in the evaluations of alternatives with respect to qualitative criteria. The aim of MCDM models is to aid DMs by managing the subjectivity inevitably present in MCDM problems and by integrating the subjective information provided by the DMs with objective measurements. Applying MCDM models thus leads to better considered, explainable, and justifiable decisions.

It should be mentioned that most MCDM methods focus on "solving" well-formulated MCDM problems. However, in practice most problems are not well-formulated. MCDM actually begins when someone feels that a particular issue matters enough to explore the potential of formal modeling (Belton and Stewart 2002). Thus the whole MCDM process actually starts by revealing all that is relevant to the problem in question. The MCDM process consists of three phases (Belton and Stewart 2002):

1. Problem identification and structuring: Before we can solve any problem we have to develop understanding of the problem, which starts by identifying the decision that has to be made, the criteria important for the decision, and the set of alternatives to be evaluated. The alternatives are defined either explicitly (a discrete list of alternatives) or implicitly by a set of constraints of the decision problem.
2. Model building and use: After the problem is identified and structured, a formal model of DMs' preferences and value judgments is built to support DMs in their search for the most preferred solution of the decision problem. Preference models consist of two main components:

   - a model of preferences in terms of individual criteria describing the relative preference (or importance) of achieving various levels of performance for individual criteria,
   - an aggregation model to combine preferences across individual criteria.

3. Development of action plans: The result provided by an MCDM model does not solve the actual decision problem. The result has to be implemented in practice by translating it into specific action plans.

In this book the model-building phase of the MCDM process is of interest. Therefore, we will assume to have an already fully defined decision problem. This means, we already have the set of criteria relevant for the problem and the set of alternatives that will be evaluated. The criteria can be structured into a hierarchy where a criterion in one level is decomposed into sub-criteria in the lower level. Thus, the criteria in the highest level may be more general (or abstract), while the criteria in the lowest level should be enough specific in order to enable the evaluation of alternatives with respect to individual criteria in the lowest level. Formally, the goal of the decision problem can be added into the highest level of the hierarchy, and the set of decision alternatives can be added into the lowest level of the hierarchy, so that the hierarchy represents the whole decision problem. We will call such a hierarchy a problem hierarchy.

An example of a problem hierarchy is shown in Fig. 1.1. This hierarchy represents the decision problem of evaluating airline service quality studied by Tsaur et al. (2002). The goal of the decision problem—evaluation of airline service quality—is stated in the highest level of the hierarchy (which is shown in the figure on the left side for ease of presentation). Five general criteria relevant for the problem were identified. These criteria are given in the second level of the hierarchy. Each of the general criteria is further broken down into two to four specific criteria given in the third level of the problem hierarchy. For example, the criterion "reliability" is broken into the criteria "professional skill of crew", "timeliness", and "safety". The criteria in the third level of the problem hierarchy are specific enough in order to enable the evaluation of decision alternatives with respect to these criteria. Finally, various decision alternatives – airline companies whose service quality is going to be evaluated – are specified in the lowest level of the problem hierarchy (shown in the figure on the right side).

MCDM models vary in terms of the type and the strength of assumptions made. This is important to keep in mind when comparing different MCDM models. For example, it is not possible to compare two MCDM models with different assumptions in terms of how well they model and aggregate decision preferences. Such models should be rather compared on the basis of the guidance and the insight into the problem provided to the DM (Belton and Stewart 2002). Contrarily, MCDM models with the same assumptions applied to the same decision problem with the same preference information provided by the DM can be compared in terms of the quality of modeling and aggregating decision preferences.

Even though it is not possible to say whether an MCDM model provides a "correct" solution, we can always distinguish between two types of MCDM models: (1) MCDM models that properly reflect the assumptions made and the preference information provided by the DM, and (2) MCDM models that do not properly reflect either the assumptions made or the preference information provided by the DM. We can say that the MCDM models from the first group provide valid results that properly represent

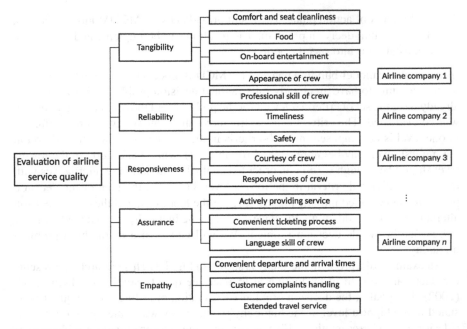

Fig. 1.1 Hierarchy of the decision problem of evaluating airline service quality

DM's preferences, while applying the MCDM models from the second group leads to invalid results that do not represent DM's preferences. The differentiation of methods in these two groups is of crucial importance in this book; many MCDM methods will be reviewed in Chap. 4, and it will be demonstrated that they belong to the second group, which means that they are not suitable to support DMs in MCDM.

MCDM is based on the concept of the relative importance of criteria that is generally represented by means of quantitative importance priorities (weights). It is important to keep in mind that the meaning of these priorities depends strongly on the MCDM model used, and understanding the meaning of the priorities in the given MCDM model is of crucial importance to correctly interpret the analysis results (Belton and Stewart 2002).

MCDM models can be classified into three broad categories (Belton and Stewart 2002):

1. Value measurement methods: Numerical scores are constructed in order to represent the degree in which one alternative is preferred to another with respect to individual criteria. The scores are then aggregated to determine the degree in which one alternative is preferred to another with respect to all criteria.
2. Outranking methods: Alternatives are compared pairwisely with respect to individual criteria in order to determine the extent to which the preference of one over another can be asserted. The preference information is then aggregated to deter-

mine the strength of evidence favoring one alternative over another with respect to all criteria.

3. Goal, aspiration, or reference level methods: Desirable/satisfactory levels are established for individual criteria, and the alternatives closest to achieving these desirable levels are searched for.

A significant part of MCDM methods is based on pairwise comparison (PC). The first PC method was developed by Thurstone (1927). Since then, PC methods have undergone a great development. PCs of objects (usually decision criteria and decision alternatives) have been widely used in many well-known MCDM methods such as Analytic Hierarchy Process (AHP), Preference Ranking Organization Method for Enrichment Evaluation (PROMETHEE), Elimination and Choice Expressing Reality (ELECTRE), and their derivatives; see, e.g., Figueira et al. (2005). AHP belongs to value measurement methods, while PROMETHEE and ELECTRE are representatives of outranking methods. The outranking methods differ from the value measurement methods in a sense that they do not use an aggregation value function. The outranking methods provide an outranking relation on the set of alternatives, while the value measurement methods provide a value for each alternative in order to produce a preference order on the set of alternatives. In this book, PC methods belonging to the group of value measurement MCDM methods are of interest, and from now on the simple expression "PC methods" will be used to refer exclusively to "PC value measurement methods".

1.1.1 Pairwise Comparison Methods

PC methods are based on the idea of comparing only two objects (decision criteria or decision alternatives) at a time, which is significantly less demanding on cognitive capabilities of DMs than comparing several objects at the same time. Considering a fully defined MCDM problem with a given problem hierarchy, such as the one given in Fig. 1.1, a set of objects in one level of the hierarchy is always compared pairwisely with respect to the superior object from the higher level of the hierarchy. In particular, the highest-level criteria are compared pairwisely with respect to the decision goal, each group of sub-criteria is compared pairwisely with respect to the superior criterion, and the set of alternatives in the lowest level of the problem hierarchy is compared pairwisely with respect to the individual lowest-level criteria. In the decision problem represented in Fig. 1.1, for example, the lowest-level criteria "professional skill of crew", "timeliness", and "safety" are compared pairwisely with respect to the superior criterion "reliability".

The most usual way to represent PCs of a finite set of objects with respect to the superior object is to structure them into a pairwise comparison matrix (PCM). A PCM of n objects $o_1, \ldots, o_n$ is a square matrix $C = \{c_{ij}\}_{i,j=1}^{n}$ where the element c_{ij} expresses the intensity of preference (or importance) of object o_i over object o_j. In the decision problem with the problem hierarchy given in Fig. 1.1, 21 PCMs would be constructed. In particular, we would construct 1 PCM of the highest-level criteria with respect to the decision goal, 5 PCMs of the lowest-level criteria with respect to the superior criteria, and 15 PCMs of alternatives with respect to each of the lowest-level criteria. For example, the PCM of the lowest-level criteria "professional skill of crew", "timeliness", and "safety" with respect to the superior criterion "reliability" would be a PCM $C = \{c_{ij}\}_{i,j=1}^{3}$ in which the PC c_{12} would express the intensity of preference of "professional skill of crew" over "timeliness", the PC c_{13} would express the intensity of preference of "professional skill of crew" over "safety", and so on.

The PCs c_{ij} and c_{ji} in a PCM $C = \{c_{ij}\}_{i,j=1}^{n}$ express intensities of preference on the same pair of objects o_i and o_j, just in the opposite order. Thus, there is a relation between the PCs c_{ij} and c_{ji}. The type of the relation depends on the type of the PCs used, i.e., on the form in which the intensity of preference of one object over another is expressed. In this book, two types of relation between the PCs of objects are of interest—multiplicative reciprocity and additive reciprocity. Multiplicative reciprocity of the related PCs is inherent to multiplicative PCMs that are used to model preference information provided in form of a preference ratio, while additive reciprocity of the related PCs is inherent to additive PCMs that are used to model preference information provided in form of a preference difference. Additive PCMs are further divided into additive PCMs with additive representation and additive PCMs with multiplicative representation. All three types of PCMs will be defined in detail in Chap. 2.

PCMs have been studied in detail by many researchers, and many different PC methods based on multiplicative and additive PCMs have been developed to support MCDM. The PC methods basically consist in the following steps:

1. constructing PCMs, i.e., comparing pairwisely objects from one level of the problem hierarchy with respect to the superior object from the upper level of the hierarchy and structuring the PCs into a PCM;
2. verifying consistency (or acceptable inconsistency) of the PCMs, i.e., verifying whether the preference information provided by the DM in form of PCs is consistent in some sense;
3. deriving priorities of objects (criteria and alternatives) from the PCMs;
4. aggregating priorities of criteria and alternatives within the problem hierarchy into final priorities of alternatives representing the final multi-criteria evaluations of alternatives.

Further extensions of PC methods have also been developed to deal with decision problems involving several DMs. Each of the above mentioned steps encounters some difficulties. In the first step, the choice of an appropriate scale for expressing the intensities of preference plays a crucial role. In the second step, the choice of a proper consistency condition compatible with the type of the preference information provided in the PCM and with the scale chosen for making PCs is very important. In the third step, it is necessary to choose such a method and a normalization condition that are compatible with the type of the preference information contained in the PCM. Finally, in the fourth step, the choice of a suitable aggregation method is of crucial importance. Unfortunately, not all of these issues are addressed properly in the literature, which may lead to wrong results and, consequently, to low-quality decisions.

1.1.2 Incomplete Pairwise Comparison Matrices

Standard PC methods require complete preference information. However, in practice, it is not always possible to obtain complete preference information from the DM, e.g., due to time or cost limitations. This problem occurs in particular when the DM is required to provide a large number of PCs. To handle this problem various methods for dealing with incomplete large-dimensional PCMs have been proposed in the literature. These methods usually consist of two main steps:

1. identifying a set of PCs that should be provided by the DM in an incomplete PCM;
2. deriving priorities of objects from an incomplete PCM—This is done either by proposing a method for deriving priorities from the obtained incomplete PCM or by proposing a method for automatically completing the missing PCs in the incomplete PCM (i.e. with no additional preference information from the DM) and then deriving priorities from the obtained complete PCM by using one of the standard methods.

Both steps encounter some challenges. In the first step, the appropriate choice of the PCs and of the number of PCs required from the DM plays a key role. In the second step, the choice of an appropriate method is of crucial importance in order to derive priorities that best represent DM's preferences and the incompleteness of preference information. However, not all of these issues are addressed properly in the literature, which may again lead to misleading results that do not reflect DM's preferences. This problem is a motivation for the second research question of the book formulated in Sect. 1.3 and afterwards answered in Chap. 5.

1.2 Fuzzy Multi-criteria Decision Making

As mentioned already in Sect. 1.1, subjectivity is present in every real-world decision problem, e.g., in the choice of the criteria relevant for the problem or in the relative importance assigned to each criterion. In addition, in most real-world decision problems the goals, problem constraints, or consequences of possible actions are not known precisely (Bellman and Zadeh 1970), which adds uncertainty into MCDM. The "traditional" MCDM methods discussed in Sect. 1.1 were not designed to deal with uncertainty in MCDM problems. Uncertainty in decision making has been modeled and analyzed by means of probabilistic theory and fuzzy set theory. The aim of probabilistic theory is to capture the stochastic nature of decision making, while fuzzy set theory aims to capture the subjectivity (imprecision) of human behavior and human thinking (Dubois and Prade 1986). Probabilities could be conveniently used, e.g., to model the expected income from various investment alternatives in investment decision problems, whereas fuzzy sets could be used, e.g., to subjectively evaluate reputation of various suppliers in supplier decision problems.

Usually, DMs express their subjective evaluations linguistically by using assessment terms such as "good", "average", "poor", etc., rather than numbers, as this way is much more natural for them. The meaning of linguistic terms then has to be modeled mathematically in MCDM models. In the "traditional" MCDM methods discussed in Sect. 1.1 the meaning of linguistic terms is modeled by "crisp" numbers. However, many researchers have agreed that crisp numbers cannot properly capture the vagueness in meaning of linguistic terms, whereas fuzzy sets, introduced by Zadeh (1965), are suitable for this purpose.

A fuzzy set is basically an extension of a "classic" set with not a sharp boundary between the elements that belong to the set and the elements that do not belong there. For example, a set of all numbers on a dice is a classic set as there is clearly a sharp boundary separating the elements belonging to the set (the integers 1, 2, 3, 4, 5, and 6) and the elements not belonging there (these are all other numbers we can imagine). Now imagine the set of red apples. Where is the boundary separating the red apples from the apples that are not red? A completely red apple belongs to the set while a completely green apple does not belong there. What about an apple that is from 99% red? Does it belong to the set? And what about an apple that is red from 30%? Does it still belong there? Let us choose as the sharp boundary 70% redness, for example. This would mean that an apple red from 71% belongs to the set of red apples, while an apple red from 69% does not belong there, which is not very intuitive. It does not seem natural to put a sharp boundary separating red apples from apples that are not red. Instead, it would be more intuitive to assign to each percentage of redness a degree of membership to the set of red apples. These degrees of membership would allow us to express that some apples belong to the set of red apples only partially, which is not possible with classic sets. The fuzzy sets will be defined later in Chap. 3.

The need for modeling the subjectivity of human behavior and human thinking in decision making resulted in the development of a new decision-making field—fuzzy MCDM. The pioneering work in fuzzy MCDM was done by Bellman and Zadeh (1970) who extended the Maxmin method to the fuzzy environment. However, the data in their method were still expressed by crisp numbers. The first proper fuzzy MCDM methods were developed by Baas and Kwakernaak (1977) and Kwakernaak (1979) who extended the Simple Additive Weighting method, and by Efstathiou and Rajkovic (1979) who proposed a fuzzy extension of the Multiple Attribute Utility Function method. Afterwards, many other approaches followed (see, e.g., Chen and Hwang (1992) for the review).

There are two main phases in fuzzy MCDM (Zimmermann 1987): (1) aggregation of the performance scores with respect to all criteria for each alternative, and (2) ranking of the alternatives according to the aggregated scores. Depending on which phase the fuzzy MCDM methods focus on, they can be classified into two categories (Zimmermann 1987):

- fuzzy MCDM methods focusing only on the first phase or on both phases,
- fuzzy MCDM methods focusing only on the second phase.

The fuzzy MCDM methods of interest in this book are the fuzzy PC methods belonging to the first category.

1.2.1 Fuzzy Pairwise Comparison Methods

PCMs (both multiplicative and additive) contain PCs of objects in terms of intensities of preference provided by DMs. These PCs are crisp numbers. However, as already mentioned above, crisp numbers cannot capture subjectivity, which is integral to human mind. Moreover, it is very natural for DMs to provide the intensities of preference in a linguistic form or directly by using fuzzy numbers rather than crisp numbers when the information about the problem is imprecise. Thus, in order to properly capture the subjectivity and imprecision of information, the PC methods have been extended to fuzzy numbers.

Most often, triangular fuzzy numbers are used for the fuzzy extension of multiplicative PCMs. Less often, trapezoidal fuzzy numbers or intervals are used for this purpose. Contrarily, unlike multiplicative PCMs, additive PCMs are usually extended to intervals rather than to fuzzy numbers. Intervals can be understood as a special case of trapezoidal fuzzy numbers where all elements from the given interval have the same degree of membership to the fuzzy set. Note that sometimes it will be explicitly distinguished between the terms "fuzzy" and "interval" in this book. Often, however, only the term "fuzzy" will be used for the simplicity, and, in such case, intervals will be understood as a particular case of fuzzy numbers, namely of trapezoidal fuzzy numbers.

The fuzzy extension of PC methods to fuzzy numbers basically consists in using fuzzy numbers at the stage of providing PCs of objects in a PCM. These fuzzy numbers can be either chosen from a predefined scale of fuzzy numbers assigned to linguistic terms expressing different intensities of preference, or entered expertly by the DM without linguistic representation. Having a fuzzy PCM (FPCM), i.e. a PCM with elements in form of fuzzy numbers, the aim is to derive fuzzy priorities of these objects that reflect properly the fuzzy preference information contained in the FPCM. As it will be shown in Chap. 4, many fuzzy PC methods proposed in the literature fail in this matter. This issue led to formulating the first research question of this book that is provided in Sect. 1.3 and afterwards answered in Chap. 4.

1.2.2 Critics of Fuzzy Extension of AHP

Saaty (2006) argued that, in AHP (one of the PC methods), "the numbers assigned to judgments are already fuzzy and making them more fuzzy does not help produce more valid outcome" (Saaty 2006, p. 457). Saaty and Tran (2007) demonstrated on several examples the invalidity of fuzzification of AHP and concluded that "one should never use fuzzy arithmetic on AHP judgment matrices" (Saaty and Tran 2007, p. 970). Clearly, when the PCs obtained from a DM are crisp there is no gain from fuzzifying them without a good reason. However, when the PCs are vague or when linguistic terms are used to express intensities of preference on pairs of compared objects, fuzzy numbers should be applied instead of crisp numbers (Krejčí et al. 2017). Krejčí et al. (2017) further showed that by neglecting the available information about the uncertainty of intensities of preferences an important part of knowledge about the decision-making problem is lost, which leads to invalid results and thus to solutions that do not represent the original preference information.

Besides Saaty and Trans critics, very harsh critics of the fuzzy extension of AHP were provided also by Zhü (2014). He heavily criticizes well-known fuzzy approaches to AHP and claims fallacy of all of them. However, Zhü's critics are based mostly on misinterpreting other researchers claims and rejecting consolidated bases of fuzzy set theory. Fedrizzi and Krejčí (2015) showed that no reliable evidence of the fallacy of AHP was provided by Zhü (2014), and thus fuzzy AHP and more in general fuzzy PC methods based on both multiplicative and additive FPCMs remain a valid MCDM research area.

It is clear that fuzzy AHP, and in general fuzzy PC methods based on multiplicative and additive FPCMs, have some critical aspects (some of them indicated by Zhü 2014) that need to be investigated and rectified. One of them is the issue regarding the reciprocity (multiplicative or additive) of the related PCs, which is an inherent property of PCMs. Reciprocity of the related PC has been neglected in multiplicative and additive FPCMs for a long time, which led to some incorrect results. Nevertheless, this drawback can be relatively easily removed by applying constrained fuzzy arithmetic instead of standard fuzzy arithmetic to the computations with fuzzy numbers (Fedrizzi and Krejčí 2015). Section 1.2.3 is devoted to clarify this point.

Another issue related to the extension of PC methods to fuzzy PC methods is the invariance of methods under permutation of objects. A PC method is said to be invariant under permutation of objects if the result of the method does not depend on the permutation of objects compared in a PCM. Even though the invariance of methods under permutation of objects was introduced as one of the axioms which "good" methods should meet (see, e.g., Fichtner 1986; Brunelli and Fedrizzi 2015), many fuzzy PC methods violate this property. Nevertheless, also the invariance of fuzzy PC methods under permutation can be easily achieved by properly applying constrained fuzzy arithmetic instead of standard fuzzy arithmetic to the fuzzy extension of the PC methods. Removing the problems regarding the reciprocity of the related PCs and the invariance of fuzzy PC methods under permutation by properly applying constrained fuzzy arithmetic is the key tool for answering the first research question of this book formulated in Sect. 1.3.

1.2.3 Constrained Fuzzy/Interval Arithmetic

Constrained fuzzy arithmetic, introduced by Klir (1997) and Klir and Pan (1998), should be applied to computations with fuzzy numbers whenever there are interactions of any type present between the fuzzy numbers. Similarly, constrained interval arithmetic, that was recently introduced by Lodwick and Jenkins (2013), should be applied to computations with intervals when interactions are present. Constrained interval arithmetic can be seen as a special case of constrained fuzzy arithmetic since intervals are special cases of trapezoidal fuzzy numbers. Therefore, all comments and discussions on constrained fuzzy arithmetic in this book apply also for interval fuzzy arithmetic if not specified otherwise. Motivation for the use of constrained fuzzy arithmetic is given in the following example. Since fuzzy numbers have not been properly defined yet (this will be done in Chap. 4), intervals will be used for explanation in the example. Moreover, trying to understand the core idea of constrained arithmetic is at the beginning significantly easier with intervals than with fuzzy numbers.

Example 1 Let us assume we have exactly one liter, i.e. 1000 ml, of beer in one bottle, and we want to share it with our friend. We pour half of the beer into a half-liter. Of course, it is highly improbable that we would manage to pour exactly 500 ml into the half-liter. Our friend has a look on the half-liter and estimates that there is for sure something between 450 and 520 ml of beer inside. What can we say about the rest of the beer in our bottle? Based on the information from our friend and on our common sense, we can say that there is for sure something between 480 and 550 ml in our bottle. How did we arrive to this conclusion? Well, if there is 450 ml of beer in our friend's half-liter, then there has to be $1000 - 450 = 550$ ml left in our bottle. In case there is 520 ml of beer in our friend's half-liter, then there has to be $1000 - 520 = 480$ ml left in our bottle. This is nothing else but the standard interval arithmetic we know; $1000 - [450, 520] = [1000 - 520, 1000 - 450] = [480, 550]$.

What happens if we pour our friend's half-liter back to the one-liter bottle? (Note that for simplicity of explanation we consider here an ideal situation in which no losses of the amount of beer are caused during the manipulation, i.e., we do not consider that there might be any liquid left in the emptied half-liter. In real case, there would actually be some very small amount of liquid left on the sides and the bottom of the half-liter after emptying it.) Answering this question is trivial, is not it? Our common sense says that we have to get again exactly one liter (i.e. 1000 ml) of beer, unless we spill some of it.

But what mathematics is behind this very simple problem? If we apply again standard interval arithmetic we know, we get

$$[450, 520] + [480, 550] = [450 + 480, 520 + 550] = [930, 1070]. \qquad (1.1)$$

So this would suggest that there is something between 930 and 1070 ml of beer. But everyone would agree that this is not possible; there has to be again exactly 1000 ml in our bottle. So what is going on here? How is it possible that we are not able to illustrate mathematically such a trivial problem?

The problem is that there are dependencies between the amounts of beer in our friend's half-liter and in our bottle. We had exactly 1000 ml of beer at the beginning, and we pour a part of it into a half-liter. Therefore, no matter what the exact amount x between 450 and 520 ml of beer in the half-liter is, we know for sure that the rest of beer in our bottle is $y = 1000 - x$ ml. Then, when pouring the half-liter back to the bottle, we still keep this relation in mind. However this relation is not reflected in the mathematical representation (1.1) of the problem; it is necessary to apply a more sophisticated approach than just standard interval arithmetic. We should correctly compute

$$z = x + y \quad \text{where}$$

$$x \in [450, 520]$$

$$y \in [480, 550]$$

$$y = 1000 - x$$

This is nothing else than constrained (in this particular case interval) arithmetic. $\triangle$

Constrained fuzzy (or interval) arithmetic is not applicable only when sharing beer with friends. It should be applied whenever there are some dependencies between operands in arithmetic operations on fuzzy numbers (or intervals). As stated at the end of Sect. 1.2.2, there are dependencies between PCs in a FPCM; in particular, reciprocal relations between the related PCs. Since reciprocity of the related PCs is an inherent property of PCMs, it is necessary to extend it properly also to FPCMs. For that it is necessary to correctly apply constrained fuzzy arithmetic.

1.3   Goal of the Book

As already mentioned in Sect. 1.2, "traditional" MCDM methods including the PC methods based on multiplicative PCMs as well as on additive PCMs (both with additive and multiplicative representation) are often criticized because of their inability to capture uncertainty stemming from subjectivity of human thinking and from incompleteness of information that are often closely related to MCDM problems. This uncertainty has an impact on the PCs provided by DMs in PCMs. In order to capture the uncertainty, the PC methods originally proposed for crisp PCMs have been extended to fuzzy numbers. The fuzzy extension often consists in simply replacing the crisp PCs in the given MCDM model by fuzzy PCs and applying standard fuzzy arithmetic to obtain the desired fuzzy priorities. However, this approach often fails in handling appropriately the uncertain preference information contained in the FPCM, which may lead to false results. Therefore, the first research question of this book is:

(1) *Based on a FPCM of objects, how should fuzzy priorities of these objects be determined so that they reflect properly all preference information available in the FPCM?*

In order to answer this research question, it is necessary to fully understand the meaning of PCs in a PCM and to identify inherent properties of PCMs. As discussed in Sect. 1.2.2, the crucial inherent property of PCMs is the reciprocity of the related PCs. The concept of reciprocity of the related PCs becomes more complex when extended to FPCMs, and handling appropriately the reciprocity property becomes of key importance in order to process correctly the preference information contained in the FPCM and to arrive to reasonable conclusions. Unfortunately, this issue is usually omitted in the literature. This leads to results (resulting fuzzy priorities of objects in particular) that are often excessively uncertain and do not reflect correctly the preference information available in the original FPCM. Another crucial property related to both PCMs and FPCMs is the invariance of methods under permutation of objects. Many fuzzy PC methods violate this property, which leads to false results. The drawbacks regarding the violation of the reciprocity property and of the invariance of methods under permutation of objects can be removed by applying constrained fuzzy arithmetic to the computations instead of standard fuzzy arithmetic. Unfortunately, the use of constrained fuzzy arithmetic is still neglected in the literature on fuzzy PC methods. In fact, I have encountered a single research paper dealing with this important topic—Enea and Piazza (2004).

The following four steps are pursued in the book in order to properly answer the research question (1):

(1.a) the well-known and in practice most often applied PC methods (based on multiplicative PCMs, additive PCMs with additive representation, and additive PCMs with multiplicative representation) dealing with the construction of PCMs, consistency verification, and priorities computation are critically reviewed;

(1.b) the existing approaches to the fuzzy extension of the PC methods reviewed within step (1.a) are critically reviewed and their drawbacks regarding the violation of the reciprocity property and of the invariance under permutation of objects are identified;

(1.c) it is demonstrated that it is necessary to apply constrained fuzzy arithmetic to the fuzzy extension of PC methods in order to obtain meaningful results reflecting the preference information contained in the original FPCM and not suffering from the drawbacks identified within step (1.b);

(1.d) a new fuzzy extension of the PC methods critically reviewed within step (1.a) is proposed by applying constrained fuzzy arithmetic so that it does not suffer from the drawbacks identified within step (1.b) and reflects properly all preference information available in FPCMs.

Note that because of the excessive extent of the topic, only the first three steps of the PC methods listed in Sect. 1.1.1 and their fuzzy extension will be addressed in this book, i.e., 1. construction of PCMs, 2. consistency verification, and 3. computation of priorities. A detailed study of the fourth step, i.e., aggregation of the priorities of alternatives and criteria into final priorities of alternatives representing final multi-criteria evaluations of the alternatives, and of the extension of all four steps to multiple DMs is left for future research. Nevertheless, some work on the fuzzy extension of aggregation methods by applying constrained fuzzy arithmetic has already been done; in particular, the fuzzy extension of the weighted arithmetic mean method based on constrained fuzzy arithmetic has been introduced by Krejčí et al. (2017).

Beside the inability to capture uncertainty, the "traditional" PC methods are also not able to cope with the situations where it is not possible or reasonable to obtain complete preference information from DMs, for example due to time or cost limitations. This problem occurs especially in the situations where large-dimensional PCMs are involved. That is why various PC methods for dealing with incomplete large-dimensional PCMs have been proposed in the literature.

When dealing with incomplete large-dimensional PCMs, compromise between reducing the number of PCs required from the DM as much as possible and obtaining reasonable priorities of objects is of paramount importance. Thus, the second research question of this book is:

(2) *How can the amount of preference information required from the DM in a large-dimensional PCM be reduced while still obtaining comparable priorities of objects?*

In order to properly answer this research question, it is necessary to first clarify what is meant by the "comparable" priorities. In the context of incomplete large-dimensional PCMs in this book, by comparable priorities of objects will be meant such priorities of objects that are close enough to the actual priorities that would be obtained from the hypothetical complete PCM, i.e., such priorities that approximate the actual priorities well enough.

The following two steps are pursued in the book in order to properly answer the research question (2):

(2.a) an efficient method for partially filling an incomplete large-dimensional PCM that minimizes the number of PCs required from the DM but provides a sufficient amount of preference information is proposed;

(2.b) a suitable method for deriving priorities from an incomplete large-dimensional PCM that reflect the incompleteness of preference information and that are "close" to the priorities obtainable from the hypothetical complete PCM is proposed.

The basic idea for answering the research question (2) is to design an algorithm based on an optimal sequential choice of the PCs that should be provided by the DM and on the concept of weak consistency. Based on the weak consistency, the missing PCs in the matrix should be replaced by intervals providing ranges for the missing preference information, thus obtaining a large-dimensional fuzzy (more precisely interval) PCM. Afterwards, methods for deriving fuzzy priorities from FPCMs that are going to be proposed in this book within step (1.d) should be applied in order to obtain interval priorities that properly represent the incompleteness of preference information contained in the original incomplete large-dimensional PCM. Note that because of the excessive extent of the topic only the problems where the DM provides PCs in the form of crisp numbers are considered in this book. The generalization of the proposed PC method to fuzzy numbers is again left for future research.

1.4 Structure of the Book

This book is divided into five chapters. This chapter, provides an introduction to the topic of the book and states the research questions and the steps pursued in the book in order to answer the research questions.

Chapter 2 provides a critical review of well-known and in real-life MCDM problems most often applied PC methods (step 1.a). Three types of PCMs are studied in this chapter—multiplicative PCMs, additive PCMs with additive representation, and additive PCMs with multiplicative representation—and transformations between the three approaches are examined.

Chapter 3 reviews basic concepts from fuzzy set theory that play a key role in this book. Trapezoidal and triangular fuzzy numbers and intervals, that are later used for the fuzzy extension of PC methods, are defined. Standard fuzzy arithmetic and constrained fuzzy arithmetic are studied in detail, and the necessity of applying constrained fuzzy arithmetic to arithmetic operations with fuzzy numbers in the presence of constraints on operands is emphasized.

Chapter 4 is the central chapter of the book that provides the answer to the research question (1). In this chapter, the fuzzy extension of the PC methods reviewed in Chap. 2 is studied. In particular, the fuzzy extensions proposed in the literature are critically reviewed, and their drawbacks regarding violation of the reciprocity of the related PCs and of the invariance under permutation of objects are identified (step 1.b). Necessity of applying constrained fuzzy arithmetic to the fuzzy extension of the PC methods is emphasized in order to remove these drawbacks (step 1.c), and a proper fuzzy extension of the PC methods is proposed afterwards (step 1.d). In the final part of the chapter, transformations between the new PC methods based on constrained fuzzy arithmetic proposed for all three types of FPCMs are studied.

In Chap. 5, findings from Chap. 4 are utilized in order to answer the research question (2), i.e., to deal with incomplete PCMs of large dimensions. In particular, an algorithm for identifying iteratively PCs that should be provided by the DM in an incomplete large-dimensional PCM (step 2.a) and a method for obtaining interval priorities from such an incomplete large-dimensional PCM (step 2.b) are proposed.

Finally, Chap. 6 contains discussion and perspectives for future research.

Chapter 2
Pairwise Comparison Matrices

Abstract This chapter provides a critical review of well-known and in real-life multi-criteria decision making problems most often applied pairwise comparison methods. Three types of pairwise comparison matrices are studied in this chapter—multiplicative pairwise comparison matrices, additive pairwise comparison matrices with additive representation, and additive pairwise comparison matrices with multiplicative representation. The focus is put on the construction of pairwise comparison matrices, definitions of consistency, and methods for deriving priorities of objects from pairwise comparison matrices. Further, the transformations between the approaches for the three different pairwise comparison matrices are studied. The chapter pays a particular attention to two key properties of the pairwise comparison matrices and the related methods—reciprocity of the related pairwise comparisons and the invariance of the pairwise comparison methods under permutation of objects.

2.1 Introduction

In MCDM models, rating of objects (alternatives, criteria) is usually required from decision makers (DMs) in order to solve MCDM problems. However, because of the cognitive limitation and incapability to compare several objects at the same time DMs often have problems with rating the objects (assigning priorities to the objects) directly. This limitation can be easily overcome by providing PCs of objects and then deriving the desired rating (priorities) of objects. This is the main concept of the PC methods briefly introduced already in Sect. 1.1.1. PC allows the DM to consider only two objects at a time, which is significantly less demanding on cognitive capabilities of the DM than considering them all. These PCs can be conveniently structured into a pairwise comparison matrix (PCM). Afterwards, an appropriate method is applied to the PCM in order to derive priorities $w_1, \ldots, w_n$ of objects representing DM's preferences.

PCs of n objects $o_1, \ldots, o_n$ are structured into a PCM $C = \left\{ c_{ij} \right\}_{i,j=1}^{n}$ as follows:

© Springer International Publishing AG, part of Springer Nature 2018
J. Krejčí, *Pairwise Comparison Matrices and their Fuzzy Extension*, Studies in Fuzziness and Soft Computing 366, https://doi.org/10.1007/978-3-319-77715-3_2

$$
C = \begin{array}{c} \\ o_1 \\ o_2 \\ \vdots \\ o_n \end{array}
\begin{array}{c} \begin{array}{cccc} o_1 & o_2 & \ldots & o_n \end{array} \\
\begin{pmatrix} c_{11} & c_{12} & \ldots & c_{1n} \\ c_{21} & c_{22} & \ldots & c_{2n} \\ \vdots & \vdots & \ddots & \vdots \\ c_{n1} & c_{n2} & \ldots & c_{nn} \end{pmatrix} \end{array}. \tag{2.1}
$$

The PC c_{ij} in the i-th row and the j-th column of the PCM C expresses the intensity of preference of object o_i over object o_j. For easier understanding, the rows and the columns in the PCM (2.1) are labeled by the names of the compared objects. This labeling is usually omitted in the literature and, for simplicity, it will be omitted also in this book.

Notice that the PCs c_{ij} and c_{ji} express the intensities of preference on the same pair of objects o_i and o_j; the PC c_{ij} expresses the intensity of preference of object o_i over object o_j, and the PC c_{ji} expresses the intensity of preference of object o_j over object o_i. Therefore, it is obvious that these PCs are in relation. The type of relation depends on the type of representation used for expressing PCs. In this book, two types of relation between the PCs c_{ij} and c_{ji} are of interest: the multiplicative-reciprocity relation $c_{ji} = \frac{1}{c_{ij}}$ and the additive-reciprocity relation $c_{ji} = 1 - c_{ij}$. Reciprocity relation is an inherent property of a PCM that results naturally from the interpretation of the PCs in a PCM. Both types of the reciprocity relation will be studied in detail in the following sections.

There exists no canonical order in which to assign to n objects the labels $o_1, \ldots, o_n$. The objects can be labeled in $n!$ different ways. By changing labeling of objects in a PCM, the preference information contained in the PCM does not change; the original PCs are only permuted accordingly. Thus, it is desirable that the priorities of objects derived from a PCM are independent of the order in which the objects are associated with the rows and the columns of the PCM (Fichtner 1986). This means that the priorities $w_1, \ldots, w_n$ of objects should not change under any permutation of the PCM C. Fichtner (1986) introduced the invariance under permutation of objects as one of the axioms which "good" methods for deriving priorities from PCMs should meet.

Being P a permutation matrix, i.e. a square matrix with exactly one entry equal to 1 in each row and column and 0 elsewhere, $C^\pi = PCP^T$ is a permutation of C associated with P. Further, let $\mathcal{C}$ denote a certain class of PCMs. Then, invariance under permutation of methods for deriving priorities of objects from PCMs can be formally defined as follows.

Definition 1 Let a method for deriving priorities $\underline{w} = (w_1, \ldots, w_n)^T$ of objects from PCMs in a certain class $\mathcal{C}$ be described by a function $f : \mathcal{C} \to \mathbb{R}^n$, i.e. $\underline{w} = f(C), C \in \mathcal{C}$. Then the method is said to be invariant under permutation of objects if

$$
f(PCP^T) = Pf(C), \qquad \forall C \in \mathcal{C} \text{ and for any permutation matrix } P.
$$

In order to obtain reliable priorities of objects from PCMs, DMs should behave rationally when providing intensities of preference, i.e., they should be consistent in their preferences. This basically means that DMs should not enter PCs into a PCM randomly without thinking carefully about their meaning but they should fully focus on this task.

Various definitions of consistency as well as inconsistency indices have been defined in the literature in order to control the consistency of PCMs. Again, it comes natural to require invariance of both the definitions of consistency and of the inconsistency indices. Brunelli and Fedrizzi (2015) even introduced the invariance under permutation of objects as one of the axioms characterizing inconsistency indices.

Definition 2 A definition of consistency for PCMs in a certain class C is said to be invariant under permutation of objects if $\forall C \in C$ the following holds:

$$C \text{ consistent } \Rightarrow PCP^T \text{ consistent for every } P,$$

$$C \text{ not consistent } \Rightarrow PCP^T \text{ not consistent for any } P,$$

where P is a permutation matrix.

Definition 3 An inconsistency index $I : C \rightarrow \mathbb{R}$ defined on a certain class C of PCMs is said to be invariant under permutation of objects if

$$I(PCP^T) = I(C), \qquad \forall C \in C \text{ and for any permutation matrix } P.$$

As already mentioned in Sect. 1.1.1, there exist two basic types of PCMs in MCDM: multiplicative PCMs and additive PCMs. The additive PCMs are further divided into two types depending on the representation used. In the following sections, all three types of PCMs are defined, and well-known and most often applied consistency conditions, inconsistency indices, and methods for deriving priorities of objects from these PCMs are reviewed.

2.2 Multiplicative Pairwise Comparison Matrices

The first bases of the theory on multiplicative PCMs were given by Saaty (1977, 1980) who introduced a complete method for supporting MCDM based on this type of PCMs called AHP. AHP covers all main steps of PC methods, from structuring the problem into a hierarchy, over the construction of multiplicative PCMs, verifying their consistency, deriving priorities of objects on different levels of the hierarchy, up to the aggregation of the priorities on different levels of the hierarchy into the final priorities of decision alternatives. In this section, the methods for constructing multiplicative PCMs, verifying their consistency, and deriving priorities of objects from multiplicative PCMs are reviewed.

2.2.1   Construction of MPCMs

Definition 4 A multiplicative pairwise comparison matrix (MPCM) of n objects $o_1, \ldots, o_n$ is a square matrix $M = \{m_{ij}\}_{i,j=1}^n$, whose elements $m_{ij}, i, j = 1, \ldots, n$, indicate the ratio of preference intensity of object o_i to that of object o_j. In other words, element m_{ij} indicates that o_i is m_{ij}-times as good as o_j. Further, a MPCM $M = \{m_{ij}\}_{i,j=1}^n$ has to be multiplicatively reciprocal, i.e.

$$m_{ij} = \frac{1}{m_{ji}}, \qquad i, j = 1, \ldots, n. \tag{2.2}$$

Example 2 The matrix

$$M = \begin{pmatrix} 1 & 2 & 4 & 7 \\ \frac{1}{2} & 1 & 2 & 5 \\ \frac{1}{4} & \frac{1}{2} & 1 & 3 \\ \frac{1}{7} & \frac{1}{5} & \frac{1}{3} & 1 \end{pmatrix} \tag{2.3}$$

represents a MPCM of four objects o_1, o_2, o_3, and o_4; the matrix clearly satisfies condition (2.2) of multiplicative reciprocity required in Definition 4.

According to Definition 4, element $m_{13} = 4$ indicates that object o_1 is 4-times as good as object o_3 (or in other words: 4-times preferred to o_3). Thus, according to the common sense, object o_3 should be $\frac{1}{4}$-times as good as object o_1 (or in other words: 4-times less preferred to o_1), i.e. $m_{31} = \frac{1}{4}$. Therefore, multiplicative reciprocity (2.2) is a very natural property of MPCMs resulting from the interpretation of its elements.

Notice that, according to the multiplicative reciprocity, the elements on the main diagonal of the MPCM M have to be equal to 1, i.e. $m_{ii} = 1, i = 1, \ldots, 4$. This requirement is in compliance with the interpretation of the PCs in the MPCM. The element m_{ii} represents the PC of object o_i with object $o_i, i = 1, \ldots, 4$, i.e. the PC of one object with itself. It is natural that o_i is 1-times as good as itself (or in other words: equally preferred), which results in $m_{ii} = 1$. $\qquad\qquad\triangle$

As shown in Example 2, multiplicative reciprocity (2.2) is an inherent property that results naturally from the interpretation of the PCs in a MPCM. Thus, it is always sufficient to provide one of the related PCs m_{ij} and $m_{ji}, i, j \in \{1, \ldots, n\}$, in a MPCM, and the second one is then determined automatically based on the multiplicative-reciprocity property (2.2).

Various other terms besides "MPCM" are used in the literature. Most often they are called just "PCMs" (see, e.g., Saaty (1977, 2006) etc.). Sometimes they are referred to as "reciprocal PCMs" or "reciprocal preference relations". In this book, the term "MPCM" is used in order to emphasize the fact that these PCMs are multiplicatively reciprocal, and to distinguish them clearly from additive PCMs (defined later in Sect. 2.3) that are additively reciprocal.

To make PCs of objects, Saaty (1977) suggested to use integer numbers between 1 and 9 and their reciprocals. This means that one object can be up to 9-times preferred

Table 2.1 Saaty's scale

Intensity of preference	Linguistic term
1	Equal preference
3	Weak preference
5	Strong preference
7	Demonstrated preference
9	Absolute preference
2, 4, 6, 8	Intermediate values between the two adjacent judgments connected by word "between"

or more important over another one. The choice of this 9-point scale was done based on the psychological experiments showing that humans are not able to compare simultaneously more than 5 to 9 objects (Miller 1956).

Each integer from the scale is also assigned a linguistic term expressing the intensity of preference (importance) of one object over another one. For example, number 1 represents equal preference and number 9 absolute preference. Thus, DMs can express their intensities of preference either numerically using the integers or linguistically using the assigned linguistic terms. Usually, Saaty's scale of integers 1, 3, 5, 7, and 9 given in Table 2.1 is used for PCs. When there is a need for more detailed PCs, a more detailed scale with intermediate values 2, 4, 6, and 8 is utilized. The intermediate values are expressed using the neighboring terms and connecting them with the word "between". For example, 2 is interpreted as "between equal and weak preference".

However, even though "being 9-times preferred" is the highest intensity of preference available in Saaty's scale to make PCs, it is not a natural maximum of intensity of preference. In fact, there is no natural maximum of number of times an object can be preferred to another one. Thus, it is quite difficult for DMs to express their intensities of preference in terms of multiples when they do not have a natural maximum value of intensity of preference available. On the other hand, number 9 in Saaty's scale is assigned the linguistic term "absolutely preferred". This may cause further confusion to DMs since "absolute preference" is naturally interpreted as the maximum possible intensity of preference. Similarly, number 3 standing for "3-times preferred" is assigned a linguistic term "weak preference". Not even this linguistic term corresponds to human intuition. Most of the DMs would probably assign "weak preference" to a number much closer to 1, which stands for equal preference, and would use the number 3 to model much stronger intensity of preference. Thus, it seems that the linguistic terms in Saaty's scale do not correspond very well to the respective numerical values that are distributed uniformly in the interval [1, 9]. Some further problems related to Saaty's scale will be mentioned in the following section. A detailed and interesting discussion on modeling linguistic terms from Saaty's scale is given by Stoklasa (2014).

Usually, either the term "intensity of preference" or the term "intensity of importance" is used in relation to PCs of objects. The choice of the particular term often
depends on the context. When comparing alternatives with respect to a given criterion, we usually use the term "preference". When comparing criteria with respect to
the goal of the decision-making problem, the term "importance" is often preferred.
For the simplicity, only one of the terms will always be used hereafter.

Saaty's discrete scale $\left\{\frac{1}{9}, \frac{1}{8}, \ldots, 1, 2, \ldots, 8, 9\right\}$ is without doubts the most commonly used scale in practice. However, in general, any scale S of real numbers
containing the neutral element 1 and such that for any element x in S also the element $\frac{1}{x}$ belongs to S can be used for the construction of MPCMs. Particular cases of
such scale are the intervals $\left[\frac{1}{\sigma}, \sigma\right]$, $\sigma > 1$, and $]0, \infty[$ (Gavalec et al. 2015).

2.2.2 Consistency of MPCMs

Multiplicative reciprocity (2.2) is an inherent property of MPCMs resulting from the
interpretation of the PCs in the matrices. However, requiring the multiplicative reciprocity of the related PCs is not sufficient to guarantee that the preference information
contained in a MPCM is reasonable and that the priorities of objects derived from
such a matrix are reliable, i.e. that they represent preferences of the DMs properly.

In order to guarantee reliability of the priorities obtainable from a MPCM, DMs
should be consistent in their preferences when entering PCs into the MPCM. Various
consistency conditions have been defined for MPCMs. A well-known and most often
applied consistency condition is the traditional multiplicative-consistency condition.

2.2.2.1 Multiplicative Consistency

Definition 5 (Saaty 1980) A MPCM $M = \{m_{ij}\}_{i,j=1}^{n}$ is said to be *multiplicatively
consistent* if it satisfies the multiplicative-transitivity property

$$m_{ij} = m_{ik} m_{kj}, \qquad i, j, k = 1, \ldots, n. \tag{2.4}$$

Definition 5 of multiplicative consistency is clearly invariant under permutation
of objects compared in the MPCM M.

Example 3 The MPCM

$$M = \begin{pmatrix} 1 & 2 & 4 & 8 \\ \frac{1}{2} & 1 & 2 & 4 \\ \frac{1}{4} & \frac{1}{2} & 1 & 2 \\ \frac{1}{8} & \frac{1}{4} & \frac{1}{2} & 1 \end{pmatrix} \tag{2.5}$$

is multiplicatively consistent according to Definition 5 since it satisfies the multiplicative-transitivity property (2.4). By permuting the MPCM M using the permutation matrix

$$P = \begin{pmatrix} 1 & 0 & 0 & 0 \\ 0 & 0 & 0 & 1 \\ 0 & 1 & 0 & 0 \\ 0 & 0 & 1 & 0 \end{pmatrix}, \tag{2.6}$$

for example, we obtain the MPCM $M^\pi = PMP^T$ in the form

$$M^\pi = \begin{pmatrix} 1 & 8 & 2 & 4 \\ \frac{1}{8} & 1 & \frac{1}{4} & \frac{1}{2} \\ \frac{1}{2} & 4 & 1 & 2 \\ \frac{1}{4} & 2 & \frac{1}{2} & 1 \end{pmatrix}, \tag{2.7}$$

which is again multiplicatively consistent according to Definition 5. $\triangle$

The following theorem provides us with alternative ways to verify multiplicative consistency of MPCMs.

Theorem 1 *For a MPCM $M = \{m_{ij}\}_{i,j=1}^n$, the following statements are equivalent:*

(i) $m_{ij} = m_{ik}m_{kj}, \quad i,j,k = 1,\ldots,n,$
(ii) $m_{ij}m_{jk}m_{ki} = 1, \quad i,j,k = 1,\ldots,n,$
(iii) $m_{ij}m_{jk}m_{ki} = m_{ik}m_{kj}m_{ji}, \quad i,j,k = 1,\ldots,n.$

Further, Saaty (1994) derived the following characterization of multiplicatively consistent MPCMs.

Proposition 1 *A MPCM $M = \{m_{ij}\}_{i,j=1}^n$ is multiplicatively consistent if and only if there exists a positive vector $\underline{w} = (w_1,\ldots,w_n)^T$ such that*

$$m_{ij} = \frac{w_i}{w_j}, \quad i,j = 1,\ldots,n. \tag{2.8}$$

The notation $\underline{w} = (w_1,\ldots,w_n)^T$ will be used hereafter to represent exclusively a priority vector associated with a MPCM.

According to Proposition 1, when a MPCM $M = \{m_{ij}\}_{i,j=1}^n$ of n objects is multiplicatively consistent, there exist priorities $w_1,\ldots,w_n$ of objects using which we can determine precisely the original PCs $m_{ij}, i,j = 1,\ldots,n$, in the MPCM M by applying the characterization (2.8). In fact, each column of a multiplicatively consistent MPCM M is a priority vector satisfying these properties. On the other hand, when the priorities $w_1,\ldots,w_n$ of objects are available, we can construct a multiplicatively consistent MPCM of objects by applying (2.8). Notice that the multiplicative reciprocity of M is always guaranteed by applying (2.8) since $m_{ji} = \frac{w_j}{w_i} = \frac{1}{\frac{w_i}{w_j}} = \frac{1}{m_{ij}}$.

Example 4 Since the MPCM (2.5) in Example 3 is multiplicatively consistent, there exists a positive priority vector $\underline{w} = (w_1, w_2, w_3, w_4)^T$ satisfying the characterization property (2.8). It is, for example, the priority vector $\underline{w} = (1, \frac{1}{2}, \frac{1}{4}, \frac{1}{8})^T$. $\triangle$

Multiplicative-consistency condition (2.4) seems to be reasonable when using numerical values for expressing the intensities of preference. When, for example, object o_i is 3-times preferred to object o_k ($m_{ik} = 3$), and object o_k is 3-times preferred to object o_j ($m_{kj} = 3$), then it is quite natural to require object o_i to be 9-times preferred to object o_j ($m_{ij} = 3 \cdot 3 = 9$). But what happens when $m_{ik} = 3$ and $m_{kj} = 5$? In order to be consistent we should write $m_{ij} = 3 \cdot 5 = 15$. However, this numerical value is out of Saaty's scale that is usually used for making PCs. Note that this problem would not occur with the scale $S =]0, \infty[$, where for any $m_{ik} \in S$ and $m_{kj} \in S$ also $m_{ij} = m_{ik} m_{kj} \in S$.

Clearly, it is very difficult or even impossible (especially for MPCMs of large dimensions) to keep multiplicative consistency (2.4) when using only the values from limited Saaty's scale and their reciprocals. Thus, Saaty (1980) defined Consistency Index *CI* to measure inconsistency of MPCMs. *CI* is given for a MPCM $M = \{m_{ij}\}_{i,j=1}^n$ as

$$CI = \frac{\lambda - n}{n - 1}, \tag{2.9}$$

where n is the number of objects compared in the matrix, and λ is the maximal eigenvalue of the matrix. *CI* defined by (2.9) is invariant under permutation of objects compared in the MPCM M (Brunelli and Fedrizzi 2015). Further, Saaty (1994) showed that when the DM is absolutely consistent in his or her judgments, i.e. when the MPCM is multiplicatively consistent according to (2.4), then $\lambda = n$, and thus $CI = 0$.

Since reaching absolute multiplicative consistency is very difficult, especially with limited Saaty's scale, some degree of inconsistency is allowed. *CI* is required to be close to 0. However, determining the valued of *CI* below which MPCMs are regarded as acceptably inconsistent and above which they are regarded as inconsistent is complicated. Thus, Saaty (1980) defined Consistency Ratio *CR* in the form

$$CR = \frac{CI}{RI}, \tag{2.10}$$

where *RI* is Random Index that is computed as an average value of *CI* of MPCMs randomly generated from the elements of Saaty's scale separately for each order n of MPCMs. Table 2.2 shows the values of *RI* (rounded to two decimal places) for MPCMs up to dimension $n = 10$ as given by Alonso and Lamata (2006).

According to Saaty (1980), a MPCM is regarded as acceptably inconsistent if $CR \leq 0.1$. A MPCM such that $CR > 0.1$ is then considered as inconsistent, and the PCs in such a matrix should be reconsidered. Nevertheless, even the relaxed requirement of $CR \leq 0.1$ is quite difficult to reach, especially for MPCMs of large dimensions.

Table 2.2 Random index RI

n	1	2	3	4	5	6	7	8	9	10
RI	0	0	0.52	0.89	1.11	1.25	1.35	1.41	1.45	1.49

Besides CR, many other inconsistency indices have been proposed in the literature for verifying an acceptable level of inconsistency of MPCMs. Some of them are reviewed, e.g., by Brunelli and Fedrizzi (2015) and by Gavalec et al. (2015).

The interpretation of the multiplicative consistency condition (2.4) does not seem very intuitive when linguistic terms from Saaty's scale are used for expressing the intensities of preference. When, for example, object o_i is weakly preferred to object o_k ($m_{ik} = 3$), and object o_k is weakly preferred to object o_j ($m_{kj} = 3$), then, according to the multiplicative consistency (2.4), it should follow that $m_{ij} = 3 \cdot 3 = 9$, which means that object o_i should be absolutely preferred to object o_j. However, absolute preference of o_i over o_j seems to be too strong in comparison to weak preferences of o_i over o_k and of o_k over o_j. Using a much smaller preference than the absolute preference would probably be more intuitive in this case. Thus, other consistency conditions respecting better the linguistic interpretation of preference intensities have been proposed in the literature. One of them is the weak-consistency condition introduced by Jandová and Talašová (2013) and by Stoklasa et al. (2013).

2.2.2.2 Weak Consistency

Jandová and Talašová (2013) and Stoklasa et al. (2013) proposed the weak-consistency condition as an intuitive minimum consistency requirement for MPCMs $M = \{m_{ij}\}_{i,j=1}^n$. The idea is to require the preference intensity m_{ij} to be at least the maximal value of the preference intensities m_{ik} and m_{kj} for each triplet of objects o_i, o_j, o_k, $i, j, k \in \{1, \ldots, n\}$.

Definition 6 (Jandová and Talašová 2013) A MPCM $M = \{m_{ij}\}_{i,j=1}^n$ is said to be weakly consistent if

$$m_{ik} > 1 \ \wedge \ m_{kj} > 1 \ \Rightarrow \ m_{ij} \geq \max\{m_{ik}, m_{kj}\},$$

$$m_{ik} = 1 \ \wedge \ m_{kj} \geq 1 \ \Rightarrow \ m_{ij} = \max\{m_{ik}, m_{kj}\}, \tag{2.11}$$

$$m_{ik} \geq 1 \ \wedge \ m_{kj} = 1 \ \Rightarrow \ m_{ij} = \max\{m_{ik}, m_{kj}\},$$

holds for $i, j, k = 1, \ldots, n$.

It is obvious that Definition 6 of weak consistency is again invariant under permutation of objects compared in the MPCM M.

The weak-consistency condition is suitable especially when linguistic terms are used for expressing the intensities of preference instead of real numbers. For example, when object o_i is weakly preferred to object o_k ($m_{ik} = 3$), and object o_k is strongly preferred to object o_j ($m_{kj} = 5$), then object o_i has to be at least strongly preferred to object o_j ($m_{ij} \geq \max\{3, 5\} = 5$). Thus, the requirement of weak consistency seems to be very natural, and it provides DMs with some space for expressing their intensities of preference—a desired tolerance. Moreover, it is very easy to control while entering PCs into the MPCM.

The weak-consistency condition was also defined in more relaxed forms by Krejčí (2017e) and by Basile and D'Apuzzo (2002). Recently, Cavallo and D'Apuzzo (2016) proposed a definition of weak consistency very similar to the weak consistency in Definition 6. In this book, the weak-consistency condition given by Definition 6 is adopted and later applied in Chap. 5 in a novel method for dealing with large-dimensional PCMs.

Jandová and Talašová (2013) derived some rules equivalent to the weak-consistency condition (2.11). They are formulated in the following theorem.

Theorem 2 (Jandová and Talašová 2013) *For a MPCM $M = \{m_{ij}\}_{i,j=1}^{n}$, the following statements are equivalent:*

(i) M is weakly consistent according to Definition 6.
(ii) For every $i, j, k = 1, \ldots, n$:

$$
\begin{aligned}
m_{ik} < 1 \wedge m_{kj} < 1 &\Rightarrow m_{ij} \leq \min\{m_{ik}, m_{kj}\}, \\
m_{ik} = 1 \wedge m_{kj} \leq 1 &\Rightarrow m_{ij} = \min\{m_{ik}, m_{kj}\}, \\
m_{ik} \leq 1 \wedge m_{kj} = 1 &\Rightarrow m_{ij} = \min\{m_{ik}, m_{kj}\}.
\end{aligned}
\tag{2.12}
$$

(iii) For every $i, j, k = 1, \ldots, n$:

$$
\begin{aligned}
1 < \tfrac{1}{m_{kj}} < m_{ik} &\Rightarrow \tfrac{1}{m_{ik}} \leq m_{ij} \leq m_{ik}, \\
1 < m_{ik} < \tfrac{1}{m_{kj}} &\Rightarrow m_{kj} \leq m_{ij} < 1, \\
1 < \tfrac{1}{m_{kj}} < m_{ik} &\Rightarrow 1 < m_{ij} \leq m_{ik}.
\end{aligned}
\tag{2.13}
$$

(iv) For every $i, j, k = 1, \ldots, n$:

$$
\begin{aligned}
\tfrac{1}{m_{kj}} < m_{ik} < 1 &\Rightarrow m_{kj} \leq m_{ij}, \\
1 < m_{kj} < \tfrac{1}{m_{ik}} &\Rightarrow m_{ik} \leq m_{ij} < 1, \\
1 < m_{kj} = \tfrac{1}{m_{ik}} &\Rightarrow \tfrac{1}{m_{kj}} \leq m_{ij} \leq m_{kj}.
\end{aligned}
\tag{2.14}
$$

Further, Jandová and Talašová (2013) derived some interesting properties of weakly consistent MPCMs. For example, every weakly consistent MPCM can be permuted in such a way that the objects compared in the MPCM are ordered from the most preferred to the least preferred. In such an ordered weakly consistent MPCM, all elements above the main diagonal are greater or equal to 1, i.e.

$m_{ij} \geq 1$, $i, j = 1, \ldots, n$, $i < j$. Moreover, the sequences of elements in the rows are non-decreasing and the sequences of elements in the columns are non-increasing.

Example 5 The MPCM

$$M = \begin{pmatrix} 1 & 2 & 5 & 9 \\ \frac{1}{2} & 1 & 4 & 7 \\ \frac{1}{5} & \frac{1}{4} & 1 & 6 \\ \frac{1}{9} & \frac{1}{7} & \frac{1}{6} & 1 \end{pmatrix} \tag{2.15}$$

is obviously weakly consistent according to Definition 6 since the sequences of the PCs in the rows are non-decreasing and the sequences of the PCs in the columns are non-increasing. △

Jandová and Talašová (2013) showed that every MPCM that is multiplicatively consistent according to Definition 5 is also weakly consistent according to Definition 6. In fact, from $m_{ik} > 1 \wedge m_{kj} > 1$ it follows that $m_{ij} = m_{ik}m_{kj} \geq \max\{m_{ik}, m_{kj}\}$. For $m_{ik} = 1 \wedge m_{kj} \geq 1$, and for $m_{ik} \geq 1 \wedge m_{kj} = 1$, we get $m_{ij} = m_{ik}m_{kj} = \max\{m_{ik}, m_{kj}\}$.

The properties of weakly consistent MPCMs reviewed in this section are utilized in the novel method for dealing with large-dimensional PCMs that is described in detail in Chap. 5.

2.2.3 Deriving Priorities from MPCMs

Based on the given MPCM $M = \{m_{ij}\}_{i,j=1}^{n}$ of n objects $o_1, \ldots, o_n$, it is endeavoured to derive the priority vector $\underline{w} = (w_1, \ldots, w_n)^T$ that would best represent the relative preference (or importance) of the objects with respect to the other objects in the set.

We know from the previous section that when a MPCM $M = \{m_{ij}\}_{i,j=1}^{n}$ is multiplicatively consistent according to (2.4), then there exists a positive priority vector $\underline{w} = (w_1, \ldots, w_n)^T$ such that $m_{ij} = \frac{w_i}{w_j}$, $i, j = 1, \ldots, n$. For example, any column of the multiplicatively consistent MPCM can by used to represent the priority vector $\underline{w} = (w_1, \ldots, w_n)^T$. When the MPCM $M = \{m_{ij}\}_{i,j=1}^{n}$ is not multiplicatively consistent, then the ratios of the priorities only estimate the PCs in the matrix, i.e.

$$m_{ij} \approx \frac{w_i}{w_j}, \quad i, j = 1, \ldots, n. \tag{2.16}$$

Priorities $w_1, \ldots, w_n$ are given on a ratio scale. Because the ratios $\frac{w_i}{w_j}$, $i, j = 1, \ldots, n$, play a key role, these ratios cannot change under any normalization of the priorities. Thus, the normalization can be done only by multiplying the priorities by a constant $c > 0$ ($\frac{c \cdot w_i}{c \cdot w_j} = \frac{w_i}{w_j}$). This means that when there exists a positive vector $\underline{w} = (w_1, \ldots, w_n)^T$ representing the priorities of objects $o_1, \ldots, o_n$, then also any vector obtained from $\underline{w}$ by the transformation

$$w_i \rightarrow c \cdot w_i, \quad i = 1, \ldots, n, \tag{2.17}$$

where $c > 0$, represents the priorities of the objects. In the literature, the normalization condition

$$\sum_{i=1}^{n} w_i = 1, \quad w_i \in [0, 1], \quad i = 1, \ldots, n, \tag{2.18}$$

is usually applied in order to reach the uniqueness. This normalization condition is applied also in this book, and for simplicity and when no confusion arises, the normalized priorities are referred to only as priorities.

Many methods have been proposed in the literature for deriving priorities of objects from MPCMs. In this book, two well-known methods-the eigenvector method and the geometric-mean method-are of interest, and only the extension of these two methods to fuzzy MPCMs will be dealt with here. For a review of other methods, see, e.g., Gavalec et al. (2015).

2.2.3.1 Eigenvector Method

The eigenvector method (EVM), originally proposed by Saaty (1977) for deriving priorities in AHP, is one of the oldest methods for deriving priorities of objects from MPCMs. According to this method, the priorities of objects are calculated as the components $w_1, \ldots, w_n$ of the normalized maximal eigenvector $\underline{w}$ corresponding to the maximal eigenvalue λ_{MAX} of the MPCM M.

Let A be a square matrix of size n. The scalar λ and the vector $\underline{w}$ of size n satisfying $A\underline{w} = \lambda\underline{w}$ are called the eigenvalue and the eigenvector of the matrix A, respectively. The set of all eigenvalues of matrix A is obtained as the solution to the equation $|A - \lambda I| = 0$ where I denotes the identity matrix of size n and $|.|$ denotes the determinant of a given matrix.

From Perron-Frobenius theorem it follows that for a positive matrix $A = \{a_{ij}\}_{i,j=1}^{n}$, i.e. $a_{ij} > 0$ for $i, j = 1, \ldots, n$, there exists a positive eigenvalue λ_{MAX} such that $|\lambda| < \lambda_{MAX}$ for any other eigenvalue λ of A. Such eigenvalue

$$\lambda_{MAX} = \max\{\lambda; |A - \lambda I| = 0\} \tag{2.19}$$

is called *the maximal eigenvalue* of A. Further, there exists a positive eigenvector

$$\underline{w}_{MAX} = (w_1, \ldots, w_n)^T: \quad A\underline{w}_{MAX} = \lambda_{MAX}\underline{w}_{MAX} \tag{2.20}$$

corresponding to λ_{MAX} called *the maximal eigenvector* of A. The maximal eigenvector $\underline{w}_{MAX}$ is unique up to a multiplicative constant. That is why normalization (2.18) is usually applied to obtain a unique solution—the normalized maximal eigenvector

$$\underline{w}_{MAX} = (w_1, \ldots, w_n)^T : \quad A\underline{w}_{MAX} = \lambda_{MAX}\underline{w}_{MAX}, \quad \sum_{i=1}^{n} w_i = 1. \qquad (2.21)$$

A MPCM $M = \{m_{ij}\}_{i,j=1}^{n}$ is a positive square matrix. Therefore, there always exists a positive maximal eigenvalue λ_{MAX} and a normalized positive maximal eigenvector $\underline{w}_{MAX}$. The components of this eigenvector represent the priorities of objects compared in the MPCM M. When a MPCM $M = \{m_{ij}\}_{i,j=1}^{n}$ is multiplicatively consistent according to (2.4), then $\lambda_{MAX} = n$, and $\underline{w}_{MAX}$ satisfies (2.8). Further, it is a well-known fact that the EVM is invariant under permutation of objects in the MPCM M (see, e.g., Fichtner (1986)).

Only the maximal eigenvalues and the corresponding maximal eigenvectors are considered in this book. Thus, for the sake of simplicity, the lower index $_{MAX}$ is omitted, and only the notation λ and $\underline{w}$ is used hereafter. Further, for later use in optimization formulas for deriving fuzzy maximal eigenvalues and normalized fuzzy maximal eigenvectors of FMPCMs in Sects. 4.2.2.4 and 4.2.3.1, it is particularly useful to denote the maximal eigenvalue λ of matrix M as $EVM_\lambda(M)$ and the normalized maximal eigenvector $\underline{w}$ of matrix M corresponding to λ as $EVM_{\underline{w}}(M)$.

Further, the following property, that is later used in Sect. 4.2.2.4, results from the Perron-Frobenius Theorem. Let $A = \{a_{ij}\}_{i,j=1}^{n}$ and $B = \{b_{ij}\}_{i,j=1}^{n}$ be two positive matrices, and let $a_{ij} \geq b_{ij}$ for $i,j = 1, \ldots, n$ and $a_{kl} > b_{kl}$ for $k, l \in \{1, \ldots, n\}$. Then, the maximal eigenvalue of A is greater than the maximal eigenvalue of B, i.e. $EVM_\lambda(A) > EVM_\lambda(B)$.

2.2.3.2 Geometric-Mean Method

The geometric mean method (GMM) is another well-known method for deriving priorities of objects from MPCMs. This method gives the same results as the logarithmic least squares method (LLSM).

LLSM utilizes the characterization property (2.16) of a MPCM $M = \{m_{ij}\}_{i,j=1}^{n}$, and it is based on the minimization of the sum of squared errors of logarithms:

$$\min \sum_{i=1}^{n-1} \sum_{j=i+1}^{n} \left(\ln m_{ij} - \ln \frac{w_i}{w_j} \right)^2 \rightarrow w_i, \quad i = 1, \ldots, n. \qquad (2.22)$$

It was shown that the optimal solution of (2.22) is always unique (up to a multiplicative constant) and can be determined by the geometric mean of the elements in the rows of the MPCM M, i.e.

$$w_i = \sqrt[n]{\prod_{j=1}^{n} m_{ij}}, \quad i = 1, \ldots, n. \qquad (2.23)$$

By employing the normalization condition (2.18), the normalized priorities can be computed directly as

$$
w_i = \frac{\sqrt[n]{\prod_{j=1}^{n} m_{ij}}}{\sum_{k=1}^{n} \sqrt[n]{\prod_{j=1}^{n} m_{kj}}}, \qquad i = 1, \ldots, n. \tag{2.24}
$$

It can be easily verified that the GMM given by formula (2.24) is invariant under permutation of objects in the MPCM M.

Furthermore, Saaty and Vargas (1984) showed that when a MPCM is multiplicatively consistent according to (2.4), then EVM and GMM (or equivalently LLSM) lead to the same results. This means that the priority vector obtained from a multiplicatively consistent MPCM by GMM satisfies (2.8) too. When a MPCM is close to multiplicative consistency, then the methods provide very similar but not identical results (Crawford and Williams 1985).

There have been many studies comparing EVM and GMM. Saaty and Vargas (1984) compared EVM, GMM, and the least squares method. They concluded that EVM is the only method that guarantees rank preservation (i.e. $m_{ik} \geq m_{jk}$ for all $k = 1, \ldots, n$ implies $w_i \geq w_j$) under inconsistency. Further, Saaty and Hu (1998) showed an illustrative example where the ranking of alternatives obtained by GMM differs from the ranking obtained by EVM. Based on this example, Saaty and Hu concluded that EVM is the only valid method for deriving priorities from MPCMs, in particular from inconsistent MPCMs. However, it is not acceptable to derive such a strong conclusion based on one illustrative example. Furthermore, showing that GMM leads to a ranking different from the one obtained by EVM does not surely demonstrate that GMM leads to rank reversal; this conclusion is based on the unfounded assumption that EVM provides the correct solution.

Crawford and Williams (1985) ran some simulations to compare the performance of EVM and GMM under different error distributions and metrics. The simulations suggest better performance of GMM in priorities estimation as well as in rank preservation. Other studies favoring GMM over EVM have been done, e.g., by Barzilai (1997), Blaquero et al. (2006), Dijkstra (2013).

2.3 Additive Pairwise Comparison Matrices

The first bases of the theory of additive PCMs were given by Orlovski (1978), Nurmi (1981), Tanino (1984), Kacprzyk (1986). In this section, the methods for constructing additive PCMs are reviewed and two types of additive PCMs are defined—additive PCMs with additive representation and additive PCMs with multiplicative representation. Afterwards, definitions of consistency and methods for verifying consistency and deriving priorities of objects from both types of additive PCMs are reviewed.

2.3.1 Construction of APCMs

Definition 7 An additive pairwise comparison matrix (APCM) of n objects $o_1, \ldots,$ o_n is a square matrix $A = \{a_{ij}\}_{i,j=1}^{n}$ whose elements $a_{ij}, i, j = 1, \ldots, n$, are defined on interval $[0, 1]$. Further, the matrix is additively reciprocal, i.e.

$$a_{ij} = 1 - a_{ji}, \quad i, j = 1, \ldots, n. \tag{2.25}$$

Unlike in the case of MPCMs, there exists no widely accepted discrete scale with assigned linguistic terms for expressing the intensities of preference for APCMs. Often, the whole interval $[0, 1]$ is used, i.e.

$$
\begin{aligned}
a_{ij} &= 1 && \text{if } o_i \text{ is absolutely preferred to } o_j, \\
a_{ij} &\in]0.5, 1[&& \text{if } o_i \text{ is preferred to } o_j, \\
a_{ij} &= 0.5 && \text{if } o_i \text{ and } o_j \text{ are equally preferred (indifferent),} \\
a_{ij} &\in]0, 0.5[&& \text{if } o_i \text{ is less preferred to } o_j, \\
a_{ij} &= 0 && \text{if } o_i \text{ is absolutely less preferred to } o_j.
\end{aligned}
\tag{2.26}
$$

Example 6 Matrix

$$A = \begin{pmatrix} 0.5 & 0.7 & 1 \\ 0.3 & 0.5 & 0.9 \\ 0 & 0.1 & 0.5 \end{pmatrix} \tag{2.27}$$

represents an APCM of three objects o_1, o_2, and o_3 as this matrix clearly satisfies condition (2.25) of additive reciprocity required in Definition 7.

According to Definition 7, element $a_{13} = 1$ indicates that object o_1 is absolutely preferred to object o_3. Thus, according to the common sense, object o_3 should be absolutely less preferred to object o_1, i.e. $a_{31} = 0$. Therefore, additive reciprocity (2.25) is a very natural property of APCMs resulting from the interpretation of its elements.

Notice that according to the additive reciprocity, the elements on the main diagonal of the APCM A have to be equal to 0.5, i.e. $a_{ii} = 0.5, i = 1, 2, 3$. This requirement is in compliance with the interpretation of the PCs in the APCM. The element a_{ii} represents the PC of object o_i with itself. It is natural that o_i is equally preferred to itself, which means $a_{ii} = 0.5, i = 1, 2, 3$. $\triangle$

As shown in Example 6, additive reciprocity (2.25) is an inherent property that results naturally from the interpretation of the PCs in an APCM. Thus, it is always sufficient to provide one of the related PCs a_{ij} and $a_{ji}, i, j \in \{1, \ldots, n\}$, in an APCM and the second one is then determined automatically based on the additive-reciprocity property (2.25).

Various other terms are used in the literature besides "APCM". Most often they are called "fuzzy preference relations" (see, e.g., Bezdek et al. (1978), Nurmi (1981), Tanino (1984), Kacprzyk (1986), Cabrerizo et al. (2014), Gavalec et al. (2015)), sometimes reciprocal relations (see, e.g., Baets et al. (2006), Fedrizzi and Brunelli (2009), Fedrizzi and Brunelli (2010)), additively reciprocal relations (see, e.g., Fedrizzi and Giove (2013)), or reciprocal preference relations (see, e.g., Chiclana et al. (2009)). In this book, the term "APCM" is used in order to emphasize the fact that these PCMs are additively reciprocal, and to make an analogy to MPCMs that are multiplicatively reciprocal.

The use of the term "fuzzy preference relation" in this book might even be confusing or misleading. Historically, fuzzy preference relations were defined as a fuzzy extension of binary preference relations (Bezdek et al. (1978), Nurmi (1981), Tanino (1984), Kacprzyk (1986)). Binary preference relation b on a finite set of objects $O = \{o_1, \ldots, o_n\}$ is defined as $b : O \times O \to \{0, 1\}$ where $b(o_i, o_j) = 1$ if o_i is preferred to o_j, and $b(o_i, o_j) = 0$ if o_j is preferred to o_i. Fuzzy preference relation $\tilde{b}$ on a finite set of objects $O = \{o_1, \ldots, o_n\}$ is then defined as a fuzzy set on the cartesian product $O \times O$ characterized by the membership function $\mu_{\tilde{b}} : O \times O \to [0, 1]$ where $\mu_{\tilde{b}}(o_i, o_j) = \alpha$ represents the degree of preference of object o_i to object o_j. (Fuzzy sets will be defined properly in Chap. 3.)

In this book, the fuzzy extension of degrees of preference (intensities of preference) is of interest. Therefore, because the intensities of preference both in MPCMs and APCMs will be later in this book modeled by fuzzy numbers instead of crisp numbers, the world "fuzzy" is reserved for describing these extended versions of PCMs, i.e., they will be called fuzzy MPCMs and fuzzy APCMs. If the term "fuzzy preference relation" was used instead of "APCM", we would have to deal with the fuzzy extension of fuzzy preference relations, i.e. "fuzzy fuzzy preference relations", which would create confusion.

Similarly as for MPCMs, the preference information contained in an APCM should be reasonable in order to ensure that the priorities of objects derived from such a matrix represent properly DMs' preferences. In order to verify whether the DMs are consistent in their preferences, it is necessary to define an appropriate consistency condition for APCMs.

Two traditional and well-known consistency conditions for APCMs are the additive-consistency condition and the multiplicative-consistency condition introduced by Tanino (1984). Often, APCMs are considered as one set of matrices for which it is possible to verify both additive and multiplicative consistency. However, this approach is not correct. Each type of consistency is strictly related to a particular interpretation of the PCs in the APCM A. In fact, before constructing an APCM A, it is necessary to choose between additive and multiplicative representation (associated with additive and multiplicative consistency, respectively), which has an impact on the values of the entries in the APCM A. This means that comparing a set of n objects pairwisely in an APCM by using additive representation, and comparing the same set of objects pairwisely in an APCM by using multiplicative representation, we obtain two APCMs with different entries. Depending on the representation used, we can then verify the associated additive or multiplicative consistency. Furthermore, it is necessary to distinguish between the two representations also when deriving priorities of objects from APCMs.

The necessity of distinguishing between APCMs with additive and multiplicative representation will be much clearer as soon as the definitions of both are provided. In Sect. 2.3.2, APCMs with additive representation will be introduced and, in Sect. 2.3.3, APCMs with multiplicative representation will be dealt with.

2.3.2 Additive Pairwise Comparison Matrices with Additive Representation

Definition 8 An APCM with additive representation (APCM-A) is an APCM $R = \{r_{ij}\}_{i,j=1}^{n}$ satisfying the additive-reciprocity property (2.25) where $r_{ij} - r_{ji}$ indicates the difference of preference intensity of object o_i and of object o_j.

2.3.2.1 Additive Consistency of APCMs-A

Definition 9 (Tanino 1984) An APCM-A $R = \{r_{ij}\}_{i,j=1}^{n}$ is said to be additively consistent if it satisfies the additive-transitivity property

$$r_{ij} = r_{ik} + r_{kj} - 0.5, \qquad i,j,k = 1,\ldots,n. \tag{2.28}$$

Definition 9 of additive consistency is invariant under permutation of objects in the APCM-A R.

Example 7 The APCM-A

$$R = \begin{pmatrix} 0.5\ 0.6\ 0.8\ 0.9 \\ 0.4\ 0.5\ 0.7\ 0.8 \\ 0.2\ 0.3\ 0.5\ 0.6 \\ 0.1\ 0.2\ 0.4\ 0.5 \end{pmatrix} \tag{2.29}$$

is additively consistent according to Definition 8 since it satisfies the additive-transitivity property (2.28). Also the permuted matrix $R^{\pi} = PRP^{T}$ obtained by applying the permutation matrix (2.6) is additively consistent. $\triangle$

The following theorem provides us with alternative ways to verify additive consistency of APCMs-A.

Theorem 3 *For an APCM-A $R = \{r_{ij}\}_{i,j=1}^{n}$, the following statements are equivalent:*

(i) *R is additively consistent according to the additive-transitivity property* (2.28),
(ii)

$$r_{ij} + r_{jk} + r_{ki} = r_{ik} + r_{kj} + r_{ji}, \qquad i,j,k = 1,\ldots,n, \qquad (2.30)$$

(iii)

$$r_{ij} + r_{jk} + r_{ki} = \frac{3}{2}, \qquad i,j,k = 1,\ldots,n. \qquad (2.31)$$

Note 1 According to Theorem 3, the additive consistency of APCMs-A can be defined by using any of the expressions (2.30) and (2.31). These expressions will be referred to later when an extension of the definition of additive consistency to fuzzy APCMs-A is dealt with.

Further, Tanino (1984) derived the following characterization of additively consistent APCMs-A.

Proposition 2 (Tanino's characterization) *An APCM-A* $R = \left\{r_{ij}\right\}_{i,j=1}^{n}$ *is additively consistent if and only if there exists a non-negative vector* $\underline{v} = (v_1,\ldots,v_n)^T$, $\left|v_i - v_j\right| \le 1, i,j = 1,\ldots,n$, *such that*

$$r_{ij} = 0.5\left(v_i - v_j + 1\right), \qquad i,j = 1,\ldots,n. \qquad (2.32)$$

The notation $\underline{v} = (v_1,\ldots,v_n)^T$ will be used hereafter to represent exclusively a priority vector associated with an APCM-A.

Proposition 2 says that when an APCM-A $R = \left\{r_{ij}\right\}_{i,j=1}^{n}$ of n objects is additively consistent, there exist priorities $v_1,\ldots,v_n$ of objects using which we can determine precisely the original PCs $r_{ij}, i,j = 1,\ldots,n$, in the APCM-A R by applying Tanino's characterization (2.32). On the other hand, when the priorities $v_1,\ldots,v_n$ of objects are given such that $|v_i - v_j| \le 1, i,j = 1,\ldots,n$, we can construct an additively consistent APCM-A of objects by applying (2.32). Notice that the additive reciprocity of R is always guaranteed by applying (2.32) since $r_{ij} = 0.5\left(v_i - v_j + 1\right) = 1 - 0.5(v_j - v_i + 1) = 1 - r_{ji}$.

Example 8 Since the APCM-A (2.29) in Example 7 is additively consistent, there exists a non-negative priority vector $\underline{v} = (v_1, v_2, v_3, v_4)^T$ satisfying the characterization property (2.32). It is, for example, the priority vector $\underline{v} = (0.8, 0.6, 0.2, 0)^T$.   △

Moreover, Tanino's characterization (2.32) is in line with the interpretation of the differences of PCs in an APCM-A R as given in Definition 8. In particular, when an APCM-A R is additively consistent according to (2.28), then the difference of PCs indicating the difference of preference intensity of o_i and of o_j corresponds to the difference of their priorities (Krejčí 2016), i.e.

$$r_{ij} - r_{ji} = v_i - v_j, \qquad i,j = 1,\ldots,n. \qquad (2.33)$$

When the APCM-A R is not additively consistent, then

$$r_{ij} - r_{ji} \approx v_i - v_j, \qquad i, j = 1, \ldots, n. \tag{2.34}$$

Example 9 From the intensity of preference $r_{12} = 0.6$ in the APCM-A (2.29) we know immediately that the difference between the priorities v_1 and v_2 is 0.2 ($v_1 - v_2 = r_{12} - r_{21} = 0.6 - 0.4 = 0.2$). Similarly also for the remaining differences between the priorities. The priority vector $\underline{v} = (0.8, 0.6, 0.2, 0)^T$ shown in Example 8 satisfies this property. $\triangle$

Similarly to the multiplicative-consistency condition for MPCMs, also the additive-consistency condition (2.28) is very difficult to keep when PCs in an APCM-A R are done using the scale $[0, 1]$. Having $r_{ik} = 0.9$ and $r_{kj} = 1$, for example, it follows that $r_{ij} = r_{ik} + r_{kj} - 0.5 = 1.4$, which exceeds the interval $[0, 1]$. Thus, it is possible to keep additive consistency only when using the intensities of preference close to the indifference value 0.5. This is not always reasonable or even possible (in particular for large-dimensional APCMs-A). Thus, other consistency conditions have been proposed in the literature. One of them is the weak-consistency condition proposed by Jandová et al. (2017).

The weak-consistency condition as introduced by Jandová et al. (2017) is applied in the same form both to APCMs-A and to APCMs with multiplicative representation (introduced later) without distinguishing among them, i.e., the weak-consistency condition is defined in general for APCMs. Therefore, this consistency condition will be reviewed later in Sect. 2.3.3.2, after introducing APCMs with multiplicative representation.

2.3.2.2 Deriving Priorities from APCMs-A

Based on the given APCM-A $R = \left\{ r_{ij} \right\}_{i,j=1}^{n}$ of n objects it is endeavoured to derive the priority vector $\underline{v} = (v_1, \ldots, v_n)^T$ that would best represent the relative preference of the objects with respect to the other objects in the set.

As discussed in the previous section, when an APCM-A $R = \left\{ r_{ij} \right\}_{i,j=1}^{n}$ is additively consistent according to (2.28), then there exist priorities $v_1, \ldots, v_n$, $|v_i - v_j| \leq 1$, $i, j = 1, \ldots, n$, such that $r_{ij} = 0.5(v_i - v_j + 1)$. When an APCM-A $R = \left\{ r_{ij} \right\}_{i,j=1}^{n}$ is not additively consistent, then the given expression only estimates the PCs in the matrix, i.e.

$$r_{ij} \approx 0.5(v_i - v_j + 1), \qquad i, j = 1, \ldots, n. \tag{2.35}$$

Fedrizzi and Brunelli (2010) proved that for an APCM-A $R = \left\{ r_{ij} \right\}_{i,j=1}^{n}$ the only vector of priorities (up to an additive constant) satisfying (2.32) is $\underline{v} = (v_1, \ldots, v_n)^T$ such that

$$v_i = \frac{2}{n} \sum_{j=1}^{n} r_{ij}, \qquad i = 1, \ldots, n. \tag{2.36}$$

Notice that the method for deriving priorities of objects from APCMs-A by formula (2.36) is again invariant under permutation of objects.

Proposition 3 *(Krejčí 2016) Given an APCM-A $R = \{r_{ij}\}_{i,j=1}^{n}$, the priorities $v_1, \ldots, v_n$ obtained from R by formula (2.36) are such that*

$$\sum_{i=1}^{n} v_i = n. \tag{2.37}$$

Proof

$$\sum_{i=1}^{n} v_i = \sum_{i=1}^{n} \frac{2}{n} \sum_{j=1}^{n} r_{ij} = \frac{2}{n} \sum_{i=1}^{n} \sum_{j=1}^{n} r_{ij} = \frac{2}{n} \left(\sum_{i=1}^{n} r_{ii} + \sum_{i=1}^{n} \sum_{\substack{j=1 \\ j \neq i}}^{n} r_{ij} \right) =$$
$$\frac{2}{n} \left(\frac{n}{2} + \frac{n(n-1)}{2} \right) = n$$

$\square$

Remark 1 Notice that the property (2.37) of the priorities given by (2.36) is independent of the additive consistency; it depends only on the additive-reciprocity condition (2.25). This means that also the sum of the priorities of objects obtained from an additively inconsistent APCM-A still equals n.

As shown by Fedrizzi and Brunelli (2010), there exist infinitely many priority vectors satisfying Proposition 2. These priority vectors can be generated from (2.36) by adding an arbitrary constant. This means that for a priority vector $\underline{v} = (v_1, \ldots, v_n)^T$ satisfying Proposition 2 also any priority vector obtained from $\underline{v}$ by the transformation

$$v_i \rightarrow v_i + c, \qquad i = 1, \ldots, n, \tag{2.38}$$

where $c \in \mathbb{R}$, satisfies Proposition 2.

Note that it is not possible to multiply the priorities (2.36) as it is done in the case of MPCMs, where the ratios of the priorities estimate the original PCs in the matrix. In the case of APCMs-A, the original PCs r_{ij} in the APCM-A $R = \{r_{ij}\}_{i,j=1}^{n}$ are estimated by the differences between the priorities v_i and v_j by means of (2.35), and thus, these differences have to remain unchanged.

In order to reach uniqueness, a normalization condition is usually applied to the priority vectors. In the case of MPCMs, the normalization condition

$$\sum_{i=1}^{n} w_i = 1, \quad w_i \in [0, 1], \ i = 1, \ldots, n, \tag{2.39}$$

is usually applied, see Sect. 2.2.3. It is worth to note that the condition (2.39) is reachable independently of the requirement of multiplicative consistency of MPCMs, i.e., even the priorities obtained from a multiplicatively inconsistent MPCM can be normalized so that they satisfy (2.39).

The normalization condition (2.39) has been applied also to the priorities obtained from APCMs-A (see, e.g., Xu (2004, 2007a); Xu and Chen (2008a, b); Wang and Li (2012); and the list of other papers provided by Fedrizzi and Brunelli (2009)). However, Fedrizzi and Brunelli (2009) showed that the normalization condition (2.39) is incompatible with Proposition 2.

Fedrizzi and Brunelli (2009) proposed a new normalization condition in the form

$$\min_{i=1,\ldots,n} v_i = 0, \quad v_i \in [0, 1], \quad i = 1, \ldots, n. \tag{2.40}$$

However, the normalization condition (2.40) is reachable only for additively consistent APCMs-A. For additively inconsistent APCMs-A, in general, the normalized priorities satisfying the condition $\min_{i=1,\ldots,n} v_i = 0$ do not satisfy the condition $v_i \in [0, 1]$, $i = 1, \ldots, n$, as illustrated on the following example.

Example 10 Let us examine the priorities obtainable from the APCM-A

$$A = \begin{pmatrix} 0.5 & 0.8 & 1 \\ 0.2 & 0.5 & 0.9 \\ 0 & 0.1 & 0.5 \end{pmatrix}, \tag{2.41}$$

which is not additively consistent; $r_{12} + r_{23} - 0.5 = 0.8 + 0.9 - 0.5 = 1.2 \neq 1 = r_{13}$. The priorities of objects obtained by formula (2.36) are in the form $v_1 = \frac{23}{15}$, $v_2 = \frac{16}{15}$, $v_3 = \frac{6}{15}$. By applying the normalization condition $\min_{i=1,\ldots,n} v_i = 0$, we obtain normalized priorities in the form $v_1 = \frac{17}{15}$, $v_2 = \frac{10}{15}$, $v_3 = 0$. Clearly, $v_1 > 1$, which violates the normalization condition $v_i \in [0, 1]$, $i = 1, \ldots, n$.   △

Proposition 4 (Krejčí 2016) *Given an APCM-A $R = \{r_{ij}\}_{i,j=1}^n$, $n \geq 3$, there exists no normalization condition of the type (2.38) for the priorities (2.36) that would guarantee the fulfillment of the property $v_i \in [0, 1]$, $i = 1, \ldots, n$.*

Proof There exist infinitely many priority vectors obtainable from (2.36) by the transformation (2.38). In order to modify the priorities so that $v_i \in [0, 1]$, $i = 1, \ldots, n$, a suitable constant c has to be added to the priorities (2.36). Further, we know that the differences between the priorities do not change by adding a constant to them; $(v_i + c) - (v_j + c) = v_i - v_j$, $i, j = 1, \ldots, n$. Clearly, the priorities (2.36) could be normalized so that $v_i + c \in [0, 1]$, $i = 1, \ldots, n$, if and only if $|v_i - v_j| \leq 1$, $i, j = 1, \ldots, n$. However, it will be shown that $|v_i - v_j| \leq 1$, $i, j = 1, \ldots, n$, is not reachable in general.

Let o_i, $i \in \{1, \ldots, n\}$, be such that it is absolutely preferred to all other objects, and let o_j, $j \in \{1, \ldots, n\}$, be such that all other objects are absolutely preferred to o_j. Then, for $n \geq 3$

$$v_i - v_j = \frac{2}{n} \sum_{k=1}^{n} r_{ik} - \frac{2}{n} \sum_{k=1}^{n} r_{jk} = \frac{2}{n} ((0.5 + n - 1) - (0.5 + 0)) = \frac{2n - 2}{n} > 1.$$

$\square$

According to Proposition 4, the property $v_i \in [0, 1]$, $i = 1, \ldots, n$, cannot be guaranteed for additively inconsistent APCMs-A under any normalization condition of the type (2.38). However, as discussed in the previous section, it is difficult or even impossible to reach additive consistency of APCMs-A in many MCDM problems, especially because of the restricted scale [0, 1] used for expressing the intensities of preference of one compared object over another. In general, the higher the dimension of an APCM-A is, the more difficult reaching the consistency is. Even when the DM is asked to reconsider his or her preferences, it does not have to lead to an additively consistent APCM-A. Therefore, in real-life applications, priorities of objects have to be often derived from additively inconsistent APCMs-A. This calls for a normalization condition applicable also to the priorities obtained from these additively inconsistent APCMs-A. Recall that for MPCMs there is such a normalization condition—(2.39).

Since the condition $v_i \in [0, 1]$, $i = 1, \ldots, n$, is not reachable for priorities obtained from additively inconsistent APCMs-A, we may weaken the normalization condition (2.39) to

$$\sum_{i=1}^{n} v_i = 1, \qquad v_i \in \mathbb{R}, \; i = 1, \ldots, n, \tag{2.42}$$

without any further constraints on the priorities. By applying this normalization condition to the priorities obtained by formulas (2.36), we derive formulas for obtaining normalized priorities from an APCM-A as

$$v_i = \frac{2}{n} \sum_{j=1}^{n} r_{ij} - \frac{n-1}{n}, \qquad i = 1, \ldots, n. \tag{2.43}$$

Notice that the method for deriving normalized priorities from an APCM-A by formula (2.43) is again invariant under permutation of objects.

Proposition 5 (Krejčí 2016) *Given an APCM-A* $R = \{r_{ij}\}_{i,j=1}^{n}$, *the priorities* $v_1, \ldots, v_n$ *obtained from R by formula (2.43) are such that*

$$\sum_{i=1}^{n} v_i = 1 \tag{2.44}$$

and

$$-1 < v_i \le 1, \qquad i = 1, \ldots, n. \tag{2.45}$$

Proof

$$\sum_{i=1}^{n} v_i = \sum_{i=1}^{n} \left(\frac{2}{n} \sum_{j=1}^{n} r_{ij} - \frac{n-1}{n} \right) = \frac{2}{n} \sum_{i=1}^{n} \sum_{j=1}^{n} r_{ij} - (n-1) = 1,$$

which proves (2.44).

The value of the priority v_i, $i \in \{1, \dots, n\}$, obtained by formula (2.43) depends only on the PCs in the i-th row of the matrix, i.e. on the intensities of preference of object o_i over the other objects. To prove the inequality $v_i \le 1$, we just need to show that the priority of object o_i will not exceed 1 even for the highest possible intensities of preference of object o_i over all other objects.

Let o_i, $i \in \{1, \dots, n\}$, be absolutely preferred to all other objects. Then,

$$v_i = \frac{2}{n} \sum_{j=1}^{n} r_{ij} - \frac{n-1}{n} = \frac{1}{n} \left(2 \sum_{j=1}^{n} r_{ij} - n + 1 \right) = \frac{1}{n} \left(2 \left(n - 1 + 0.5 \right) - n + 1 \right) = 1.$$

Similarly, to prove the inequality $-1 < v_i$ we just need to show that the priority of object o_i will be greater than -1 even for the lowest possible intensities of preference of object o_i over all other objects. Let o_i, $i \in \{1, \dots, n\}$, be absolutely preferred by all other objects. Then,

$$v_i = \frac{2}{n} \sum_{j=1}^{n} r_{ij} - \frac{n-1}{n} = \frac{1}{n} \left(2 \left(0 + 0.5 \right) - n + 1 \right) = \frac{2 - n}{n} = -1 + \frac{2}{n} > -1$$

$\square$

Example 11 The priority vector obtainable from the APCM-A (2.41) in Example 10 by the formulas (2.43) is in the form $\underline{v} = (\frac{13}{15}, \frac{6}{15}, \frac{-4}{15})^T$. This priority vector clearly satisfies both normalization properties (2.44) and (2.45). $\triangle$

Liu et al. (2012b) showed that the normalization condition $\sum_{i=1}^{n} v_i = 1$, $v_i \in [0, 1]$, $i = 1, \dots, n$, is reachable for the priorities (2.43) obtainable from an additively consistent APCM-A $R = \{r_{ij}\}_{i,j=1}^{n}$, if and only if

$$\min_{1 \le i \le n} \sum_{k=1}^{n} r_{ik} \ge \frac{n-1}{2}.$$

According to Proposition 5, some of the priorities normalized according to the normalization condition (2.42) can be negative, i.e. $v_i < 0$, $i \in \{1, \dots, n\}$. To avoid these situations, Meng et al. (2016) proposed "normalization" of the priorities (2.43) as follows:

$$v_i = \max\left\{0, \frac{2}{n}\sum_{j=1}^{n} r_{ij} - \frac{n-1}{n}\right\}, \qquad i, = 1, \ldots, n.$$

However, such "normalization" of the priorities, similarly to the inappropriate normalization condition (2.39), distorts the preference information contained in APCMs-A, and thus it is not appropriate. Moreover, the possible negativity of some of the priorities normalized according to (2.42) is not a problem at all because the scale on which the priorities are given is an interval scale; the differences between the priorities are meaningful. For example the normalized priorities $v_1 = \frac{13}{15}, v_2 = \frac{6}{15}, v_3 = \frac{-4}{15}$ obtained in Example 11 from the APCM-A (2.41) by the formula (2.43) tell us, for example, that $r_{23} - r_{32}$ is estimated as $v_2 - v_3 = \frac{2}{3}$ or that r_{23} is estimated as $0.5 + 0.5\frac{2}{3} = \frac{5}{6}$.

Remark 2 It is necessary to mention that also a more general characterization than Tanino's characterization (2.32) has appeared in the literature (see, e.g., Xu et al. (2009, 2014b), Liu et al. (2012b)):

$$r_{ij} = 0.5 + \beta(v_i - v_j), \qquad \beta \geq \max_{i=1,\ldots,n}\left\{\frac{n}{2} - \sum_{j=1}^{n} r_{ij}\right\} > 0 \qquad (2.46)$$

together with priorities

$$v_i = \frac{1}{n\beta}\sum_{j=1}^{n} r_{ij} - \frac{1}{2\beta} + \frac{1}{n} \qquad (2.47)$$

satisfying this characterization and normalization condition $\sum_{i=1}^{n} v_i = 1$, $v_i \in [0, 1]$. More particularly, Xu et al. (2009) proposed to set $\beta = \frac{n}{2}$, while Xu et al. (2011) and Hu et al. (2014) assumed $\beta = \frac{n-1}{2}$. It is true that by assuming the characterization (2.46) the obtained normalized priorities (2.47) are always non-negative and constrained to interval [0, 1]. However, the priorities miss an intuitive interpretation. In particular, $r_{ij} - r_{ji} = 0.5 + \beta(v_i - v_j) - 0.5 - \beta(v_j - v_i) = 2\beta(v_i - v_j)$, which would mean that the difference of priorities gives us $\frac{1}{2\beta}$-th of the difference between the related PCs in the APCM-A. This is quite difficult to interpret. Particularly, for $\beta = \frac{n}{2}$ we obtain $v_i - v_j = \frac{1}{n}(r_{ij} - r_{ji})$, and for $\beta = \frac{n-1}{2}$ we obtain $v_i - v_j = \frac{1}{n-1}(r_{ij} - r_{ji})$. Notice that for $\beta = \frac{1}{2}$ the characterization (2.46) equals to Tanino's characterization (2.32), and the corresponding priorities (2.47) equal to priorities (2.43) with a clear and intuitive interpretation $v_i - v_j = r_{ij} - r_{ji}$.

2.3.3 Additive Pairwise Comparison Matrices with Multiplicative Representation

Definition 10 An APCM with multiplicative representation (APCM-M) is an APCM $Q = \{q_{ij}\}_{i,j=1}^{n}$, $q_{ij} \in]0, 1[$, satisfying the additive-reciprocity property (2.25) where $\frac{q_{ij}}{q_{ji}}$ indicates the ratio of preference intensity of object o_i to that of object o_j, i.e. o_i is $\frac{q_{ij}}{q_{ji}}$-times as good as o_j.

The requirement of $q_{ij} \in]0, 1[$, $i, j = 1, \ldots, n$, in the definition of an APCM-M means that none of the objects compared in the APCM-M can be absolutely preferred to another one. This requirement is necessary in order not to divide by 0 in the ratios $\frac{q_{ij}}{q_{ji}}$. Because of the additive-reciprocity property of APCMs, the constraint $q_{ij} \in]0, 1[$, $i, j = 1, \ldots, n$, can be equivalently written as $q_{ij} > 0$, $i, j = 1, \ldots, n$.

2.3.3.1 Multiplicative Consistency of APCMs-M

Definition 11 (Tanino 1984) An APCM-M $Q = \{q_{ij}\}_{i,j=1}^{n}$ is said to be multiplicatively consistent if it satisfies the multiplicative-transitivity property

$$\frac{q_{ij}}{q_{ji}} = \frac{q_{ik}}{q_{ki}} \frac{q_{kj}}{q_{jk}}, \qquad i, j, k = 1, \ldots, n. \tag{2.48}$$

Definition 11 of multiplicative consistency is invariant under permutation of objects in the APCM-M Q.

Example 12 The APCM-M

$$Q = \begin{pmatrix} \frac{1}{2} & \frac{3}{5} & \frac{3}{5} & \frac{3}{4} \\ \frac{2}{5} & \frac{1}{2} & \frac{1}{2} & \frac{2}{3} \\ \frac{2}{5} & \frac{1}{2} & \frac{1}{2} & \frac{2}{3} \\ \frac{1}{4} & \frac{1}{3} & \frac{1}{3} & \frac{1}{2} \end{pmatrix} \tag{2.49}$$

is multiplicatively consistent according to Definition 11 since it satisfies the multiplicative-transitivity property (2.48). Furthermore, also the permuted matrix $Q^\pi = PQP^T$ obtained by using the permutation matrix (2.6) is multiplicatively consistent. $\triangle$

The following theorem provides us with alternative ways to verify multiplicative consistency of APCMs-M.

Theorem 4 *For an APCM-M $Q = \{q_{ij}\}_{i,j=1}^{n}$, $q_{ij} > 0$, $i, j = 1, \ldots, n$, the following statements are equivalent:*

(i) Q *is multiplicatively consistent according to Tanino's multiplicative-transitivity property (2.48),*

(ii)

$$q_{ij}q_{jk}q_{ki} = q_{ik}q_{kj}q_{ji}, \qquad i, j, k = 1, \ldots, n, \tag{2.50}$$

(iii)

$$\frac{q_{ij}}{q_{ji}}\frac{q_{jk}}{q_{kj}}\frac{q_{ki}}{q_{ik}} = 1, \qquad i, j, k = 1, \ldots, n, \tag{2.51}$$

(iv)

$$\frac{q_{ij}}{q_{ji}}\frac{q_{jk}}{q_{kj}}\frac{q_{ki}}{q_{ik}} = \frac{q_{ik}}{q_{ki}}\frac{q_{kj}}{q_{jk}}\frac{q_{ji}}{q_{ij}}, \qquad i, j, k = 1, \ldots, n, \tag{2.52}$$

(v)

$$q_{ij} = \frac{q_{ik}q_{kj}}{q_{ik}q_{kj} + (1 - q_{ik})(1 - q_{kj})} \qquad i, j, k = 1, \ldots, n. \tag{2.53}$$

Further, Tanino (1984) derived the following characterization of multiplicatively consistent APCMs-M.

Proposition 6 (Tanino's characterization) *An APCM-M* $Q = \{q_{ij}\}_{i,j=1}^{n}, q_{ij} > 0, i, j = 1, \ldots, n,$ *is multiplicatively consistent if and only if there exists a positive vector* $\underline{u} = (u_1, \ldots, u_n)^T$ *such that*

$$q_{ij} = \frac{u_i}{u_i + u_j}, \qquad i, j = 1, \ldots, n. \tag{2.54}$$

The notation $\underline{u} = (u_1, \ldots, u_n)^T$ will be used hereafter to represent exclusively a priority vector associated with an APCM-M.

Proposition 6 says that when an APCM-M $Q = \{q_{ij}\}_{i,j=1}^{n}$ of n objects is multiplicatively consistent then there exist non-negative priorities $u_1, \ldots, u_n$ of objects using which we can determine precisely the original PCs q_{ij} in the APCM-M Q by applying Tanino's characterization (2.54). On the other hand, when the non-negative priorities $u_1, \ldots, u_n$ of objects are known, we can construct a multiplicatively consistent APCM-M of objects by applying (2.54). Notice that the additive reciprocity of Q is always guaranteed by applying (2.54) since $q_{ij} = \frac{u_i}{u_i + u_j} = \frac{1}{1 + \frac{u_j}{u_i}} = 1 - \frac{\frac{u_j}{u_i}}{1 + \frac{u_j}{u_i}} = 1 - \frac{u_j}{u_j + u_i} = 1 - q_{ji}.$

Example 13 Since the APCM-M (2.49) in Example 12 is multiplicatively consistent, there exists a positive priority vector $\underline{u} = (u_1, u_2, u_3, u_4)^T$ satisfying Tanino's characterization property (2.54). It is, for example, the priority vector $\underline{u} = (1, \frac{2}{3}, \frac{2}{3}, \frac{1}{3})^T.$ △

Moreover, Tanino's characterization (2.54) is in line with the interpretation of the ratios of PCs in an APCM-M Q as given Definition 10. In particular, when an

APCM-M Q is multiplicatively consistent according to (2.48), then the ratio of PCs indicating the ratio of preference intensity of o_i to that of o_j corresponds to the ratio of their priorities, i.e.

$$\frac{q_{ij}}{q_{ji}} = \frac{u_i}{u_j}, \quad i, j = 1, \ldots, n. \tag{2.55}$$

When the APCM-M Q is not multiplicatively consistent, then

$$\frac{q_{ij}}{q_{ji}} \approx \frac{u_i}{u_j}, \quad i, j = 1, \ldots, n. \tag{2.56}$$

Example 14 From the intensity of preference $q_{12} = \frac{3}{5}$ in the APCM-M (2.49) we know immediately that the ratio of the priorities u_1 and u_2 is $\frac{3}{2}$ ($\frac{u_1}{u_2} = \frac{q_{12}}{q_{21}} = \frac{3}{5}\frac{5}{2} = \frac{3}{2}$). Similarly also for the remaining differences between the priorities. The priority vector $\underline{u} = (1, \frac{2}{3}, \frac{2}{3}, \frac{1}{3})^T$ shown in Example 13 satisfies this property. $\triangle$

Unlike in the case of the additive-consistency condition (2.28) for APCMs-A, the interval $]0, 1[$ can never be exceeded when trying to keep the multiplicative consistency (2.48) for APCMs-M. That is why APCMs-M become of more and more interest to researchers. However, despite this advantage, the interpretation of the multiplicative consistency is not very intuitive for DMs. Further, DMs have difficulties using the open interval $]0, 1[$ for expressing the intensities of preferences since this scale has no minimum and maximum.

Various other consistency conditions with more intuitive interpretation have been introduced. One of them is the weak-consistency condition proposed by Jandová et al. (2017) which was already mentioned in relation to APCMs-A. Since the weak-consistency condition is defined for APCMs without distinguishing between APCMs-A and APCMs-M, it is reviewed separately in the following section.

2.3.3.2 Weak Consistency of APCMs

Analogously to the weak consistency for MPCMs defined by Jandová and Talašová (2013), also the weak-consistency condition introduced in this section for APCMs is based on the properties that are intuitively supposed to hold.

Definition 12 (Jandová et al. 2017) An APCM $A = \{a_{ij}\}_{i,j=1}^{n}$ is said to be weakly consistent if

$$
\begin{aligned}
a_{ik} > 0.5 \ \wedge \ a_{kj} > 0.5 &\Rightarrow a_{ij} \geq \max\{a_{ik}, a_{kj}\}, \\
a_{ik} = 0.5 \ \wedge \ a_{kj} \geq 0.5 &\Rightarrow a_{ij} = \max\{a_{ik}, a_{kj}\}, \\
a_{ik} \geq 0.5 \ \wedge \ a_{kj} = 0.5 &\Rightarrow a_{ij} = \max\{a_{ik}, a_{kj}\},
\end{aligned} \tag{2.57}
$$

holds for $i, j, k = 1, \ldots, n$.

It is easy to verify that Definition 12 of weak consistency is invariant under permutation of objects in the APCM A.

Definition 12 does not distinguish between APCMs-A and APCMs-M; for both types of APCMs, the weak-consistency condition is defined in the same form.

Similarly as for the weak-consistency condition (2.11) for MPCMs, weak-consistency condition (2.57) for APCMs provides an intuitive minimum consistency requirement for APCMs. For example, when object o_i is preferred to object o_k with intensity 0.8, and object o_k is preferred to object o_j with intensity 0.6, then object o_i has to be preferred to object o_j with intensity at least 0.8 ($a_{ij} \geq \max\{0.8, 0.6\} = 0.8$). Thus, the requirement of weak consistency is very intuitive, it provides DMs with some space for expressing their intensities of preference, and it is very easy to control while entering PCs into the APCM.

In this book, the weak-consistency condition given by Definition 12 is adopted and later applied in Chap. 5 in a novel method for dealing with large-dimensional PCMs. However, it is worth to note that the weak consistency given by Definition 12 is not the only weak version of consistency for APCMs. For example, even a more relaxed form of the weak-consistency condition (2.11) for APCMs, strong stochastic transitivity, was introduced already half a century ago by Luce and Suppes (1965).

In the following theorem, rules equivalent to the weakly-consistency condition (2.57) are formulated.

Theorem 5 (Jandová et al. 2017) *For an APCM $A = \{a_{ij}\}_{i,j=1}^{n}$, the following statements are equivalent:*

(i) A is weakly consistent according to Definition 12.
(ii) For every $i, j, k = 1, \dots, n$:

$$
\begin{aligned}
a_{ik} < 0.5 \ \wedge \ a_{kj} < 0.5 \ &\Rightarrow \ a_{ij} \leq \min\{a_{ik}, a_{kj}\}, \\
a_{ik} = 0.5 \ \wedge \ a_{kj} \leq 0.5 \ &\Rightarrow \ a_{ij} = \min\{a_{ik}, a_{kj}\}, \\
a_{ik} \leq 0.5 \ \wedge \ a_{kj} = 0.5 \ &\Rightarrow \ a_{ij} = \min\{a_{ik}, a_{kj}\}.
\end{aligned}
\tag{2.58}
$$

(iii) For every $i, j, k = 1, \dots, n$:

$$
\begin{aligned}
0.5 < 1 - a_{kj} < a_{ik} \ &\Rightarrow \ 1 - a_{ik} \leq a_{ij} \leq a_{ik}, \\
0.5 < a_{ik} < 1 - a_{kj} \ &\Rightarrow \ a_{kj} \leq a_{ij} < 0.5, \\
0.5 < 1 - a_{kj} < a_{ik} \ &\Rightarrow \ 0.5 < a_{ij} \leq a_{ik}.
\end{aligned}
\tag{2.59}
$$

(iv) For every $i, j, k = 1, \dots, n$:

$$
\begin{aligned}
1 - a_{kj} < a_{ik} < 1 \ &\Rightarrow \ a_{kj} \leq a_{ij}, \\
0.5 < a_{kj} < 1 - a_{ik} \ &\Rightarrow \ a_{ik} \leq a_{ij} < 0.5, \\
0.5 < a_{kj} = 1 - a_{ik} \ &\Rightarrow \ 1 - a_{kj} \leq a_{ij} \leq a_{kj}.
\end{aligned}
\tag{2.60}
$$

Similarly to Definition 6 of weak consistency for MPCMs, it is possible to derive some interesting properties for APCMs weakly consistent according to Definition 12. Every weakly consistent APCM $A = \{a_{ij}\}_{i,j=1}^{n}$ can be permuted in such a way that the objects compared in the APCM are ordered from the most preferred to the least preferred. In such an ordered weakly consistent APCM all the elements above the main diagonal are greater or equal to 0.5, i.e. $a_{ij} \geq 0.5$, $i,j = 1, \ldots, n$, $i < j$. Further, the sequences of elements in the rows are non-decreasing and the sequences of elements in the columns are non-increasing.

The following propositions show that the weak-consistency condition is weaker than both the additive-consistency condition for APCMs-A and the multiplicative-consistency condition for APCMs-M.

Proposition 7 *An APCM-A $R = \{r_{ij}\}_{i,j=1}^{n}$ additively consistent according to Definition 9 is also weakly consistent according to Definition 12.*

Proof For $r_{ik} > 0.5 \wedge r_{kj} > 0.5$, we get immediately $r_{ij} = r_{ik} + r_{kj} - 0.5 > \max\{r_{ik}, r_{kj}\}$. For $r_{ik} = 0.5 \wedge r_{kj} \geq 0.5$ we get $r_{ij} = r_{ik} + r_{kj} - 0.5 = r_{kj} \geq \max\{r_{ik}, r_{kj}\}$, and analogously for $r_{ijk} \geq 0.5 \wedge r_{kj} = 0.5$. □

Proposition 8 *An APCM-M $Q = \{q_{ij}\}_{i,j=1}^{n}$ multiplicatively consistent according to Definition 11 is also weakly consistent according to Definition 12.*

Proof It is convenient to consider the multiplicative-consistency condition in the form (2.53) for the proof. The proof for $q_{ik} = 0.5 \wedge q_{kj} \geq 0.5$ (respectively for $q_{ik} \geq 0. \wedge q_{kj} = 0.5$) is trivial:

$$q_{ij} = \frac{q_{ik}q_{kj}}{q_{ik}q_{kj} + (1 - q_{ik})(1 - q_{kj})} = \frac{0.5 \cdot q_{kj}}{0.5 \cdot q_{kj} + 0.5(1 - q_{kj})} = q_{kj} = \max\{q_{ik}, q_{kj}\}.$$

The validity for $q_{ik} > 0.5 \wedge q_{kj} > 0.5$ is demonstrated by contradiction. Without the loss of generality, let us assume $q_{ik} \geq q_{kj} > 0.5$, and suppose the proposition is false, i.e. $q_{ij} < \max\{q_{ik}, q_{kj}\}$. It follows that $q_{ij} < q_{ik}$. Thus, we have

$$\frac{q_{ik}q_{kj}}{q_{ik}q_{kj} + (1 - q_{ik})(1 - q_{kj})} < q_{ik}$$
$$\Downarrow$$
$$\frac{q_{kj}}{q_{ik}q_{kj} + (1 - q_{ik})(1 - q_{kj})} < 1$$
$$\Downarrow$$
$$q_{kj} < 2q_{ik}q_{kj} + 1 - q_{ik} - q_{kj}$$
$$\Downarrow$$
$$q_{ik} - 1 < q_{kj}(2q_{ik} - 2)$$
$$\Downarrow$$
$$q_{kj} < 0.5$$

which is in contradiction with the assumption $q_{kj} > 0.5$. □

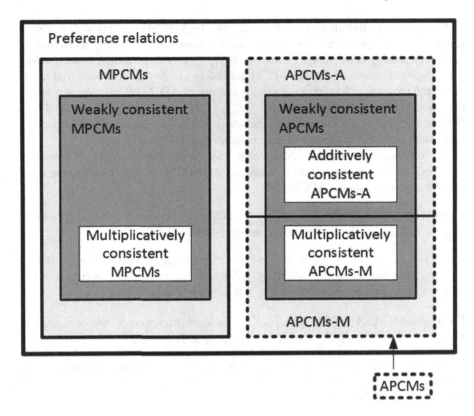

Fig. 2.1 Relations between MPCMs, APCMs-A, APCMs-M, and the associated definitions of consistency

The properties of weakly consistent APCMs described in this section are utilized in the new method for dealing with large-dimensional PCMs that is described in detail in Chapter 5.

The relations between MPCMs, APCMs-A, APCMs-M, and the associated definitions of consistency reviewed in Sects. 2.2 and 2.3 are represented by a diagram in Fig. 2.1.

2.3.3.3   Deriving Priorities from APCMs-M

As discussed in the previous section, when an APCM-M $Q = \{q_{ij}\}_{i,j=1}^{n}$ is multiplicatively consistent according to (2.48), then there exist positive priorities $u_1, \ldots, u_n$ such that $q_{ij} = \frac{u_i}{u_i+u_j}$. When the APCM-M $Q = \{q_{ij}\}_{i,j=1}^{n}$ is not multiplicatively consistent, then the given expression only estimates the PCs in the matrix, i.e.

$$q_{ij} \approx \frac{u_i}{u_i + u_j}, \qquad i,j = 1, \ldots, n. \tag{2.61}$$

Fedrizzi and Brunelli (2010) proved that, for an APCM-M $Q = \{q_{ij}\}_{i,j=1}^n$, the only vector of priorities (up to a multiplicative constant) satisfying (2.54) is $\underline{u} = (u_1, \ldots, u_n)^T$ such that

$$u_i = \sqrt[n]{\prod_{j=1}^n \frac{q_{ij}}{q_{ji}}}, \qquad i = 1, \ldots, n. \tag{2.62}$$

This method is again invariant under permutation of objects in APCMs-M.
Further, as pointed out by Xia and Xu (2011), the priorities (2.62) are such that $\prod_{i=1}^n u_i = 1$.
Because the priorities (2.62) can be multiplied by any positive constant, it is possible to apply the normalization condition (2.18), similarly as in the case of the priorities obtained from a MPCM. Thus, the normalized priorities can be computed from an APCM-M $Q = \{q_{ij}\}_{i,j=1}^n$, directly as

$$u_i = \frac{\sqrt[n]{\prod_{j=1}^n \frac{q_{ij}}{q_{ji}}}}{\sum_{k=1}^n \sqrt[n]{\prod_{j=1}^n \frac{q_{kj}}{q_{jk}}}}, \qquad i = 1, \ldots, n. \tag{2.63}$$

Remark 3 Even though the representation of PCs in APCMs-M is not very intuitive for DMs, they have a great advantage over APCMs-A regarding the normalization of the priorities. In particular, the priorities obtained from these matrices can be normalized by using the widely accepted normalization condition (2.18) with the normalized priorities still lying in the interval $]0, 1[$. As discussed in the previous section, this normalization condition is unreachable for priorities obtainable from APCMs-A.

2.4 Transformations Between MPCMs and APCMs

In this section, transformations between MPCMs, APCMs-A, and APCMs-M, and between the priorities obtainable from these PCMs are reviewed and discussed.

2.4.1 Transformations Between MPCMs and APCMs-A

It is a well-known fact that MPCMs and APCMs-A are equivalent. Fedrizzi (1990) showed that a MPCM $M = \{m_{ij}\}_{i,j=1}^{n}$ can be transformed into an APCM-A $R = \{r_{ij}\}_{i,j=1}^{n}$ by applying the transformation formula

$$r_{ij} = \frac{1}{2}\left(1 + \log_9 m_{ij}\right), \qquad i,j = 1,\dots,n. \tag{2.64}$$

The values in interval $[\frac{1}{9}, 9]$ (Saaty's scale) are transformed into values in interval $[0, 1]$, the multiplicative reciprocity (2.2) is transformed into the additive reciprocity (2.25), and the multiplicative consistency (2.4) is transformed into the additive consistency (2.28) by the formula (2.64) (Fedrizzi and Brunelli 2010).

Analogously, the inverse transformation formula

$$m_{ij} = 9^{2r_{ij}-1}, \qquad i,j = 1,\dots,n, \tag{2.65}$$

can be used to transform an APCM-A $R = \{r_{ij}\}_{i,j=1}^{n}$ into a MPCM $M = \{m_{ij}\}_{i,j=1}^{n}$ with all its relevant properties.

Further, the following theorem is valid for the transformation of the weak-consistency condition.

Theorem 6 *Let $M = \{m_{ij}\}_{i,j=1}^{n}$ be a MPCM weakly consistent according to Definition 6. Then the APCM-A $R = \{r_{ij}\}_{i,j=1}^{n}$ obtained from M by transformations (2.64) is weakly consistent according to Definition 12.*

Proof It is sufficient to show that when the weak-consistency condition (2.11) is valid for a MPCM M, then the weak-consistency condition (2.57) is valid for the APCM-A obtained from M by the transformation (2.64). By substituting (2.65) into the first part of (2.11), we obtain

$$9^{2r_{ik}-1} > 1 \wedge 9^{2r_{kj}-1} > 1 \Rightarrow 9^{2r_{ij}-1} \geq \max\left\{9^{2r_{ij}-1}, 9^{2r_{jk}-1}\right\}$$
$$\Downarrow$$
$$2r_{ik} - 1 > 0 \wedge 2r_{kj} - 1 > 0 \Rightarrow 2r_{ij} - 1 \geq \max\left\{2r_{ik} - 1, 2r_{kj} - 1\right\}$$
$$\Downarrow$$
$$r_{ik} > 0.5 \wedge r_{kj} > 0.5 \Rightarrow r_{ij} \geq \left\{r_{ik}, r_{kj}\right\},$$

which is the first part of (2.57). Analogously, for the other two parts of (2.11) we obtain the other two parts of (2.57).   □

Corollary 1 *Let $R = \{r_{ij}\}_{i,j=1}^{n}$ be an APCM-A weakly consistent according to Definition 12. Then the MPCM $M = \{m_{ij}\}_{i,j=1}^{n}$ obtained from R by transformations (2.65) is weakly consistent according to Definition 6.*

Remark 4 The validity of Corollary 1 follows immediately from Theorem 6 by utilizing properties of an inverse function. Note that this form of representing the results is used in the whole section. This means that the transformation of a particular property is formulated in a theorem and proved only in one direction. Afterwards, each such theorem is followed by a corollary showing the transformation of the property in the opposite direction without providing the proof.

Fedrizzi and Brunelli (2010) showed that priorities $w_1, \ldots, w_n$ of objects obtained from a MPCM M by formulas (2.23) can be transformed into the priorities $v_1, \ldots, v_n$ obtainable from the corresponding APCM-A R by (2.36) by using the transformation formula

$$v_i = 1 + \log_9 w_i, \quad i = 1, \ldots, n. \tag{2.66}$$

The inverse transformation formula is

$$w_i = 9^{v_i - 1}, \quad i = 1, \ldots, n. \tag{2.67}$$

Furthermore, Fedrizzi and Brunelli (2010) demonstrated that, $\underline{w} = (w_1, \ldots, w_n)^T$ being a priority vector representing priorities of objects compared in a MPCM, the vector $\underline{v} = (v_1, \ldots, w_n)^T$ obtained as

$$v_i = c + \log_9 w_i, \quad c \in \mathbb{R}, \quad i = 1, \ldots, n, \tag{2.68}$$

is a priority vector representing the priorities of objects compared in the corresponding APCM-A. The inverse transformation is

$$w_i = 9^{v_i - c}, \quad c \in \mathbb{R}, \quad i = 1, \ldots, n. \tag{2.69}$$

Note that it is also possible to derive a relation between the normalized priorities (2.24) obtainable from a MPCM and the normalized priorities (2.43) obtainable from the corresponding APCM-A. In particular, the normalized priorities (2.24) can be expressed as

$$w_i = \frac{\sqrt[n]{\prod_{j=1}^{n} m_{ij}}}{\sum_{k=1}^{n} \sqrt[n]{\prod_{j=1}^{n} m_{kj}}} = \frac{\sqrt[n]{\prod_{j=1}^{n} 9^{2r_{ij}-1}}}{\sum_{k=1}^{n} \sqrt[n]{\prod_{j=1}^{n} 9^{2r_{kj}-1}}} = \frac{9^{\frac{1}{n}\sum_{j=1}^{n}(2r_{ij}-1)}}{\sum_{k=1}^{n} 9^{\frac{1}{n}\sum_{j=1}^{n}(2r_{kj}-1)}} =$$

$$\frac{9^{\frac{2}{n}\sum_{j=1}^{n} r_{ij} - \frac{n-1}{n} - \frac{1}{n}}}{\sum_{k=1}^{n} 9^{\frac{2}{n}\sum_{j=1}^{n} r_{kj} - \frac{n-1}{n} - \frac{1}{n}}} \overset{(2.43)}{=} \frac{9^{v_i - \frac{1}{n}}}{\sum_{k=1}^{n} 9^{v_k - \frac{1}{n}}}. \tag{2.70}$$

Therefore, in order to obtain the normalized priority w_i of object o_i from the MPCM, the normalized priorities $v_j, j = 1, \ldots, n$, of all objects are necessary. This is due to the fact that the normalized priority w_i obtained by the formula (2.24) depends on all PCs in the MPCM, while the normalized priority v_i obtained by the formula (2.43) depends only on the PCs in the i-th row of the corresponding APCM-A.

In fact, as proved in Proposition 3, the sum of the non-normalized priorities (2.36) obtainable from an APCM-A always equals n, i.e., the sum of the priorities is independent of the PCs in the APCM-A. Thus, in order to normalize the priorities (2.36), the constant $-\frac{n-1}{n}$ is always added, i.e.,

$$v_i \rightarrow v_i - \frac{n-1}{n}, \qquad i = 1, \ldots, n.$$

Contrarily, the sum of the non-normalized priorities (2.23) obtainable from a MPCM is not constant; it depends on the values of the PCs in the MPCM. Therefore, in order to normalize the priorities obtainable from a MPCM, we have to divide them by their sum, i.e.,

$$w_i \rightarrow \frac{w_i}{\sum\limits_{k=1}^{n} w_k}, \qquad i = 1, \ldots, n.$$

Thus, it is not possible to derive a general formula that would transform a normalized priority w_i obtained by the formula (2.24) into the corresponding normalized priority v_i obtained by the formula (2.43) and vice versa. In other words, w_i is not a function of the single value v_i, but, on the contrary, it depends on the values $v_1, \ldots, v_n$, as highlighted in the expression (2.70). Nevertheless, having a particular MPCM and the corresponding APCM-A, there always exists a constant c such that the normalized priorities (2.24) can be transformed into the normalized priorities (2.43) by the transformation formula (2.68) or, in the opposite direction, by the transformation formula (2.69).

Example 15 Let us consider the APCM-A

$$R = \begin{pmatrix} 0.5 & 0.6 & 0.7 & 0.9 \\ 0.4 & 0.5 & 0.6 & 0.8 \\ 0.3 & 0.4 & 0.5 & 0.7 \\ 0.1 & 0.2 & 0.3 & 0.5 \end{pmatrix}. \tag{2.71}$$

The non-normalized priority vector obtained from the APCM-A R by the formula (2.36) is

$$\underline{v} = (1.35, 1.15, 0.95, 0.55)^T. \tag{2.72}$$

The MPCM M obtained from R by the transformation formula (2.65) is

$$M = \begin{pmatrix} 1 & 1.5518 & 2.4082 & 5.7995 \\ \frac{1}{1.5518} & 1 & 1.5518 & 3.7372 \\ \frac{1}{2.4082} & \frac{1}{1.5518} & 1 & 2.4082 \\ \frac{1}{5.7995} & \frac{1}{3.7372} & \frac{1}{2.4082} & 1 \end{pmatrix}. \tag{2.73}$$

The non-normalized priority vector obtained from the MPCM M by the formula (2.23) is

$$\underline{w} = (2.1577, 1.3904, 0.8960, 0.3720)^T. \tag{2.74}$$

The same priority vector would be obtained also by applying the transformation (2.67) to the priority vector (2.72).

Further, the normalized priority vector obtained from the APCM-A R by the formula (2.43) is $\underline{v} = (0.6, 0.4, 0.2, -0.2)^T$, and the normalized priority vector obtained from the MPCM M by the formula (2.24) is $\underline{w} = (0.4480, 0.2887, 0.1860, 0.0772)^T$. These normalized priority vectors can be transformed one into the other by using the transformations (2.68) and (2.69) with the constant $c = 0.9654$.

Notice that the APCM-A (2.71) is weakly consistent according to Definition 12. Thus, according to Corollary 1, the MPCM (2.73) is weakly consistent according to Definition 6. △

2.4.2 Transformations Between MPCMs and APCMs-M

Chiclana et al. (1998) showed that a MPCM M can be transformed into an APCM-M Q by applying the transformation formula

$$q_{ij} = \frac{m_{ij}}{1 + m_{ij}}, \quad i, j = 1, \ldots, n. \tag{2.75}$$

The values in interval $[\frac{1}{9}, 9]$ (Saaty's scale) are transformed into values in interval $[\frac{1}{10}, \frac{9}{10}] \subset]0, 1[$, the multiplicative reciprocity (2.2) transforms to the additive reciprocity (2.25), and the multiplicative consistency (2.4) transforms to the multiplicative consistency (2.48).

Analogously, the inverse transformation formula

$$m_{ij} = \frac{q_{ij}}{q_{ji}}, \quad i, j = 1, \ldots, n, \tag{2.76}$$

transforms an APCM-M $Q = \{q_{ij}\}_{i,j=1}^n$, $q_{ij} \in [\frac{1}{10}, \frac{9}{10}]$, to a MPCM $M = \{m_{ij}\}_{i,j=1}^n$, $m_{ij} \in [\frac{1}{9}, 9]$, with all relevant properties. Note that there is no transformation formula that would transform the interval $[\frac{1}{9}, 9]$ to the open interval $]0, 1[$ and vice versa.

Further, the following theorem is valid for the transformation of the weak-consistency condition.

Theorem 7 *Let $M = \{m_{ij}\}_{i,j=1}^{n}$ be a MPCM weakly consistent according to Definition 6. Then the APCM-M $Q = \{q_{ij}\}_{i,j=1}^{n}$ obtained from M by transformations (2.75) is weakly consistent according to Definition 12.*

Proof It is sufficient to show that when the weak-consistency condition (2.11) is valid for a MPCM M, then the weak-consistency condition (2.57) is valid for the APCM-M obtained from M by the transformation (2.75). By substituting (2.76) into the first part of (2.11), we obtain

$$\frac{q_{ik}}{q_{ki}} > 1 \wedge \frac{q_{kj}}{q_{jk}} > 1 \Rightarrow \frac{q_{ij}}{q_{ji}} \geq \max\left\{\frac{q_{ik}}{q_{ki}}, \frac{q_{kj}}{q_{jk}}\right\}.$$

Further,

$$\frac{q_{ik}}{q_{ki}} > 1 \Leftrightarrow \frac{q_{ik}}{1 - q_{ik}} > 1 \Leftrightarrow q_{ik} > 1 - q_{ik} \Leftrightarrow q_{ik} > 0.5,$$

and similarly we obtain $q_{kj} > 0.5$. Then,

$$\frac{q_{ij}}{q_{ji}} \geq \max\left\{\frac{q_{ik}}{q_{ki}}, \frac{q_{kj}}{q_{jk}}\right\} \Leftrightarrow \frac{1}{q_{ji}} - 1 \geq \max\left\{\frac{1}{q_{ki}} - 1, \frac{1}{q_{jk}} - 1\right\} \Leftrightarrow$$

$$\frac{1}{q_{ji}} \geq \max\left\{\frac{1}{q_{ki}}, \frac{1}{q_{jk}}\right\} \Leftrightarrow q_{ji} \leq \min\left\{q_{ki}, q_{jk}\right\} \Leftrightarrow$$

$$1 - q_{ij} \leq \min\left\{1 - q_{ik}, 1 - q_{kj}\right\} \Leftrightarrow q_{ij} \geq \max\left\{q_{ik}, q_{kj}\right\}.$$

Thus, we obtain $q_{ik} > 0.5 \wedge q_{kj} > 0.5 \Rightarrow q_{ij} \geq \max\left\{q_{ik}, q_{kj}\right\}$, which is the first part of (2.57). Analogously, from the remaining two parts of (2.11) we obtain the remaining two parts of (2.57). $\square$

Corollary 2 *Let $Q = \{q_{ij}\}_{i,j=1}^{n}$ be an APCM-M weakly consistent according to Definition 12. Then the MPCM $M = \{m_{ij}\}_{i,j=1}^{n}$ obtained from Q by transformations (2.76) is weakly consistent according to Definition 6.*

Fedrizzi and Brunelli (2010) showed that the priorities $w_1, \ldots, w_n$ of objects obtained from a MPCM M by formulas (2.23) and the priorities $u_1, \ldots, u_n$ of objects obtainable from the corresponding APCM-M Q by formulas (2.62) are identical, i.e.,

$$w_i = u_i, \qquad i = 1, \ldots, n. \tag{2.77}$$

Furthermore, Fedrizzi and Brunelli (2010) demonstrated that, $\underline{u} = (u_1, \ldots, u_n)^T$ being a priority vector representing the priorities of objects compared in an APCM-M, the vector $\underline{w} = (w_1, \ldots, w_n)^T$ obtained as

$$w_i = c \cdot u_i, \ c > 0, \qquad i = 1, \ldots, n, \tag{2.78}$$

is a priority vector representing the priorities of objects compared in the corresponding MPCM. The inverse transformation is then

$$u_i = \frac{1}{c} \cdot w_i, \ c > 0, \qquad i = 1, \ldots, n. \tag{2.79}$$

Note that in this case it is also possible to derive a direct relation between the normalized priorities (2.24) obtainable from a MPCM and the normalized priorities (2.63) obtainable from the corresponding APCM-M. In particular,

$$w_i = \frac{\sqrt[n]{\prod\limits_{j=1}^{n} m_{ij}}}{\sum\limits_{k=1}^{n} \sqrt[n]{\prod\limits_{j=1}^{n} m_{kj}}} \overset{(2.76)}{=} \frac{\sqrt[n]{\prod\limits_{j=1}^{n} \frac{q_{ij}}{q_{ji}}}}{\sum\limits_{k=1}^{n} \sqrt[n]{\prod\limits_{j=1}^{n} \frac{q_{kj}}{q_{jk}}}} = u_i.$$

This means that the normalized priorities (2.24) and (2.63) obtained from the MPCM and from the corresponding APCM-M, respectively, are identical. This simple relation between the normalized priority vectors $\underline{w}$ and $\underline{u}$ was possible to obtain only because the priority vectors are normalized in the same way;

$$w_i \rightarrow \frac{w_i}{\sum\limits_{k=1}^{n} w_k}, \qquad u_i \rightarrow \frac{u_i}{\sum\limits_{k=1}^{n} u_k}, \qquad i = 1, \ldots, n.$$

Example 16 Let us consider the MPCM M given by (2.73). The APCM-M obtained from M by the transformation formula (2.75) is

$$Q = \begin{pmatrix} 0.5 & 0.6081 & 0.7066 & 0.8529 \\ 0.3919 & 0.5 & 0.6081 & 0.7889 \\ 0.2934 & 0.3919 & 0.5 & 0.7066 \\ 0.1471 & 0.2111 & 0.2934 & 0.5 \end{pmatrix}. \tag{2.80}$$

The non-normalized priority vector obtained from the APCM-M Q by the formula (2.62) is

$$\underline{u} = (2.1577, 1.3904, 0.8960, 0.3720)^T, \tag{2.81}$$

i.e. it is identical to the priority vector (2.74). This result is in line with the transformation formula (2.77).

Notice that the APCM-M (2.80) is weakly consistent according to Definition 12. This conclusion follows also from Corollary 2.   △

2.4.3 Transformations Between APCMs-A and APCMs-M

Since there exist transformations between MPCMs and APCMs-A and between
MPCMs and APCMs-M, it is clear that there exist also transformations between
APCMs-A and APCMs-M. These transformations can be derived directly by composing the corresponding formulas from the previous two sections as specified in the
following theorems.

Theorem 8 *An APCM-A $R = \{r_{ij}\}_{i,j=1}^{n}$ can be transformed into an APCM-M $Q = \{q_{ij}\}_{i,j=1}^{n}$ with $q_{ij} \in [\frac{1}{10}, \frac{9}{10}], i, j = 1, \ldots, n$, by transformation formula*

$$q_{ij} = \frac{9^{2r_{ij}-1}}{1 + 9^{2r_{ij}-1}}, \qquad i, j = 1, \ldots, n. \tag{2.82}$$

Proof Because the transformation formula (2.65) transforms an APCM-A into a
MPCM, and (2.75) transforms a MPCM into an APCM-M, then the composition of
these formulas transforms an APCM-A into an APCM-M. By composing (2.65) and
(2.75) we immediately obtain (2.82).   □

Corollary 3 *An APCM-M $Q = \{q_{ij}\}_{i,j=1}^{n}$ with $q_{ij} \in [\frac{1}{10}, \frac{9}{10}], i, j = 1, \ldots, n$, can be transformed into an APCM-A $R = \{r_{ij}\}_{i,j=1}^{n}$ by transformation formula*

$$r_{ij} = \frac{1}{2}\left(1 + \log_9 \frac{q_{ij}}{q_{ji}}\right), \qquad i, j = 1, \ldots, n. \tag{2.83}$$

Theorem 9 *Let $R = \{r_{ij}\}_{i,j=1}^{n}$ be an APCM-A additively consistent according to Definition 9. Then the APCM-M $Q = \{q_{ij}\}_{i,j=1}^{n}$ obtained from R by transformations (2.82) is multiplicatively consistent according to Definition 11.*

Proof Because the transformation formula (2.65) transforms additive consistency
(2.28) of an APCM-A into multiplicative consistency (2.4) of the corresponding
MPCM, and (2.75) transforms multiplicative consistency (2.4) of a MPCM into
multiplicative consistency (2.48) of the corresponding APCM-M, then the composition (2.82) of these formulas transforms additive consistency of an APCM-A into
multiplicative consistency of the corresponding APCM-M.   □

Corollary 4 *Let $Q = \{q_{ij}\}_{i,j=1}^{n}$ be an APCM-M multiplicatively consistent according to Definition 11. Then the APCM-A $R = \{r_{ij}\}_{i,j=1}^{n}$ obtained from Q by transformations (2.83) is additively consistent according to Definition 9.*

Further, the following theorem is valid for the transformation of the weak-
consistency condition.

Theorem 10 *Let $R = \{r_{ij}\}_{i,j=1}^{n}$ be an APCM-A weakly consistent according to Definition 12. Then also the APCM-M $Q = \{q_{ij}\}_{i,j=1}^{n}$ obtained from R by transformations (2.82) is weakly consistent according to Definition 12.*

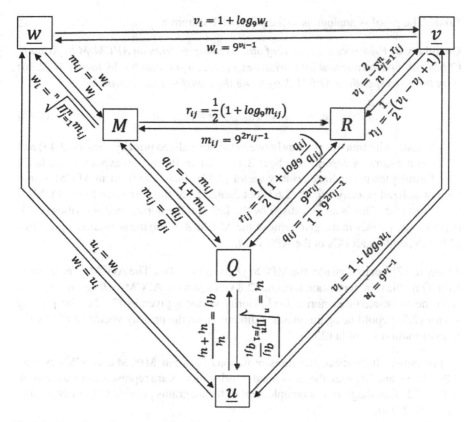

Fig. 2.2 Transformations between MPCMs, APCMs-A, and APCMs-M

Proof Because the transformation formula (2.65) transforms the weak consistency (2.57) of an APCM-A into the weak consistency (2.11) of the corresponding MPCM, and (2.75) transforms weak consistency (2.11) of a MPCM into weak consistency (2.57) of the corresponding APCM-M, then the composition (2.82) of these formulas transforms weak consistency of an APCM-A into weak consistency of the corresponding APCM-M. □

Corollary 5 *Let $Q = \{q_{ij}\}_{i,j=1}^{n}$ be an APCM-M weakly consistent according to Definition 12. Then also the APCM-A $R = \{r_{ij}\}_{i,j=1}^{n}$ obtained from Q by transformations (2.83) is weakly consistent according to Definition 12.*

Theorem 11 *Priorities $v_1, \ldots, v_n$ of objects obtained from an APCM-A by formula (2.36) can be transformed into priorities $u_1, \ldots, u_n$ obtainable by formulas (2.62) from the corresponding APCM-M by using the transformation formula*

$$u_i = 9^{v_i-1}, \quad i = 1, \ldots, n. \tag{2.84}$$

Proof The proof is analogous to the proof of Theorem 8.   □

Corollary 6 *Priorities $u_1, \ldots, u_n$ of objects obtained from an APCM-M by formula (2.62) can be transformed into priorities $v_1, \ldots, v_n$ obtainable by formulas (2.36) from the corresponding APCM-A by using the transformation formula*

$$v_i = 1 + \log_9 u_i, \quad i = 1, \ldots, n. \tag{2.85}$$

The issue with transformations between the normalized priority vectors (2.43) and (2.63) is the same as described in Sect. 2.4.1. That is, there is no explicit formula for transforming the normalized priority vector (2.43) obtainable from an APCM-A into the normalized priority vector (2.63) obtainable from the corresponding APCM-M and vice versa. This is again caused by the fact that the normalized priorities (2.43) depend only on PCs in the given row of the APCM-A while the normalized priorities (2.63) depend on all PCs in the APCM-M.

Example 17 Let us consider the MPCM Q given by (2.80). The APCM-A obtainable from Q by the transformation formula (2.83) is again the APCM-A R given by (2.71) with the associated non-normalized priority vector $\underline{v}$ given as (2.72). The priority vector (2.72) could be again obtained directly from the priority vector (2.81) by the transformation formula (2.85).   △

For better illustration, the transformations between MPCMs, APCMs-A, and APCMs-M, and between the associated priority vectors are represented by a diagram in Fig. 2.2. This diagram is a complement of the diagrams provided by (Fedrizzi and Brunelli, 2010).

Chapter 3
Fuzzy Set Theory

Abstract This chapter reviews concepts from fuzzy set theory indispensable for the fuzzy extension of th e multi-criteria decision making methods based on pairwise comparison matrices. Trapezoidal and triangular fuzzy numbers and intervals, which are most often used for the fuzzy extension of pairwise comparison methods, are introduced here, and normalization of fuzzy vectors is studied. Standard fuzzy arithmetic, that is usually used for the fuzzy extension of pairwise comparison methods, is reviewed, and the difference between applying standard fuzzy arithmetic and simplified standard fuzzy arithmetic on computations with fuzzy numbers is illustrated graphically on numerical examples. It is shown on an example that standard fuzzy arithmetic is not able to cope with problems in which there are interactions between the operands (even when these problems seem to be very simple and their solutions are intuitive) and that constrained fuzzy arithmetic needs to be applied instead. Because there are interactions between the operands in arithmetic operations conducted in fuzzy pairwise comparison methods, it results to be necessary to apply constrained fuzzy arithmetic instead of standard fuzzy arithmetic in these methods. Therefore, this section provides a detailed introduction to constrained fuzzy arithmetic complemented by illustrative examples.

3.1 Introduction to Fuzzy Sets

The traditional "crisp" set theory is based on the concept of "crisp set". A crisp set is defined in such a way that there exists a precise unambiguous distinction between the elements that belong to the crisp set and those that do not belong to the crisp set. A crisp set S on a given universe U can be defined in three basic ways (Klir and Yuan 1995):

- Set S is defined by listing all its members, i.e., $S = \{m_1, \ldots, m_k\}$ denotes the set S whose members are $m_1, \ldots, m_k$. Only finite sets can be defined in this way.
- Set S is defined by formulating a property that is met by all its members, i.e., $S = \{x; f(x)\}$ denotes the set S on the universe U whose members $x \in U$ satisfy the property f.

© Springer International Publishing AG, part of Springer Nature 2018
J. Krejčí, *Pairwise Comparison Matrices and their Fuzzy Extension*, Studies in Fuzziness and Soft Computing 366, https://doi.org/10.1007/978-3-319-77715-3_3

- Set S is defined by providing a characteristic function $\chi_S : U \to \{0, 1\}$ declaring which elements of the universe U are members of the set and which are not as

$$\chi_S(x) = \begin{cases} 1 & \text{for } x \in S, \\ 0 & \text{for } x \notin S. \end{cases}$$

Example 18 The set of all even numbers on a dice can be defined as $S_1 = \{2, 4, 6\}$. The set of men who are at least 190 cm tall can be defined as $S_2 = \{x; x \geq 190\}$, where $x \in U_2$ is actually the height of a man in centimeters. The sets S_1 and S_2 are defined unambiguously; there exists a clear distinction between the elements that belong to the sets and those that do not belong there. For example, the number 1 from the universe $U_1 = \{1, 2, \ldots, 6\}$ of all numbers on a dice clearly does not belong to the set $S_1 = \{2, 4, 6\}$, i.e. $1 \notin S_1$, while the number 4 clearly belongs there, i.e. $4 \in S_1$. Similarly, a man who is 160 cm tall obviously does not belong the set of men tall at least 190 cm, i.e. $160 \notin S_2$, while a man tall 195 cm obviously belongs there, i.e. $195 \in S_2$.   △

Example 19 Now let us assume we want to define a set of tall men, i.e. without explicitly defining their height in cm. First of all, the meaning of the adjective "tall" depends on the context. Do we want to define the set of tall men on the universe of basketball players, horse racers, Vietnamese, Norwegians, ...? Further, the meaning of "a tall man" is perceived subjectively by every evaluator. An evaluator who is 200 cm tall will probably not consider a 185 cm tall man as tall while an evaluator that is only 150 cm tall will probably do. Since the meaning of the adjective "tall" is vague, even for a particular universe of men and for a particular evaluator, it is very difficult to define the set of tall men. Let us say, I—a 167 cm tall woman—am the evaluator, and I attempt to define the meaning of "a tall European man". A man over 190 cm is definitely tall for me, while a man under 170 cm is not tall. But what about the men between 170 and 190 cm tall? Are they tall or not? It is difficult to draw a line above which I perceive a European man as tall and below which I perceive a European man as not tall. For example, if I chose 180 cm as the border, a man of 181 cm would be considered tall, while a man of 179 cm would be considered not tall. This is against our intuition.

The traditional "crisp" set theory is clearly not able to deal with this paradox. Thus other tools are needed. It feels very natural to describe the European men between 170 and 190 cm as "tall in some degree", i.e. by using "a partial degree of membership". This is the concept of fuzzy set theory.   △

Fuzzy set theory was initiated by Zadeh (1965, 1975a, b, c). Zadeh (1965) introduced a fuzzy set as a generalization of a crisp set with not precise boundaries.

Definition 13 Let U be a nonempty universe. A fuzzy set $\tilde{S}$ on U is characterized by its membership function $\mu_{\tilde{S}}(x)$ which associates to each element $x \in U$ a real number in the interval $[0, 1]$, i.e. $\mu_{\tilde{S}} : U \to [0, 1]$. $\mu_{\tilde{S}}(x)$ is called the degree of membership of the element x to the fuzzy set $\tilde{S}$.

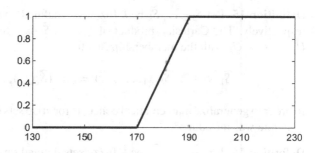

Fig. 3.1 Membership function of the fuzzy set "tall European man"

The membership function $\mu_{\tilde{S}} : U \to [0, 1]$ is a generalization of the characterization function $\chi_S : U \to \{0, 1\}$ that besides the values 1 and 0 representing the total membership and the total non-membership, respectively, allows the values between 0 and 1 for expressing degrees of membership to the fuzzy set $\tilde{S}$.

Example 20 By applying the concept of fuzzy sets, we can now easily define the meaning of "a tall European man" discussed in Example 19. One may define the fuzzy set $\tilde{S}$ of tall European men by the membership function

$$\mu_{\tilde{S}}(x) = \begin{cases} 0 & \text{for } x \leq 170, \\ \frac{x-170}{20} & \text{for } 170 < x < 190, \\ 1 & \text{for } x \geq 190. \end{cases} \tag{3.1}$$

The membership function is graphically represented in Fig. 3.1. However, as mentioned in Example 19, the actual membership function can differ for every evaluator. △

Note 2 For simplicity, $\tilde{S}(x)$ will be used hereafter to denote the membership function of a fuzzy set $\tilde{S}$ instead of $\mu_{\tilde{S}}(x)$.

A fuzzy set $\tilde{S}$ is defined uniquely by its membership function $\tilde{S}(x) : U \to [0, 1]$. Besides the membership function, also other characteristics of fuzzy sets are used to describe them.

Definition 14 Let $\tilde{S}$ be a fuzzy set defined on the universe U. The set $Core\ \tilde{S} := \{x \in U; \tilde{S}(x) = 1\}$ denotes the core of $\tilde{S}$, the set $Supp\ \tilde{S} := \{x \in U; \tilde{S}(x) > 0\}$ denotes the support of $\tilde{S}$, and the set $\tilde{S}_{(\alpha)} := \{x \in U; \tilde{S}(x) \geq \alpha\}$, $\alpha \in]0, 1]$, denotes the α-cut of $\tilde{S}$. Fuzzy set $\tilde{S}$ is said to be a normal fuzzy set if $\exists x \in U : \tilde{S}(x) = 1$, i.e. if $Core\ \tilde{S} \neq \emptyset$.

The set of all fuzzy sets defined on $\mathbb{R}$ is denoted $\mathcal{F}(\mathbb{R})$.

Example 21 Let us consider the fuzzy set $\tilde{S}$ defined by (3.1) and graphically represented in Fig. 3.1. The core of $\tilde{S}$ is $Core\ \tilde{S} := [190, \infty[$, the support of $\tilde{S}$ is $Supp\ \tilde{S} :=]170, \infty[$, and the 0.5-cut of $\tilde{S}$ is $\tilde{S}_{(0.5)} := [180, \infty[$. Moreover, $\tilde{S}$ is a normal fuzzy set. △

Definition 15 Let $\widetilde{S}_1, \ldots, \widetilde{S}_k$ be k fuzzy sets defined on the universes $U_1, \ldots, U_k$, respectively. The Cartesian product of $\widetilde{S}_1, \ldots, \widetilde{S}_k$ is a fuzzy set $\widetilde{S}_1 \times \cdots \times \widetilde{S}_k$ on $U_1 \times \cdots \times U_k$ with the membership function

$$\widetilde{S}_1 \times \cdots \times \widetilde{S}_k(x_1, \ldots, x_k) = \min\left\{\widetilde{S}_1(x_1), \ldots, \widetilde{S}_k(x_k)\right\}. \tag{3.2}$$

In order to generalize mathematical concepts for crisp sets to fuzzy sets, the extension principle was introduced.

Definition 16 Let $\widetilde{S}_1, \ldots, \widetilde{S}_k$ be k fuzzy sets defined on the universes $U_1, \ldots, U_k$, respectively, and let $U = U_1 \times \cdots \times U_k$. Further, let f be a mapping from the universe U to the universe V, $y = f(x_1, \ldots, x_k)$. The extension principle defines the membership function of a fuzzy set $\widetilde{S}$ on V as

$$\widetilde{S}(y) = \begin{cases} \sup\left\{\min\left\{\widetilde{S}_1(x_1), \ldots, \widetilde{S}_k(x_k)\right\}; (x_1, \ldots, x_k) \in U : y = f(x_1, \ldots, x_k)\right\} \\ \qquad\qquad \text{if } \{(x_1, \ldots, x_k) \in U : y = f(x_1, \ldots, x_k)\} \neq \emptyset, \\ 0, \qquad\qquad \text{otherwise.} \end{cases}$$

$$\tag{3.3}$$

Various types of fuzzy sets have been defined in the literature. Fuzzy numbers-a particular type of the fuzzy sets defined on $\mathbb{R}$-proved to be of a particular significance. "They should capture our intuitive conceptions of approximate numbers or intervals, such as "numbers that are close to a given real number" or "numbers that are around a given interval of real numbers". Such concepts are essential for characterizing states of fuzzy variables and, consequently, play an important role in many applications, including fuzzy control, decision making, approximate reasoning, optimization, and statistics with imprecise probabilities" (Klir and Yuan (1995), p. 97).

Definition 17 A fuzzy set $\widetilde{n}$ on $\mathbb{R}$ is said to be a fuzzy number if it satisfies the following properties:

(i) $\widetilde{n}$ is a normal fuzzy set, i.e. $\exists x \in \mathbb{R} : \widetilde{n}(x) = 1$;
(ii) the α-cuts $\widetilde{n}_{(\alpha)}$ are closed intervals for every $\alpha \in]0, 1]$;
(iii) the support of $\widetilde{n}$ is bounded, i.e. $\exists r_1, r_2 \in \mathbb{R} : Supp\, \widetilde{n} \subseteq [r_1, r_2]$.

A fuzzy number $\widetilde{n}$ is said to be positive if $\exists r_1, r_2 \in \mathbb{R}^+ : Supp\, \widetilde{n} \subseteq [r_1, r_2]$. The set of all fuzzy numbers is denoted by $\mathcal{F}_N(\mathbb{R})$, and the set of all positive fuzzy numbers is denoted by $\mathcal{F}_N(\mathbb{R}^+)$.

Example 22 Let us consider the fuzzy set

$$\widetilde{n}(x) = \begin{cases} 0 & \text{for } x \leq 3 \text{ and } x \geq 9, \\ \frac{x-3}{2} & \text{for } 3 < x < 5, \\ \frac{9-x}{3} & \text{for } 6 < x < 9, \\ 1 & \text{for } 5 \leq x \leq 6, \end{cases} \tag{3.4}$$

that is graphically represented in Fig. 3.2.

Fig. 3.2 Fuzzy set $\widetilde{n}$ given by (3.4)

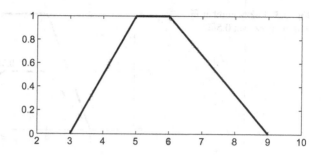

Since $\widetilde{n}(x) = 1$ for $x \in [5, 6]$, $\widetilde{n}$ is a normal fuzzy set. Further, the α-cuts $\widetilde{n}_{(\alpha)}$ are closed intervals for every $\alpha \in]0, 1]$, in particular $\widetilde{n}_{(\alpha)} = [3+2\alpha, 9-3\alpha] \forall \alpha \in]0, 1]$. Finally, since $Supp\ \widetilde{n} =]3, 9[\subseteq [3, 9]$, the support of $\widetilde{n}$ is bounded. Thus, $\widetilde{n}$ given by (3.4) satisfies Definition 17 of a fuzzy number. Moreover, $\widetilde{n}$ is a positive fuzzy number as $Supp\ \widetilde{n} \subseteq [3, 9]$ and $3, 9 \in \mathbb{R}^+$. $\triangle$

3.2 Alpha-Cut Representation

In the previous section, a definition of α-cuts of a fuzzy set was provided. In this Section, α-cuts will be reviewed in more detail since they are of a particular usefulness in fuzzy set theory. For the sake of integrity of this section, the definition of α-cuts will be recalled here once again.

In the literature, two basic types of α-cuts are used.

Definition 18 Let $\widetilde{S}$ be a fuzzy set on a nonempty universe U. Then, the set $\widetilde{S}_{(\alpha)} = \{x \in U; \widetilde{S}(x) \geq \alpha\}$ for $\alpha \in]0, 1]$ is called (weak) α-cut (or (weak) α-level set) of $\widetilde{S}$. The set $\widetilde{S}_{(\alpha)}^{>} = \{x \in U; \widetilde{S}(x) > \alpha\}$ for $\alpha \in [0, 1[$ is called strong α-cut (or strong α-level set) of $\widetilde{S}$.

When the membership function $\widetilde{S}(x)$ of $\widetilde{S}$ is continuous, the distinction between the (weak) α-cuts and the strong α-cuts is not necessary in applications. In the following, only the (weak) α-cuts are considered.

Remark 5 Notice that α-cut of $\widetilde{S}$ is not defined for $\alpha = 0$. However, for later use, it is convenient to define 0-cut of $\widetilde{S}$ as the closure[1] of the support of $\widetilde{S}$, $\widetilde{S}_{(0)} = Cl(Supp\,\widetilde{S})$.

Theorem 12 *Let $\widetilde{S}$ be a fuzzy set on a nonempty universe U. Then its membership function $\widetilde{S}(x)$ is given as $\widetilde{S}(x) = \sup \{\alpha; x \in \widetilde{S}_{(\alpha)}, \alpha \in [0, 1]\}$, $\forall x \in U$.*

Definition 19 Let $\widetilde{S}$ be a fuzzy set on a nonempty universe U. Then the α-multiple of $\widetilde{S}$ is a fuzzy set $\alpha\widetilde{S}$ on U with the membership function $(\alpha\widetilde{S})(x) = \alpha \cdot \widetilde{S}(x)$, $\forall x \in U$.

---

[1]The closure of interval $\overline{u}$ is the smallest closed interval containing $\overline{u}$; e.g. $Cl(]1, 3[) = [1, 3]$.

Fig. 3.3 Fuzzy number $\widetilde{m}$ and the fuzzy set $0.5\widetilde{m}$

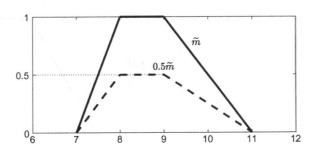

Fig. 3.4 Union of fuzzy sets $\widetilde{n}$ and $0.5\widetilde{m}$

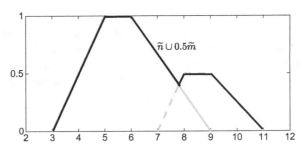

Example 23 Let us consider the fuzzy number $\widetilde{m}$ given in Fig. 3.3. The 0.5-multiple of $\widetilde{m}$ is then a fuzzy set $0.5\widetilde{m}$ illustrated in Fig. 3.3.   △

Definition 20 Let $\widetilde{S}_i, i = 1, \ldots, k$, be fuzzy sets on a nonempty universe U. Then the union of $\widetilde{S}_i, i = 1, \ldots, k$, is a fuzzy set $\widetilde{S} = \bigcup_{i=1}^{k} \widetilde{S}_i$ on U with the membership function $\widetilde{S}(x) = \max\left\{\widetilde{S}_1(x), \ldots, \widetilde{S}_k(x)\right\}, \forall x \in U$.

Example 24 Let us consider the fuzzy number $\widetilde{n}$ defined by (3.4) and the fuzzy set $0.5\widetilde{m}$ given in Fig. 3.3. The union of these two fuzzy sets is illustrated in Fig. 3.4. △

Using Definition 18 of α-cuts of a fuzzy set and Definitions 19 and 20 of the α-multiple and of the union of fuzzy sets, respectively, it is possible to derive another representation of fuzzy sets.

Theorem 13 *Let $\widetilde{S}$ be a fuzzy set on a nonempty universe U. Then*

$$\widetilde{S} = \bigcup_{\alpha=0}^{1} \alpha \widetilde{S}_{(\alpha)}. \tag{3.5}$$

The α-cut representation (3.5) is particularly convenient for fuzzy numbers, for which it can be defined easily by providing two functions.

Theorem 14 *Let $\widetilde{n} \in \mathcal{F}_N(\mathbb{R})$ and let $\widetilde{n}_{(\alpha)} = [n_{(\alpha)}^L, n_{(\alpha)}^U], \alpha \in [0, 1]$. Fuzzy number $\widetilde{n}$ can be determined uniquely by two functions $n^-, n^+ : [0, 1] \to \mathbb{R}$ defining the lower and upper boundary values $n_{(\alpha)}^L, n_{(\alpha)}^U$ of the α-cuts $\widetilde{n}_{(\alpha)}, \alpha \in [0, 1]$, of $\widetilde{n}$ satisfying*

Fig. 3.5 Functions n^- and n^+ for the fuzzy number $\tilde{n}$

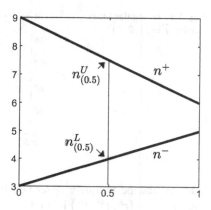

(i) $n^- : n^L_{(\alpha)} = n^-(\alpha)$ *is a bounded monotonic non-decreasing left-continuous function for* $\alpha \in]0, 1]$ *and right-continuous for* $\alpha = 0$;
(ii) $n^+ : n^U_{(\alpha)} = n^+(\alpha)$ *is a bounded monotonic non-increasing left-continuous function for* $\alpha \in]0, 1]$ *and right-continuous for* $\alpha = 0$;
(iii) $\forall \alpha \in [0, 1] : n^L_{(\alpha)} \leq n^U_{(\alpha)}$.

Example 25 Let us consider the fuzzy number $\tilde{n}$ defined by (3.4). The functions n^- and n^+ and the boundary values $n^L_{(0.5)}$ and $n^U_{(0.5)}$ of the 0.5-cut $\tilde{n}_{(0.5)}$ of $\tilde{n}$ are represented in Fig. 3.5. △

3.3 Trapezoidal Fuzzy Numbers

In the fuzzy extension of MCDM methods, and in particular in the fuzzy extension of the MCDM methods based on PCMs, special types of fuzzy numbers are usually used; namely, triangular and trapezoidal fuzzy numbers, and intervals. All three types of fuzzy numbers are introduced in this section.

Triangular fuzzy numbers were introduced as a special case of fuzzy numbers by Laarhoven and Pedrycz (1983).

Definition 21 A triangular fuzzy number $\tilde{t}$ is a fuzzy number whose membership function is given as

$$\tilde{t}(x) = \begin{cases} \frac{x-t^L}{t^M-t^L}, & t^L < x < t^M, \\ 1, & x = t^M, \\ \frac{t^U-x}{t^U-t^M}, & t^M < x < t^U, \\ 0, & \text{otherwise}, \end{cases} \tag{3.6}$$

where t^L and t^U are called the lower and upper boundary values of the triangular fuzzy number $\tilde{t}$, and t^M is called the middle value of $\tilde{t}$. Every triangular fuzzy

Fig. 3.6 Triangular fuzzy number $\widetilde{t} = (1, 2, 4)$

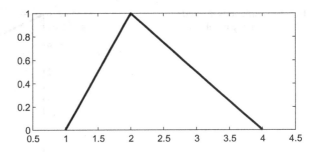

number can be uniquely described by a triplet of these representing values; the notation $\widetilde{t} = \left(t^L, t^M, t^U\right)$ is used.

A triangular fuzzy number $\widetilde{t} = \left(t^L, t^M, t^U\right)$ is positive if $t^L > 0$. The core of $\widetilde{t} = \left(t^L, t^M, t^U\right)$ is the singleton set $Core\ \widetilde{t} = \left\{t^M\right\}$, and the support is an open interval $Supp\ \widetilde{t} =]t^L, t^U[$. The α-cut, $\alpha \in [0, 1]$, of the triangular fuzzy number $\widetilde{t} = \left(t^L, t^M, t^U\right)$ is a closed interval $\widetilde{t}_{(\alpha)} = [t^L_{(\alpha)}, t^U_{(\alpha)}]$, where $t^L_{(\alpha)} = \alpha t^M + (1 - \alpha)t^L$, $t^U_{(\alpha)} = \alpha t^M + (1 - \alpha)t^U$.

Example 26 Triangular fuzzy number $\widetilde{t} = (1, 2, 4)$, given in Fig. 3.6, is positive, its core is the singleton set $Core\ \widetilde{t} = \{2\}$, and its support is the open interval $Supp\ \widetilde{t} =]1, 4[$. The α-cuts, $\alpha \in [0, 1]$, are closed intervals $\widetilde{t}_{(\alpha)} = [t^L_{(\alpha)}, t^U_{(\alpha)}]$ such that $t^L_{(\alpha)} = 1 + \alpha$, $t^U_{(\alpha)} = 4 - 2\alpha$.   △

A couple of years later, trapezoidal fuzzy numbers were introduced by Buckley (1985b) even though they got their name later.

Definition 22 A trapezoidal fuzzy number $\widetilde{z}$ is a fuzzy number whose membership function is given as

$$\widetilde{z}(x) = \begin{cases} \frac{x - z^\alpha}{z^\beta - z^\alpha}, & z^\alpha < x < z^\beta, \\ 1, & z^\beta \leq x \leq z^\gamma, \\ \frac{z^\delta - x}{z^\delta - z^\gamma}, & z^\gamma < x < z^\delta, \\ 0, & \text{otherwise.} \end{cases} \tag{3.7}$$

Every trapezoidal fuzzy number can be uniquely described by a quadruple of its representing values; the notation $\widetilde{z} = \left(z^\alpha, z^\beta, z^\gamma, z^\delta\right)$ is used.

A trapezoidal fuzzy number $\widetilde{z} = \left(z^\alpha, z^\beta, z^\gamma, z^\delta\right)$ is positive if $z^\alpha > 0$. The core of $\widetilde{z} = \left(z^\alpha, z^\beta, z^\gamma, z^\delta\right)$ is a closed interval $Core\ \widetilde{z} = \left[z^\beta, z^\gamma\right]$, and the support is an open interval $Supp\ \widetilde{z} =]z^\alpha, z^\delta[$. The α-cut, $\alpha \in [0, 1]$, of the trapezoidal fuzzy number $\widetilde{z} = \left(z^\alpha, z^\beta, z^\gamma, z^\delta\right)$ is a closed interval $\widetilde{z}_{(\alpha)} = [z^L_{(\alpha)}, z^U_{(\alpha)}]$, where

$$z^L_{(\alpha)} = \alpha z^\beta + (1 - \alpha)z^\alpha, \quad z^U_{(\alpha)} = \alpha z^\gamma + (1 - \alpha)z^\delta. \tag{3.8}$$

Fig. 3.7 Trapezoidal fuzzy number $\widetilde{z} = (-2.5, -2, 0, 1)$

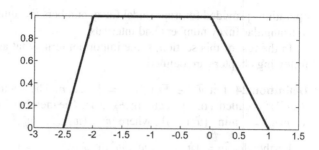

Example 27 Trapezoidal fuzzy number $\widetilde{z} = (-2.5, -2, 0, 1)$, given in Fig. 3.7, is not positive since $z^\alpha = -2.5 \not> 0$. The core of $\widetilde{z}$ is the closed interval $Core\, \widetilde{z} = [-2, 0]$ and its support is the open interval $Supp \widetilde{z} =]-2.5, 1[$. The α-cuts, $\alpha \in [0, 1]$, are closed intervals $\widetilde{z}_{(\alpha)} = [z^L_{(\alpha)}, z^U_{(\alpha)}]$ such that $z^L_{(\alpha)} = -2.5 + 0.5\alpha$, $z^U_{(\alpha)} = 1 - \alpha$.
$\triangle$

Even intervals can be understood and dealt with as fuzzy numbers.

Definition 23 An interval $\overline{v}$ is a fuzzy number whose membership function is given as

$$\overline{v}(x) = \begin{cases} 1, & v^L \le x \le v^U, \\ 0, & \text{otherwise.} \end{cases} \tag{3.9}$$

Every interval can be uniquely described by the pair of its lower and upper boundary values; the notation $\overline{v} = [v^L, v^U]$ is used.

An interval $\overline{v} = [v^L, v^U]$ is positive if $v^L > 0$. The core and the support of $\overline{v} = [v^L, v^U]$ are obviously identical to the interval $\overline{v}$, i.e. $Core\, \overline{v} = Supp\, \overline{v} = [v^L, v^U]$. Also all the α-cuts, $\alpha \in [0, 1]$, of the interval $\overline{v} = [v^L, v^U]$ are identical to the interval, i.e. $\overline{v}_{(\alpha)} = [v^L, v^U]$.

Remark 6 Notice that the interval $\overline{v}$ in Definition 23 is denoted by $^-$ instead of $^\sim$, which is usual for fuzzy numbers. That is because the notation $^-$ is more common than $^\sim$ to refer to intervals in the literature. Thus, the same notation was opted for in this book. Nevertheless, this notation does not change anything on the fact that intervals can be looked at as a particular case of fuzzy numbers.

Not only are triangular fuzzy numbers, trapezoidal fuzzy numbers, and intervals special cases of fuzzy numbers, but triangular fuzzy numbers and intervals are also special cases of trapezoidal fuzzy numbers. In particular, a triangular fuzzy number $\widetilde{t} = (t^L, t^M, t^U)$ can be written in the form $\widetilde{t} = (t^L, t^M, t^M, t^U)$ which satisfies Definition 22 of a trapezoidal fuzzy number, but the membership function given by (3.7) still has the form of a triangular fuzzy number. Analogously, an interval $\overline{v} = [v^L, v^U]$ can be written in the form $\widetilde{v} = (v^L, v^L, v^U, v^U)$ which satisfies Definition 22 of a trapezoidal fuzzy number, but the membership function given by (3.7) still has the form of an interval. Thus all concepts, definitions, and arithmetic operations provided for trapezoidal fuzzy numbers are automatically applicable also to triangular fuzzy numbers and intervals.

In the rest of this section, some important terms that are going to be used in the following chapters are defined.

Definition 24 Let $\widetilde{w}_i \in \mathcal{F}_N(\mathbb{R})$, $i = 1, \ldots, n$. The vector $\underline{\widetilde{w}} = (\widetilde{w}_1, \ldots, \widetilde{w}_n)^T \in \mathcal{F}_N(\mathbb{R})^n$ is called a fuzzy vector in $\mathcal{F}_N(\mathbb{R})^n$. The membership function of $\underline{\widetilde{w}}$ is defined as $\underline{\widetilde{w}}(\underline{w}) = \min_{i=1,\ldots,n} \{\widetilde{w}_i(w_i)\}$, where $\underline{w} = (w_1, \ldots, w_n)^T$.

Further, let $\widetilde{w}_i = (w_i^\alpha, w_i^\beta, w_i^\gamma, w_i^\delta) \in \mathcal{F}_N(\mathbb{R})$, $i = 1, \ldots, n$, be trapezoidal fuzzy numbers. Then the fuzzy vector $\underline{\widetilde{w}} = (\widetilde{w}_1, \ldots, \widetilde{w}_n)^T \in \mathcal{F}_N(\mathbb{R})^n$ can be written as $\underline{\widetilde{w}} = (\underline{w}^\alpha, \underline{w}^\beta, \underline{w}^\gamma, \underline{w}^\delta)$, where $\underline{w}^\alpha = (w_1^\alpha, \ldots, w_n^\alpha)^T$, $\underline{w}^\beta = (w_1^\beta, \ldots, w_n^\beta)^T$, $\underline{w}^\gamma = (w_1^\gamma, \ldots, w_n^\gamma)^T$, and $\underline{w}^\delta = (w_1^\delta, \ldots, w_n^\delta)^T$ are called the representing vectors of the fuzzy vector $\underline{\widetilde{w}}$.

Definition 25 Let $\widetilde{m}_{ij} \in \mathcal{F}_N(\mathbb{R})$, $i, j = 1, \ldots, n$. Then $\widetilde{M} = \{\widetilde{m}_{ij}\}_{i,j=1}^n \in \mathcal{F}_N(\mathbb{R})^{n^2}$ is called a fuzzy matrix in $\mathcal{F}_N(\mathbb{R})^{n^2}$. The membership function of $\widetilde{M}$ is defined as $\widetilde{M}(M) = \min_{i,j=1,\ldots,n} \{\widetilde{m}_{ij}(m_{ij})\}$, where $M = \{m_{ij}\}_{i,j=1}^n$.

In order to defuzzify fuzzy numbers, the center-of-area defuzzification method (Takagi and Sugeno 1985), sometimes called also the center-of-gravity method or the centroid method, is often used because of its computational simplicity and well accepted results.

Definition 26 Let $\widetilde{z} = (z^\alpha, z^\beta, z^\gamma, z^\delta)$ be a trapezoidal fuzzy number. The center of area $COA_{\widetilde{z}}$ of $\widetilde{z}$ is defined as

$$COA_{\widetilde{z}} = \frac{1}{3} \frac{(z^\delta)^2 + (z^\gamma)^2 - (z^\beta)^2 - (z^\alpha)^2 + z^\delta z^\gamma - z^\beta z^\alpha}{z^\delta + z^\gamma - z^\beta - z^\alpha}. \tag{3.10}$$

Note 3 Note that for a triangular fuzzy number $\widetilde{t} = (t^L, t^M, t^U)$, the formula (3.10) is reduced to

$$COA_{\widetilde{t}} = \frac{t^L + t^M + t^U}{3},$$

and for an interval $\overline{v} = [v^L, v^U]$, it is reduced to

$$COA_{\overline{v}} = \frac{v^L + v^U}{2}.$$

Example 28 Centers of area of the triangular fuzzy number $\widetilde{t} = (1, 2, 4)$ and of the trapezoidal fuzzy number $\widetilde{z} = (-2.5, -2, 0, 1)$ are $COA_{\widetilde{t}} = \frac{7}{3}$ and $COA_{\widetilde{z}} = -\frac{19}{22}$, respectively.   △

In the following, Ruspini's fuzzy partition (often called also a fuzzy scale) is introduced as it is particularly suitable for modeling the meaning of linguistic terms from a predefined scale used for comparing objects pairwisely.

Definition 27 (Ruspini 1969) A set of fuzzy numbers $\tilde{n}_1, \ldots, \tilde{n}_k, k > 1$, defined on interval $[a, b]$ is called Ruspini's fuzzy partition of $[a, b]$, if $\tilde{n}_i \neq \emptyset, i = 1, \ldots, k$, and $\sum_{i=1}^{k} \tilde{n}_i(x) = 1, \forall x \in [a, b]$.

For trapezoidal fuzzy numbers, the following proposition is valid.

Proposition 9 *A set of trapezoidal fuzzy numbers* $\tilde{z}_i = \left(z_i^\alpha, z_i^\beta, z_i^\gamma, z_i^\delta \right), i = 1, \ldots, k$, *defined on interval* $[a, b]$ *and numbered in the conformity with their linear ordering forms Ruspini's fuzzy partition of interval* $[a, b]$ *if and only if*

$$
\begin{aligned}
z_1^\alpha &= z_1^\beta = a, \\
z_{i-1}^\gamma &= z_i^\alpha, && i = 2, \ldots, k, \\
z_{i-1}^\delta &= z_i^\beta, && i = 2, \ldots, k, \\
z_k^\gamma &= z_k^\delta = b.
\end{aligned}
\tag{3.11}
$$

Proof Interval $[a, b]$ can be written as the union $[a, b] = \bigcup_{i=1}^{k} [z_i^\beta, z_i^\gamma] \cup \bigcup_{i=2}^{k}]z_i^\alpha, z_i^\beta[$. For $x \in [z_i^\beta, z_i^\gamma], i \in \{1, \ldots, k\}$, it holds that $\tilde{z}_i(x) = 1$ and $\tilde{z}_j(x) = 0$ for $j = 1, \ldots, k, j \neq i$. Therefore, $\sum_{j=1}^{k} \tilde{z}_j(x) = \tilde{z}_i(x) = 1$. Further, for $x \in]z_i^\alpha, z_i^\beta[=]z_{i-1}^\gamma, z_{i-1}^\delta[, i \in \{2, \ldots, k\}$, by applying (3.6), we obtain $\tilde{z}_i(x) = \frac{x - z_i^\alpha}{z_i^\beta - z_i^\alpha}$, $\tilde{z}_{i-1}(x) = \frac{z_{i-1}^\delta - x}{z_{i-1}^\delta - z_{i-1}^\gamma} = \frac{z_i^\beta - x}{z_i^\beta - z_i^\alpha}$, and $\tilde{z}_j(x) = 0$ for $j = 1, \ldots, k, \ j \neq i, i-1$. Therefore, for $x \in]z_i^\alpha, z_i^\beta[, i \in \{2, \ldots, k\}$, the equation $\sum_{j=1}^{k} \tilde{z}_j(x) = \tilde{z}_{i-1}(x) + \tilde{z}_i(x) = \frac{z_i^\beta - x}{z_i^\beta - z_i^\alpha} + \frac{x - z_i^\alpha}{z_i^\beta - z_i^\alpha} = \frac{z_i^\beta - x + x - z_i^\alpha}{z_i^\beta - z_i^\alpha} = 1$ holds, which completes the proof. $\square$

Note 4 Proposition 9 is a generalization of the proposition regarding Ruspini's fuzzy partition for triangular fuzzy numbers which was provided and proved by Krejčí (2017e).

Example 29 Trapezoidal fuzzy numbers $\tilde{z}_1 = (0, 0, 1, 2), \tilde{z}_2 = (1, 2, 3, 4), \tilde{z}_3 = (3, 4, 5, 5)$, given in Fig. 3.8, form Ruspini's fuzzy partition of interval $[0, 5]$. $\triangle$

In Chap. 4, a fuzzy extension of Saaty's scale given in Table 2.1 will be proposed in such a way that the fuzzy numbers modeling the meaning of the linguistic terms from the scale and their reciprocals form Ruspini's fuzzy partition of the given interval.

As discussed in chap. 2, the priorities of objects derived from a PCM are usually normalized. Most often, the normalization condition (2.18), $\sum_{i=1}^{n} w_i = 1, w_i \in [0, 1], i = 1, \ldots, n$, is utilized; in particular for the priorities obtained from MPCMs and APCMs-M by the methods reviewed in the previous chapter. Recall that it was shown in Sect. 2.3.2.2 that the normalization condition (2.18) is not compatible with

Fig. 3.8 Ruspini's fuzzy
partition of interval [0, 5]

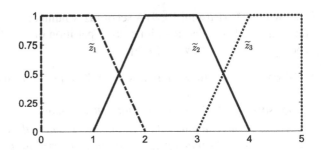

Tanino's characterization (2.32) for the priorities obtained from APCMs-A, and thus
a weaker version of the normalization condition (2.18) is needed in this case.

When extending MCDM methods based on PCMs to fuzzy numbers it is necessary,
besides other issues, to handle properly the fuzzy extension of the normalization
condition (2.18). Many definitions of normalized interval and fuzzy vectors have
been proposed in the literature; see, e.g., Chang and Lee (1995), Jiménez et al.
(2003), Wang and Elhag (2006), Sevastjanov et al. (2012). In this book, the approach
of Wang and Elhag (2006) is considered. The reason for choosing this approach will
be clarified in Chap. 4 after introducing fuzzy PCMs and methods for deriving fuzzy
priorities of objects from them.

Wang and Elhag (2006) provided a definition of the normalized interval vector
and they extended this definition to general fuzzy numbers given by means of their
α-cuts.

Definition 28 (Wang and Elhag 2006) Let $\overline{w}_i = [w_i^L, w_i^U], i = 1, \ldots, n$, be intervals, $\overline{w}_i \subseteq [0, 1]$, $i = 1, \ldots, n$, and let

$$N_{\underline{w}} = \left\{ \underline{w} = (w_1, \ldots, w_n)^T; \ \sum_{i=1}^{n} w_i = 1, \ w_i^L \leq w_i \leq w_i^U, \ i = 1, \ldots, n \right\}$$
$$(3.12)$$

be a set of normalized vectors constructed from the intervals. The intervals $\overline{w}_1, \ldots, \overline{w}_n$
are said to be normalized if

(i) there exists at least one normalized vector $\underline{w} = (w_1, \ldots, w_n)^T$ in $N_{\underline{w}}$;
(ii) the lower and upper boundary values w_i^L and w_i^U of $\overline{w}_i, i = 1, \ldots, n$, are
     attainable in $N_{\underline{w}}$.

The vector $\underline{\overline{w}} = (\overline{w}_1, \ldots, \overline{w}_n)^T$ of normalized intervals will be called a normalized interval vector.

Remark 7 As already pointed out by Pavlačka (2014), the validity of the condition
(ii) in Definition 28 automatically implies the validity of the condition (i). This means
that the condition (i) could be omitted in Definition 28.

According to the condition (ii), for every lower boundary value $w_i^L, i \in$
$\{1, \ldots, n\}$, there have to exist values $w_j \in [w_j^L, w_j^U], \ j = 1, \ldots, n, j \neq i,$

such that $w_i^L + \sum_{\substack{j=1 \\ j \neq i}}^n w_j = 1$, and analogously, for every upper boundary value $w_i^U, i \in \{1, \ldots, n\}$, there have to exist values $w_j \in [w_j^L, w_j^U]$, $j = 1, \ldots, n, j \neq i$, such that $w_i^U + \sum_{\substack{j=1 \\ j \neq i}}^n w_j = 1$.

Wang and Elhag (2006) formulated the following theorem to verify whether a set of intervals is normalized.

Theorem 15 *(Wang and Elhag 2006) Let $\overline{w}_i = [w_i^L, w_i^U], i = 1, \ldots, n$, be intervals, $\overline{w}_i \subseteq [0, 1]$, $i = 1, \ldots, n$. Then the intervals are normalized according to Definition 28 if and only if the inequalities*

$$w_i^L + \sum_{\substack{j=1 \\ j \neq i}}^n w_j^U \geq 1, \quad w_i^U + \sum_{\substack{j=1 \\ j \neq i}}^n w_j^L \leq 1 \qquad (3.13)$$

are satisfied for all $i = 1, \ldots, n$.

Remark 8 From the inequalities (3.13) it is obvious that a set of intervals $\overline{w}_i = [w_i^L, w_i^U]$, $i = 1, \ldots, n$, is normalized according to Definition 28 if and only if for any $w_i \in [w_i^L, w_i^U], i \in \{1, \ldots, n\}$, there exist $w_j \in [w_j^L, w_j^U], j = 1, \ldots, n, j \neq i$, such that $\sum_{k=1}^n w_k = 1$. In other words, any value w_i from the interval $\overline{w}_i = [w_i^L, w_i^U], i \in \{1, \ldots, n\}$, is a part of a normalized vector $\underline{w} = (w_1, \ldots, w_n)^T$ belonging to the set (3.12) of normalized vectors constructed from the intervals.

Based on Definition 28 and Remark 8, the definition of a normalized interval vector can be extended intuitively to a definition of a normalized trapezoidal fuzzy vector.

Definition 29 Let $\widetilde{w}_i = (w_i^\alpha, w_i^\beta, w_1^\gamma, w_1^\delta), i = 1, \ldots, n$, be trapezoidal fuzzy numbers, $\widetilde{w}_i \subseteq [0, 1]$, $i = 1, \ldots, n$. The trapezoidal fuzzy numbers $\widetilde{w}_1, \ldots, \widetilde{w}_n$ are said to be normalized if

$$\forall w_{i\alpha} \in \widetilde{w}_{i(\alpha)} \; \exists w_{j\alpha} \in \widetilde{w}_{j(\alpha)}, j = 1, \ldots, n, j \neq i : \; w_{i\alpha} + \sum_{\substack{j=1 \\ j \neq i}}^n w_{j\alpha} = 1 \qquad (3.14)$$

for all $\alpha \in [0, 1]$ and $i = 1, \ldots, n$.

The vector $\underline{\widetilde{w}} = (\widetilde{w}_1, \ldots, \widetilde{w}_n)^T \in \mathcal{F}_N(\mathbb{R})^n$ of normalized trapezoidal fuzzy numbers will be called the normalized fuzzy vector.

Theorem 15 can be extended to trapezoidal fuzzy numbers as follows.

Theorem 16 *Let $\widetilde{w}_i = (w_i^\alpha, w_i^\beta, w_i^\gamma, w_i^\delta), i = 1, \ldots, n$, be trapezoidal fuzzy numbers, $\widetilde{w}_i \subseteq [0, 1]$, $i = 1, \ldots, n$. Then the trapezoidal fuzzy numbers are normalized according to Definition 29 if and only if the inequalities*

$$w_i^\alpha + \sum_{\substack{j=1 \\ j \neq i}}^n w_j^\delta \geq 1, \quad w_i^\delta + \sum_{\substack{j=1 \\ j \neq i}}^n w_j^\alpha \leq 1, \quad w_i^\beta + \sum_{\substack{j=1 \\ j \neq i}}^n w_j^\gamma \geq 1, \quad w_i^\gamma + \sum_{\substack{j=1 \\ j \neq i}}^n w_j^\beta \leq 1$$

(3.15)

are satisfied for all $i = 1, \ldots, n$.

Proof First, let us show that (3.14) implies (3.15). For $\alpha = 0$ it follows from (3.14) that for $w_i^\alpha, i \in \{1, \ldots, n\}$, $\exists w_j \in \left[w_j^\alpha, w_j^\delta\right], j = 1, \ldots, n, j \neq i$: $w_i^\alpha + \sum_{\substack{j=1 \\ j \neq i}}^n w_j = 1$. Because $w_j^\delta \geq w_j$, then clearly $w_i^\alpha + \sum_{\substack{j=1 \\ j \neq i}}^n w_j^\delta \geq 1$. Similarly, for $w_i^\delta, i \in \{1, \ldots, n\}$, $\exists w_j \in \left[w_j^\alpha, w_j^\delta\right], j = 1, \ldots, n, j \neq i$: $w_i^\delta + \sum_{\substack{j=1 \\ j \neq i}}^n w_j = 1$. Because $w_j^\alpha \leq w_j$, then clearly $w_i^\delta + \sum_{\substack{j=1 \\ j \neq i}}^n w_j^\alpha \leq 1$. Analogously, for $\alpha = 1$, the inequalities $w_i^\beta + \sum_{\substack{j=1 \\ j \neq i}}^n w_j^\gamma \geq 1$ and $w_i^\gamma + \sum_{\substack{j=1 \\ j \neq i}}^n w_j^\beta \leq 1$ are derived from (3.14).

Now, let us show that (3.15) implies (3.14). From the inequalities $w_i^\delta + \sum_{\substack{j=1 \\ j \neq i}}^n w_j^\alpha \leq 1$ and $w_i^\alpha + \sum_{\substack{j=1 \\ j \neq i}}^n w_j^\delta \geq 1$, the inequalities $w_i + \sum_{\substack{j=1 \\ j \neq i}}^n w_j^\alpha \leq 1$ and $w_i + \sum_{\substack{j=1 \\ j \neq i}}^n w_j^\delta \geq 1$ follow $\forall w_i \in \left[w_i^\alpha, w_i^\delta\right]$. Therefore, $\exists w_j \in \left[w_j^\alpha, w_j^\delta\right], j = 1, \ldots, n, j \neq i$: $w_i + \sum_{\substack{j=1 \\ j \neq i}}^n w_j = 1$, which implies (3.14) for $\alpha = 0$. Analogously, from the inequalities $w_i^\gamma + \sum_{\substack{j=1 \\ j \neq i}}^n w_j^\beta \leq 1$ and $w_i^\beta + \sum_{\substack{j=1 \\ j \neq i}}^n w_j^\gamma \geq 1$, the inequalities $w_i + \sum_{\substack{j=1 \\ j \neq i}}^n w_j^\beta \leq 1$ and $w_i + \sum_{\substack{j=1 \\ j \neq i}}^n w_j^\gamma \geq 1$ follow $\forall w_i \in \left[w_i^\beta, w_i^\gamma\right]$. Therefore, $\exists w_j \in \left[w_j^\beta, w_j^\gamma\right], j = 1, \ldots, n, j \neq$: $w_i + \sum_{\substack{j=1 \\ j \neq i}}^n w_j = 1$, which implies (3.14) for $\alpha = 1$.

The proof of the validity of (3.14) for $\alpha \in]0, 1[$ is analogous; it is sufficient to show that the inequalities (3.15) hold also for the α-cuts $\widetilde{w}_{i(\alpha)} = \left[w_{i(\alpha)}^L, w_{i(\alpha)}^U\right]$ of the trapezoidal fuzzy numbers $\widetilde{w}_i = (w_i^\alpha, w_i^\beta, w_i^\gamma, w_i^\delta), i \in \{1, \ldots, n\}$, i.e.

$$w_{i(\alpha)}^L + \sum_{\substack{j=1 \\ j \neq i}}^n w_{j(\alpha)}^U \geq 1, \quad w_{i(\alpha)}^U + \sum_{\substack{j=1 \\ j \neq i}}^n w_{j(\alpha)}^L \leq 1.$$

(3.16)

Then it is enough to take the α-cuts $\widetilde{w}_{i(\alpha)} = \left[w_{i(\alpha)}^L, w_{i(\alpha)}^U\right]$ of $\widetilde{w}_i, i = 1, \ldots, n$, for $\left[w_i^\alpha, w_i^\delta\right]$ in the above part of the proof.

Using the definition (3.8) of α-cuts and formulas (3.15), we obtain

$$w_{i(\alpha)}^U + \sum_{\substack{j=1 \\ j \neq i}}^n w_{j(\alpha)}^L = \alpha w_i^\gamma + (1 - \alpha)w_i^\delta + \sum_{\substack{j=1 \\ j \neq i}}^n \left[\alpha w_j^\beta + (1 - \alpha)w_j^\alpha\right] =$$

Fig. 3.9 Normalized
triangular fuzzy numbers

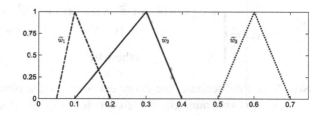

$$\alpha \left[w_i^{\gamma} + \sum_{\substack{j=1 \\ j \neq i}}^{n} w_j^{\beta} \right] + (1-\alpha) \left[w_i^{\delta} + \sum_{\substack{j=1 \\ j \neq i}}^{n} w_j^{\alpha} \right] \leq \alpha + (1-\alpha) = 1$$

and analogously the inequality $w_{i(\alpha)}^L + \sum_{\substack{j=1 \\ j \neq i}}^{n} w_{j(\alpha)}^U \geq 1$ could be demonstrated. $\square$

Remark 9 Note that for normalized triangular fuzzy numbers $\widetilde{w}_i = (w_i^L, w_i^M, w_i^U)$, $i = 1, \ldots, n$, the inequalities (3.15) are reduced to

$$w_i^L + \sum_{\substack{j=1 \\ j \neq i}}^{n} w_j^U \geq 1, \quad w_i^U + \sum_{\substack{j=1 \\ j \neq i}}^{n} w_j^L \leq 1, \quad \sum_{i=1}^{n} w_i^M = 1. \qquad (3.17)$$

Example 30 Triangular fuzzy numbers $\widetilde{w}_1 = (0.05, 0.1, 0.2)$, $\widetilde{w}_2 = (0.1, 0.3, 0.4)$, $\widetilde{w}_3 = (0.5, 0.6, 0.7)$, given in Fig. 3.9, are normalized. $\triangle$

3.4 Standard Fuzzy Arithmetic

In Definition 16, the extension principle for fuzzy sets was formulated. The extension principle (3.3) allows us to define arithmetic operations on fuzzy numbers. In this section, standard fuzzy arithmetic is defined by using both the membership functions and the α-cuts of fuzzy numbers. Afterwards, simplified standard fuzzy arithmetic is defined for trapezoidal fuzzy numbers.

There are four basic binary arithmetic operations on crisp numbers - addition, subtraction, multiplication, and division $(+, -, \cdot, /)$. Let $*$ denote one of them.

Definition 30 Let $\widetilde{n}_1, \widetilde{n}_2 \in \mathcal{F}_N(\mathbb{R})$ and let $* : \mathbb{R}^2 \to \mathbb{R}$ be a binary arithmetic operation, $z = x * y$. Then the extension of the arithmetic operation $*$ to fuzzy numbers, $* : \mathcal{F}_N(\mathbb{R})^2 \to \mathcal{F}_N(\mathbb{R})$, is defined as $\widetilde{n} = \widetilde{n}_1 * \widetilde{n}_2$ with the membership function

$$\tilde{n}(z) = \begin{cases} \sup \left\{ \min \left\{ \tilde{n}_1(x), \tilde{n}_2(y) \right\}; \ (x, y) \in \mathbb{R}^2 : z = x * y \right\} \\ \qquad \text{if } \left\{ (x, y) \in \mathbb{R}^2; \ z = x * y \right\} \neq \emptyset, \\ 0, \qquad\qquad\qquad \text{otherwise.} \end{cases} \quad (3.18)$$

Note 5 In the literature, the fuzzy extension of the binary arithmetic operations $+, -, \cdot, /$ to fuzzy numbers is often denoted as $\oplus, \ominus, \odot, \oslash$. Nevertheless, for simplicity of notation, the standard notation $+, -, \cdot, /$ will be used in this book for arithmetic operations defined on fuzzy numbers. Thus, for $\tilde{n}_1, \tilde{n}_2 \in \mathcal{F}_N(\mathbb{R})$, the notation $\tilde{n}_1 + \tilde{n}_2, \tilde{n}_1 - \tilde{n}_2, \tilde{n}_1 \cdot \tilde{n}_2$ (or simply just $\tilde{n}_1 \tilde{n}_2$), and $\tilde{n}_1 / \tilde{n}_2$ (or $\frac{\tilde{n}_1}{\tilde{n}_2}$) will be used.

Note 6 We know that division by 0 is not defined. Similarly, this limitation holds also for division of fuzzy numbers. In other words, $\tilde{n}_2 \in \mathcal{F}_N(\mathbb{R})$ in Definition 30 has to be such that $0 \notin Cl(Supp\, \tilde{n}_2)$ when division $\tilde{n}_1 / \tilde{n}_2$ is performed. Analogously, in the rest of this chapter, whenever arithmetic operation $* \in \{+, -, \cdot, /\}$ is considered on $\tilde{n}_1$ and $\tilde{n}_2$, it is automatically assumed that $0 \notin Cl(Supp\, \tilde{n}_2)$ for the case $* = /$.

Besides the four binary arithmetic operations, also the fuzzy extension of the p-th power of the variable is needed for the fuzzy extension of the methods reviewed in Chap. 2.

Definition 31 Let $\tilde{n} \in \mathcal{F}_N(\mathbb{R})$ and let $(.)^p : \mathbb{R} \to \mathbb{R}$, $p \in \mathbb{R}$, be the p-th power of the variable, $y = x^p$. Then the extension of $(.)^p$ to fuzzy numbers, $(.)^p : \mathcal{F}_N(\mathbb{R}) \to \mathcal{F}_N(\mathbb{R})$, is defined as $\tilde{m} = (\tilde{n})^p$ with the membership function

$$\tilde{m}(y) = \begin{cases} \sup \left\{ \tilde{n}(x); \ x \in \mathbb{R} : y = x^p \right\} & \text{if } \{x \in \mathbb{R}; \ y = x^p\} \neq \emptyset, \\ 0, & \text{otherwise.} \end{cases} \quad (3.19)$$

In particular, definitions of the reciprocal $\frac{1}{\tilde{n}}$ and of the k-th root $\sqrt[k]{\tilde{n}}$, $k \in \mathbb{N}$, of fuzzy number $\tilde{n} \in \mathcal{F}(\mathbb{R}^+)$ are needed. Note that the reciprocal $\frac{1}{\tilde{n}}$ is obtained by substituting $p = -1$ in Definition 31, and the k-th root $\sqrt[k]{\tilde{n}}$ is obtained by substituting $p = \frac{1}{k}$, $k \in \mathbb{N}$, in Definition 31.

In Sect. 3.2, it was shown that $\tilde{n} \in \mathcal{F}_N(\mathbb{R})$ can be represented uniquely by its α-cuts; $\tilde{n} = \bigcup_{\alpha=0}^{1} \alpha[n_{(\alpha)}^L, n_{(\alpha)}^U]$. This representation allows for an alternative definition of the fuzzy extension of arithmetic operations based on standard fuzzy arithmetic.

Definition 32 Let $\tilde{n}_1, \tilde{n}_2 \in \mathcal{F}_N(\mathbb{R})$ be given by their α-cuts as $\tilde{n}_1 = \bigcup_{\alpha=0}^{1} \alpha[n_{1(\alpha)}^L, n_{1(\alpha)}^U]$, $\tilde{n}_2 = \bigcup_{\alpha=0}^{1} \alpha[n_{2(\alpha)}^L, n_{2(\alpha)}^U]$. Further, let $* : \mathbb{R}^2 \to \mathbb{R}$ be a binary arithmetic operation, $z = x * y$. Then the extension of the arithmetic operation $*$ to fuzzy numbers, $* : \mathcal{F}_N(\mathbb{R})^2 \to \mathcal{F}_N(\mathbb{R})$, is defined as $\tilde{n} = \tilde{n}_1 * \tilde{n}_2$ with the α-cut representation $\tilde{n} = \bigcup_{\alpha=0}^{1} \alpha[n_{(\alpha)}^L, n_{(\alpha)}^U]$:

$$n^L_{(\alpha)} = \min \left\{ x * y; \ x \in [n^L_{1(\alpha)}, n^U_{1(\alpha)}], \ y \in [n^L_{2(\alpha)}, n^U_{2(\alpha)}] \right\},$$

$$n^U_{(\alpha)} = \max \left\{ x * y; \ x \in [n^L_{1(\alpha)}, n^U_{1(\alpha)}], \ y \in [n^L_{2(\alpha)}, n^U_{2(\alpha)}] \right\}. \tag{3.20}$$

Definition 33 Let $\tilde{n} \in \mathcal{F}_N(\mathbb{R})$ be given by its α-cuts as $\tilde{n} = \bigcup^1_{\alpha=0} \alpha[n^L_{(\alpha)}, n^U_{(\alpha)}]$. Further, let $(.)^p : \mathbb{R} \rightarrow \mathbb{R}$ be the p-th power of the variable, $y = x^p$. Then the extension of $(.)^p$ to fuzzy numbers, $(.)^p : \mathcal{F}_N(\mathbb{R}) \rightarrow \mathcal{F}_N(\mathbb{R})$, is defined as $\tilde{m} = (\tilde{n})^p$ with the α-cut representation $\tilde{m} = \bigcup^1_{\alpha=0} \alpha[m^L_{(\alpha)}, m^U_{(\alpha)}]$:

$$m^L_{(\alpha)} = \min \left\{ x^p; \ x \in [n^L_{(\alpha)}, n^U_{(\alpha)}] \right\},$$

$$m^U_{(\alpha)} = \max \left\{ x^p; \ x \in [n^L_{(\alpha)}, n^U_{(\alpha)}] \right\}. \tag{3.21}$$

As already mentioned in Sect. 3.3, triangular and trapezoidal fuzzy numbers and intervals are most often used for the fuzzy extension of MCDM methods based on PCMs. This class of fuzzy numbers is used for the fuzzy extension also in this book. Thus, the arithmetic operations are going to be introduced in detail for this particular class of fuzzy numbers. The α-cut representation of the fuzzy extension of arithmetic operations given by Definitions 32 and 33 is going to be used for this purpose as it is more convenient than the membership representation given by Definitions 30 and 31.

Let $\tilde{c} = \left(c^\alpha, c^\beta, c^\gamma, c^\delta\right)$ and $\tilde{d} = \left(d^\alpha, d^\beta, d^\gamma, d^\delta\right)$ be two trapezoidal fuzzy numbers. The sum of $\tilde{c}$ and $\tilde{d}$ obtained by applying the extension principle (3.20) is again a trapezoidal fuzzy number given as

$$\tilde{c} + \tilde{d} = \left(c^\alpha + d^\alpha, c^\beta + d^\beta, c^\gamma + d^\gamma, c^\delta + d^\delta\right), \tag{3.22}$$

and their difference is a trapezoidal fuzzy number given as

$$\tilde{c} - \tilde{d} = \left(c^\alpha - d^\delta, c^\beta - d^\gamma, c^\gamma - d^\beta, c^\delta - d^\alpha\right). \tag{3.23}$$

Unlike the sum and the difference, the product and the quotient of two trapezoidal fuzzy numbers as well as the reciprocal and the k-th root of a trapezoidal fuzzy number are not trapezoidal fuzzy numbers anymore when extension principle (3.20) is applied. Analogously, also the product and the quotient of two triangular fuzzy numbers as well as the reciprocal and the k-th root of a triangular fuzzy number are not in general triangular fuzzy numbers any more. However, for the sake of computational simplicity, it is a common practice in fuzzy MCDM based on FPCMs to approximate the results of these arithmetic operations by trapezoidal and triangular fuzzy numbers, respectively. Usually, authors do not even mention that simplified standard fuzzy arithmetic is used in their papers instead of standard fuzzy arithmetic.

According to simplified standard fuzzy arithmfetic, arithmetic operations are performed only on the representing values of trapezoidal fuzzy numbers, thus obtaining representing values of the resulting trapezoidal fuzzy numbers. This means that the product and the quotient of two trapezoidal fuzzy numbers $\tilde{c} = \left(c^\alpha, c^\beta, c^\gamma, c^\delta\right)$ and $\tilde{d} = \left(d^\alpha, d^\beta, d^\gamma, d^\delta\right)$ are trapezoidal fuzzy numbers given as $\tilde{p} = \tilde{c} \cdot \tilde{d} = \left(p^\alpha, p^\beta, p^\gamma, p^\delta\right)$ where

$$
\begin{aligned}
p^\alpha &= \min\left\{c^\alpha \cdot d^\alpha, c^\alpha \cdot d^\delta, c^\delta \cdot d^\alpha, c^\delta \cdot d^\delta\right\}, \\
p^\beta &= \min\left\{c^\beta \cdot d^\beta, c^\beta \cdot d^\gamma, c^\gamma \cdot d^\beta, c^\gamma \cdot d^\gamma\right\}, \\
p^\gamma &= \max\left\{c^\beta \cdot d^\beta, c^\beta \cdot d^\gamma, c^\gamma \cdot d^\beta, c^\gamma \cdot d^\gamma\right\}, \\
p^\delta &= \max\left\{c^\alpha \cdot d^\alpha, c^\alpha \cdot d^\delta, c^\delta \cdot d^\alpha, c^\delta \cdot d^\delta\right\},
\end{aligned}
\tag{3.24}
$$

and $\tilde{q} = \frac{\tilde{c}}{\tilde{d}} = \left(q^\alpha, q^\beta, q^\gamma, q^\delta\right), 0 \notin \left[d^\alpha, d^\delta\right]$, where

$$
\begin{aligned}
q^\alpha &= \min\left\{\frac{c^\alpha}{d^\alpha}, \frac{c^\alpha}{d^\delta}, \frac{c^\delta}{d^\alpha}, \frac{c^\delta}{d^\delta}\right\}, \\
q^\beta &= \min\left\{\frac{c^\beta}{d^\beta}, \frac{c^\beta}{d^\gamma}, \frac{c^\gamma}{d^\beta}, \frac{c^\gamma}{d^\gamma}\right\}, \\
q^\gamma &= \max\left\{\frac{c^\beta}{d^\beta}, \frac{c^\beta}{d^\gamma}, \frac{c^\gamma}{d^\beta}, \frac{c^\gamma}{d^\gamma}\right\}, \\
q^\delta &= \max\left\{\frac{c^\alpha}{d^\alpha}, \frac{c^\alpha}{d^\delta}, \frac{c^\delta}{d^\alpha}, \frac{c^\delta}{d^\delta}\right\},
\end{aligned}
\tag{3.25}
$$

respectively. Analogously, the reciprocal of a trapezoidal fuzzy number $\tilde{c} = \left(c^\alpha, c^\beta, c^\gamma, c^\delta\right)$, $c^\alpha > 0$, is a trapezoidal fuzzy number

$$
\frac{1}{\tilde{c}} = \left(\frac{1}{c^\delta}, \frac{1}{c^\gamma}, \frac{1}{c^\beta}, \frac{1}{c^\alpha}\right),
\tag{3.26}
$$

and the k-th root of $\tilde{c} = \left(c^\alpha, c^\beta, c^\gamma, c^\delta\right)$, $c^\alpha > 0$, is a trapezoidal fuzzy number

$$
\sqrt[k]{\tilde{c}} = \left(\sqrt[k]{c^\alpha}, \sqrt[k]{c^\beta}, \sqrt[k]{c^\gamma}, \sqrt[k]{c^\delta}\right).
\tag{3.27}
$$

In the fuzzy extension of the MCDM methods based on MPCMs and APCMs reviewed in Chap. 2, only positive fuzzy numbers are present. This enables us to further simplify the formulas (3.24) and (3.25). For $\tilde{c} = \left(c^\alpha, c^\beta, c^\gamma, c^\delta\right) \in \mathcal{F}_N(\mathbb{R}^+)$ and $\tilde{d} = \left(d^\alpha, d^\beta, d^\gamma, d^\delta\right) \in \mathcal{F}_N(\mathbb{R}^+)$, the formulas (3.24) and (3.25) are simplified to

$$
\tilde{c} \cdot \tilde{d} = \left(c^\alpha \cdot d^\alpha, c^\beta \cdot d^\beta, c^\gamma \cdot d^\gamma, c^\delta \cdot d^\delta\right)
\tag{3.28}
$$

and

$$
\frac{\tilde{c}}{\tilde{d}} = \left(\frac{c^\alpha}{d^\delta}, \frac{c^\beta}{d^\gamma}, \frac{c^\gamma}{d^\beta}, \frac{c^\delta}{d^\alpha}\right),
\tag{3.29}
$$

respectively.

The product, the quotient, the reciprocal, and the k-th root of trapezoidal fuzzy numbers (3.24)–(3.27), respectively, obtained by simplified standard fuzzy arithmetic have the same representing values as the results of these arithmetic operations obtained by properly applying extension principles (3.20) and (3.21), respectively, in standard fuzzy arithmetic. That means that the support and the core of the results of these arithmetic operations are determined correctly by applying simplified standard fuzzy arithmetic, and the left and right sides of the resulting fuzzy numbers are approximated by linear functions. This approximation is generally accepted as sufficient in the literature on the fuzzy extension of PC methods.

The following example is provided in order to demonstrate better the difference between the standard fuzzy arithmetic and the simplified standard fuzzy arithmetic.

Example 31 Let us consider the trapezoidal fuzzy number $\widetilde{c} = (2, 3, 4, 6)$ and the triangular fuzzy number $\widetilde{d} = (1, 2.5, 3)$. Clearly, $\widetilde{d}$ can be written as a trapezoidal fuzzy number in the form $\widetilde{d} = (1, 2.5, 2.5, 3)$. By applying the formulas (3.22), (3.23), and (3.26)–(3.29) based on simplified standard fuzzy arithmetic to the computation with the trapezoidal fuzzy numbers $\widetilde{c}$ and $\widetilde{d}$, we obtain the trapezoidal fuzzy numbers

$$\widetilde{c} + \widetilde{d} = (3, 5.5, 6.5, 9), \quad \widetilde{c} - \widetilde{d} = (-1, 0.5, 1.5, 5),$$

$$\widetilde{c} \cdot \widetilde{d} = (2, 7.5, 10, 18), \quad \frac{\widetilde{c}}{\widetilde{d}} = \left(\tfrac{2}{3}, \tfrac{6}{5}, \tfrac{8}{5}, 6\right),$$

$$\frac{1}{\widetilde{c}} = \left(\frac{1}{6}, \frac{1}{4}, \frac{1}{3}, \frac{1}{2}\right), \quad \sqrt[2]{\widetilde{c}} = (\sqrt[2]{2}, \sqrt[2]{3}, 2, \sqrt[2]{6}).$$

The resulting trapezoidal fuzzy numbers are represented in Fig. 3.10 together with the actual results of the arithmetic operations given by the extension principles (3.20) and (3.21). The actual results of the arithmetic operations are given by a dotted line and their trapezoidal approximations are given by a solid line. Notice that the sum and the difference of the trapezoidal fuzzy numbers $\widetilde{c}$ and $\widetilde{d}$ obtained by using standard fuzzy arithmetic are again trapezoidal fuzzy numbers. In fact, as it is obvious from Fig. 3.10, the results obtained by using standard fuzzy arithmetic coincide with the results obtained by using simplified standard fuzzy arithmetic. Contrarily, the results of other four arithmetic operations obtained by applying standard fuzzy arithmetic are not trapezoidal fuzzy numbers anymore. Nevertheless, as it is obvious from Fig. 3.10, the trapezoidal approximations of these results obtained by applying simplified standard fuzzy arithmetic have the same support and the same core as the actual results obtained by applying standard fuzzy arithmetic.   △

Since intervals are a particular case of trapezoidal fuzzy numbers, the arithmetic operations with intervals are performed according to the formulas (3.22)–(3.27) as well. Recall that interval $\overline{c} = [c^L, c^U]$ can be easily written as trapezoidal fuzzy number $\widetilde{c} = (c^L, c^L, c^U, c^U)$. However, intervals are also a particular class of crisp sets on $\mathbb{R}$ with a well-defined interval arithmetic for performing arithmetic operations on intervals. The interval arithmetic allows us to perform arithmetic operations on intervals in a much simpler way than the fuzzy arithmetic does. Nevertheless, both interval arithmetic and standard fuzzy arithmetic applied to intervals provide the

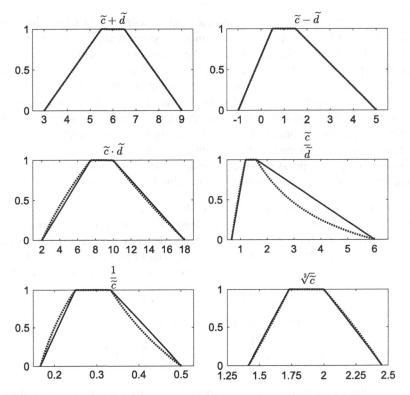

Fig. 3.10 Arithmetic operations with trapezoidal fuzzy numbers

same results. Unlike for the case of triangular and trapezoidal fuzzy numbers, the results of arithmetic operations with intervals are again intervals. Thus, there is no need for defining simplified standard fuzzy (interval) arithmetic for intervals. Let $\bar{u} = [u^L, u^U]$ and $\bar{v} = [v^L, v^U]$ be two positive intervals, i.e. $u^L > 0$, $v^L > 0$. Then the arithmetic operations are defined by using standard interval arithmetic as

$$\bar{u} + \bar{v} = [u^L + v^L, u^U + v^U], \tag{3.30}$$

$$\bar{u} - \bar{v} = [u^L - v^U, u^U - v^L], \tag{3.31}$$

$$\bar{u} \cdot \bar{v} = [u^L \cdot v^L, u^U \cdot v^U], \tag{3.32}$$

$$\frac{\bar{u}}{\bar{v}} = \left[\frac{u^L}{v^U}, \frac{u^U}{v^L} \right], \tag{3.33}$$

$$\frac{1}{\bar{u}} = \left[\frac{1}{u^U}, \frac{1}{u^L} \right], \tag{3.34}$$

$$\sqrt[k]{\tilde{u}} = \left[\sqrt[k]{u^L}, \sqrt[k]{u^U} \right]. \tag{3.35}$$

The extension of single arithmetic operations on fuzzy numbers to functions combining the arithmetic operations is straightforward.

Definition 34 Let $\tilde{n}_i \in \mathcal{F}_N(\mathbb{R}), i = 1, \ldots, k$, and let $f : \mathbb{R}^k \to \mathbb{R}$ be a function defined by means of any combination of arithmetic operations, $z = f(x_1, \ldots, x_k)$. Then the extension of the function f to fuzzy numbers, $f : \mathcal{F}_N(\mathbb{R})^k \to \mathcal{F}_N(\mathbb{R})$, is defined as $\tilde{n} = f(\tilde{n}_1, \ldots, \tilde{n}_k)$ with the membership function

$$\tilde{n}(z) = \begin{cases} \sup \left\{ \min \{\tilde{n}_1(x_1), \ldots, \tilde{n}_k(x_k)\}; \ (x_1, \ldots, x_k) \in \mathbb{R}^k : z = f(x_1, \ldots, x_k) \right\} \\ \qquad \text{if } \left\{ (x_1, \ldots, x_k) \in \mathbb{R}^k; \ z = f(x_1, \ldots, x_k) \right\} \neq \emptyset, \\ 0, \qquad \text{otherwise.} \end{cases} \tag{3.36}$$

Definition 35 Let $\tilde{n}_i \in \mathcal{F}_N(\mathbb{R})$ be given by their α-cuts as $\tilde{n}_i = \bigcup_{\alpha=0}^{1} \alpha[n_{i(\alpha)}^L, n_{i(\alpha)}^U]$, $i = 1, \ldots, k$. Further, let $f : \mathbb{R}^k \to \mathbb{R}$ be a function defined by means of any combination of arithmetic operations, $z = f(x_1, \ldots, x_k)$. Then the extension of the function f to fuzzy numbers, $f : \mathcal{F}_N(\mathbb{R})^k \to \mathcal{F}_N(\mathbb{R})$, is defined as $\tilde{n} = f(\tilde{n}_1, \ldots, \tilde{n}_k)$ with the α-cut representation $\tilde{n} = \bigcup_{\alpha=0}^{1} \alpha[n_{(\alpha)}^L, n_{(\alpha)}^U]$:

$$\begin{aligned} n_{(\alpha)}^L &= \min \left\{ f(x_1, \ldots, x_k); \ x_i \in [n_{i(\alpha)}^L, n_{i(\alpha)}^U], \ i = 1, \ldots, k \right\}, \\ n_{(\alpha)}^U &= \max \left\{ f(x_1, \ldots, x_k); \ x_i \in [n_{i(\alpha)}^L, n_{i(\alpha)}^U], \ i = 1, \ldots, k \right\}. \end{aligned} \tag{3.37}$$

The result of applying standard fuzzy arithmetic in Definitions 34 and 35 to the computations with trapezoidal fuzzy numbers can be approximated according to simplified standard fuzzy arithmetic by a trapezoidal fuzzy number $\tilde{n} = (n^\alpha, n^\beta, n^\gamma, n^\delta)$:

$$\begin{aligned} n^\alpha &= \min \left\{ f(x_1, \ldots, x_k); \ x_i \in [n_i^\alpha, n_i^\delta], \ i = 1, \ldots, k \right\}, \\ n^\beta &= \min \left\{ f(x_1, \ldots, x_k); \ x_i \in [n_i^\beta, n_i^\gamma], \ i = 1, \ldots, k \right\}, \\ n^\gamma &= \max \left\{ f(x_1, \ldots, x_k); \ x_i \in [n_i^\beta, n_i^\gamma], \ i = 1, \ldots, k \right\}, \\ n^\delta &= \max \left\{ f(x_1, \ldots, x_k); \ x_i \in [n_i^\alpha, n_i^\delta], \ i = 1, \ldots, k \right\}. \end{aligned} \tag{3.38}$$

3.5 Constrained Fuzzy Arithmetic

Constrained fuzzy arithmetic was introduced by Klir (1997) and Klir and Pan (1998) to handle correctly arithmetic operations on fuzzy numbers in the presence of constraints on operands. "When arithmetic operations are performed on real numbers, they follow unique rules that are independent of what is represented by the numbers involved. That is, the result of each particular arithmetic operation on real numbers depends only on the numbers involved and not on the entities represented by the numbers. As it is well known, the validity of this simple principle is also tacitly assumed in the usual interval arithmetic as well as fuzzy arithmetic" (Klir (1997), p. 167). Klir (1997) and Klir and Pan (1998) argued that this principle is valid neither for interval arithmetic nor for fuzzy arithmetic. Contrarily to standard arithmetic on real numbers, interval arithmetic and fuzzy arithmetic depend not only on the intervals or fuzzy numbers involved, but also on their meanings that impose constraints on the operands. These constraints often appear when different operands represent the same linguistic variable or when there are any relations among the operands.

Example 32 Let us consider function $f(x) = x - x$. Clearly, $f(x) = 0$ for any $x \in \mathbb{R}$ since both operands in the subtraction are the same.

Now let us examine the extension of the function f to fuzzy numbers $\tilde{n} \in \mathcal{F}_N(\mathbb{R})$. According to standard fuzzy arithmetic (in particular Definition 30) the fuzzy extension of the function f is given as $\tilde{m} = f(\tilde{n}) = \tilde{n} - \tilde{n}$ with the membership function

$$
\tilde{m}(z) = \begin{cases} \sup \left\{ \min \left\{ \tilde{n}(x), \tilde{n}(y) \right\} \right\}; \ (x, y) \in \mathbb{R}^2 : z = x - y \\ \quad \text{if } \left\{ (x, y) \in \mathbb{R}^2; \ z = x - y \right\} \neq \emptyset, \\ 0, \qquad\qquad \text{otherwise.} \end{cases} \tag{3.39}
$$

For trapezoidal fuzzy numbers $\tilde{n} = (n^\alpha, n^\beta, n^\gamma, n^\delta) \in \mathcal{F}_N(\mathbb{R})$ in particular, we obtain

$$
\tilde{m} = f(\tilde{n}, \tilde{n}) = \tilde{n} - \tilde{n} = (n^\alpha - n^\delta, n^\beta - n^\gamma, n^\gamma - n^\beta, n^\delta - n^\alpha) \neq 0. \tag{3.40}
$$

Let us consider a trivial example: We have a bottle of water and we drink it all. How much is left? Nothing, right? How did we arrive to this simple solution? The problem can be solved by using the function f defined above. When there is 1 l of water and we drink it all then there is $f(1) = 1 - 1 = 0$ l left. When there is 650 ml and we drink it all then there is again $f(650) = 650 - 650 = 0$ ml left.

Now let us consider we do not know precisely the amount of water in the bottle. Let us say there is about 1 l, which can be described by the triangular fuzzy number $\tilde{n} = (0.95, 1, 1.05)$, for example. How much will be now left when we drink it all? The common sense again suggests that nothing should be left. However, when we apply standard fuzzy arithmetic to find the solution of this simple problem, we get into difficulties. By using the formula (3.40) we obtain $f(\tilde{n}) = \tilde{n} - \tilde{n} = (0.95 - 1.05, 1 - 1, 1.05 - 0.95) = (-0.1, 0, 0.1)$. So according to the standard fuzzy arithmetic there should be something between -0.1 l and 0.1 l left. This conclusion is obviously nonsensical.

The problem is that we did not consider the meaning of the operands in the subtraction. From the definition of the problem, it is clear that the operands of the subtraction are in interaction. In particular, only the same values of operands are admissible; whatever the amount of water in the bottle is, we drink this exact amount. This constraint has to be considered in the computations. △

Klir (1997) and Klir and Pan (1998) examined various types of constraints on operands in constrained fuzzy arithmetic, in particular equality, inequality, and probabilistic constraints. In this book, only equality constraints are of interest.

Definition 36 Let $\tilde{n}_1, \tilde{n}_2 \in \mathcal{F}_N(\mathbb{R})$ and let $* : \mathbb{R}^2 \to \mathbb{R}$ be a binary arithmetic operation, $z = x * y$. Further, let the equality constraint $g(x, y) = 0$ express an interaction between the operands. Then, considering the constraint $g(x, y) = 0$ on the values of the operands, the extension of the arithmetic operation $*$ to fuzzy numbers, $* : \mathcal{F}_N(\mathbb{R})^2 \to \mathcal{F}(\mathbb{R})$, is defined as $\tilde{n} = \tilde{n}_1 * \tilde{n}_2$ with the membership function

$$\tilde{n}(z) = \begin{cases} \sup \left\{ \min \left\{ \tilde{n}_1(x), \tilde{n}_2(y) \right\}; \ (x, y) \in \mathbb{R}^2 : z = x * y, \ g(x, y) = 0 \right\} \\ \qquad \text{if } \left\{ (x, y) \in \mathbb{R}^2; \ z = x * y, \ g(x, y) = 0 \right\} \neq \emptyset, \quad (3.41) \\ 0, \qquad \text{otherwise.} \end{cases}$$

Note that the result of constrained fuzzy arithmetic given by the extension principle (3.41) on fuzzy numbers is a fuzzy set but not a fuzzy number in general. This is caused by the presence of the interaction constraint $g(x, y) = 0$. To guarantee that the result of constrained fuzzy arithmetic applied to fuzzy numbers is again a fuzzy number, i.e. that $*$ in Definition 36 is such that $* : \mathcal{F}_N(\mathbb{R})^2 \to \mathcal{F}_N(\mathbb{R})$, further requirements on the constraint $g(x, y) = 0$ have to be imposed.

Some constraining requirements were given by Klir and Pan (1998). In this book, only a particular type of equality constraint $g(x, y) = 0$ is needed to extend appropriately the formulas defined for PCMs in Chap. 2 to fuzzy PCMs. The equality constraints of the type $g(x, y) = 0$ that are going to be applied in Chap. 4 are such that $G = \left\{ (x, y) \in \mathbb{R}^2; \ g(x, y) = 0 \right\}$ is a connected set.[2] Being G a connected set, it is sufficient to require

$$\left\{ (x, y) \in \mathbb{R}^2; \ \tilde{n}_1(x) = 1, \ \tilde{n}_2(y) = 1, \ g(x, y) = 0 \right\} \neq \emptyset$$

to guarantee that the result of the constrained fuzzy arithmetic on fuzzy numbers given by the extension principle (3.41) is again a fuzzy number.

Analogously to standard fuzzy arithmetic, the α-cut representation of fuzzy numbers can be conveniently used also to define constrained fuzzy arithmetic.

Definition 37 Let $\tilde{n}_1, \tilde{n}_2 \in \mathcal{F}_N(\mathbb{R})$ be given by their α-cuts as $\tilde{n}_1 = \bigcup_{\alpha=0}^{1} \alpha[n_{1(\alpha)}^L, n_{1(\alpha)}^U]$, $\tilde{n}_2 = \bigcup_{\alpha=0}^{1} \alpha[n_{2(\alpha)}^L, n_{2(\alpha)}^U]$. Further, let $* : \mathbb{R}^2 \to \mathbb{R}$ be a binary arithmetic operation, $z = x * y$, and let the equality constraint $g(x, y) = 0$ express an interaction between the operands. If $G = \left\{ (x, y) \in \mathbb{R}^2; \ g(x, y) = 0 \right\}$ is a connected set and

---

[2] Set G is connected if it cannot be divided into two disjoint closed sets.

if $\{(x, y) \in \mathbb{R}^2; \, \tilde{n}_1(x) = 1, \, \tilde{n}_2(y) = 1, \, g(x, y) = 0\} \neq \emptyset$, then the extension of the arithmetic operation $*$ to fuzzy numbers, $* : \mathcal{F}_N(\mathbb{R})^2 \to \mathcal{F}_N(\mathbb{R})$, is defined as $\tilde{n} = \tilde{n}_1 * \tilde{n}_2$ with the α-cut representation $\tilde{n} = \bigcup_{\alpha=0}^{1} \alpha[n_{(\alpha)}^L, n_{(\alpha)}^U]$:

$$
\begin{aligned}
n_{(\alpha)}^L &= \min \left\{ x * y; \ x \in [n_{1(\alpha)}^L, n_{1(\alpha)}^U], \ y \in [n_{2(\alpha)}^L, n_{2(\alpha)}^U], \ g(x, y) = 0 \right\}, \\
n_{(\alpha)}^U &= \max \left\{ x * y; \ x \in [n_{1(\alpha)}^L, n_{1(\alpha)}^U], \ y \in [n_{2(\alpha)}^L, n_{2(\alpha)}^U], \ g(x, y) = 0 \right\}.
\end{aligned}
\tag{3.42}
$$

As already mentioned in the previous section, simplified standard fuzzy arithmetic is commonly used in fuzzy PC methods in order to keep the computational procedure simple. The results of arithmetic operations with triangular or trapezoidal fuzzy numbers are thus still triangular or trapezoidal fuzzy numbers, respectively, whose supports and cores correspond to the supports and cores of the actual results of the arithmetic operations determined precisely by applying extension principles (3.41) and (3.42). Recall that there is no need for simplified fuzzy arithmetic to perform arithmetic operations on intervals since the results are always intervals.

In order to be consistent with this approach, it is necessary to apply the simplified version of fuzzy arithmetic also when there appear any constraints on the operands in the computational procedure. This basically means that we want to reflect the constraints given on operands in the outcome of the computation, but, at the same time, we want to approximate the outcome by a triangular or trapezoidal fuzzy number, respectively, to keep the computational procedure as simple as possible still obtaining reliable results. Simplified constrained fuzzy arithmetic-a combination of simplified fuzzy arithmetic and constrained fuzzy arithmetic-is thus needed.

According to the simplified constrained fuzzy arithmetic, for two trapezoidal fuzzy numbers $\tilde{n}_1 = (n_1^\alpha, n_1^\beta, n_1^\gamma, n_1^\delta), \tilde{n}_2 = (n_2^\alpha, n_2^\beta, n_2^\gamma, n_2^\delta) \in \mathcal{F}_N(\mathbb{R})$, only the representing values $n^\alpha, n^\beta, n^\gamma, n^\delta$ of the resulting fuzzy number $\tilde{n} = \tilde{n}_1 * \tilde{n}_2$ are computed. However, unlike in the case of simplified standard fuzzy arithmetic, the representing values of $\tilde{n}$ are not obtained by performing the arithmetic operation $*$ on the representing values of the fuzzy numbers $\tilde{n}_1, \tilde{n}_2$. The formulas for performing arithmetic operations on trapezoidal fuzzy numbers based on the simplified constrained fuzzy arithmetic are more complex than the formulas based on the simplified standard fuzzy arithmetic; solving optimization problems is necessary to obtain the representing values of the resulting trapezoidal fuzzy number $\tilde{n} = (n^\alpha, n^\beta, n^\gamma, n^\delta)$.

Let $* : \mathbb{R}^2 \to \mathbb{R}$ be an arithmetic operation, $z = x * y$, and let $g(x, y) = 0$ represent a constraint imposed on the operands, $G = \{(x, y) \in \mathbb{R}^2; \, g(x, y) = 0\}$ being a connected set. Further, let $\tilde{n}_1 = (n_1^\alpha, n_1^\beta, n_1^\gamma, n_1^\delta), \tilde{n}_2 = (n_2^\alpha, n_2^\beta, n_2^\gamma, n_2^\delta) \in \mathcal{F}_N(\mathbb{R})$ be trapezoidal fuzzy numbers such that $\{(x, y) \in \mathbb{R}^2; \, \tilde{n}_1(x) = 1, \, \tilde{n}_2(y) = 1, \, g(x, y) = 0\} \neq \emptyset$. Then the fuzzy extension of the arithmetic operation $*$ to trapezoidal fuzzy numbers $\tilde{n}_1$ and $\tilde{n}_2$ based on the simplified constrained fuzzy arithmetic is defined as $\tilde{n} = \tilde{n}_1 * \tilde{n}_2$ with the representation $\tilde{n} = (n^\alpha, n^\beta, n^\gamma, n^\delta)$:

Fig. 3.11 Product of $\widetilde{c}$ and $\widetilde{d}$ obtained by standard fuzzy arithmetic and by simplified standard fuzzy arithmetic

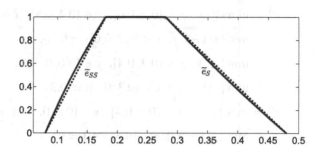

$$n^\alpha = \min\left\{x * y;\ x \in [n_1^\alpha, n_1^\delta],\ y \in [n_2^\alpha, n_2^\delta],\ g(x, y) = 0\right\},$$

$$n^\beta = \min\left\{x * y;\ x \in [n_1^\beta, n_1^\gamma],\ y \in [n_2^\beta, n_2^\gamma],\ g(x, y) = 0\right\},$$

$$n^\gamma = \max\left\{x * y;\ x \in [n_1^\beta, n_1^\gamma],\ y \in [n_2^\beta, n_2^\gamma],\ g(x, y) = 0\right\},$$

$$n^\delta = \max\left\{x * y;\ x \in [n_1^\alpha, n_1^\delta],\ y \in [n_2^\alpha, n_2^\delta],\ g(x, y) = 0\right\}.$$

(3.43)

Difference between standard and constrained fuzzy arithmetic is demonstrated on the following illustrative example.

Example 33 Let us consider trapezoidal fuzzy number $\widetilde{c} = (0.2, 0.3, 0.4, 0.6)$ and let us compute trapezoidal fuzzy number $\widetilde{d}$ as $\widetilde{d} = 1 - \widetilde{c} = (0.4, 0.6, 0.7, 0.8)$. Clearly, trapezoidal fuzzy numbers $\widetilde{c}$ and $\widetilde{d}$ are in relation; to any value $x \in Cl(Supp\ \widetilde{c})$ corresponds a value $y \in Cl(Supp\ \widetilde{d})$ such that $y = 1 - x$.

Let us compute the product $\widetilde{e} = \widetilde{c} \cdot \widetilde{d}$. By applying the simplified standard fuzzy arithmetic, and in particular formula (3.28), we obtain trapezoidal fuzzy number $\widetilde{e}_{SS} = (0.08, 0.18, 0.28, 0.48)$. In Fig. 3.11, you can compare this trapezoidal approximation with the actual outcome $\widetilde{e}_S$ of the multiplication obtainable by applying standard fuzzy arithmetic (3.20).

The relation $y = 1 - x$ between the operands represented by trapezoidal fuzzy numbers $\widetilde{c}$ and $\widetilde{d}$ was not taken into account when computing their product by using the formulas (3.28) and (3.20). Thus, the obtained results $\widetilde{e}_{SS}$ and $\widetilde{e}_S$ are both imprecise, too vague. In order to eliminate the excessive vagueness, it is necessary to apply properly the constrained fuzzy arithmetic (3.42) or, alternatively, the simplified constrained fuzzy arithmetic (3.43).

First, let us verify the requirements for the constraint on the operands. The constraint $y = 1 - x$ can be written as $g(x, y) = x + y - 1 = 0$. The set $G = \left\{(x, y) \in \mathbb{R}^2;\ x + y - 1 = 0\right\}$ is clearly connected and $\{(x, y) \in \mathbb{R}^2;\ \widetilde{c}(x) = 1,\ \widetilde{d}(y) = 1, x + y - 1 = 0\} = \{(x, 1 - x) \in [0.3, 0.4] \times [0.6, 0.7]\} \neq \emptyset$. This guarantees that the result of the constrained fuzzy arithmetic applied to $\widetilde{c}$ and $\widetilde{d}$ is again a fuzzy number. By applying (3.43) with the function $f : \mathbb{R}^2 \to \mathbb{R}$ in the form $f(x, y) = x \cdot y$ and the constraint $g(x, y) = 0$ in the form $x + y - 1 = 0$, we obtain trapezoidal fuzzy number $\widetilde{e}_{SC} = (e^\alpha, e^\beta, e^\gamma, e^\delta)$:

$$e^\alpha = \min\{x \cdot y; \ x \in [0.2, 0.6], \ y \in [0.4, 0.8], \ x + y - 1 = 0\} =$$
$$\min\{x(1 - x); \ x \in [0.2, 0.6]\} = 0.16,$$

$$e^\beta = \min\{x \cdot y; \ x \in [0.3, 0.4], \ y \in [0.6, 0.7], \ x + y - 1 = 0\} =$$
$$\min\{x(1 - x); \ x \in [0.3, 0.4]\} = 0.21,$$

$$e^\gamma = \max\{x \cdot y; \ x \in [0.3, 0.4], \ y \in [0.6, 0.7], \ x + y - 1 = 0\} = \quad\quad (3.44)$$
$$\max\{x(1 - x); \ x \in [0.3, 0.4]\} = 0.24,$$

$$e^\delta = \max\{x \cdot y; \ x \in [0.2, 0.6], \ y \in [0.4, 0.8], \ x + y - 1 = 0\} =$$
$$\max\{x(1 - x); \ x \in [0.2, 0.6]\} = 0.25.$$

Again, trapezoidal approximation $\widetilde{e}_{SC}$ is displayed in Fig. 3.12 together with the actual outcome $\widetilde{e}_C$ of the constrained fuzzy arithmetic performed by applying (3.42).

The product $\widetilde{e}_{SC} = (0.16, 0.21, 0.24, 0.25)$ obtained by simplified constrained fuzzy arithmetic is significantly less vague than the product $\widetilde{e}_{SS} = (0.08, 0.18, 0.28, 0.48)$ obtained by simplified standard fuzzy arithmetic. The difference in vagueness of both trapezoidal fuzzy numbers is even more noticeable from graphical representation, see Fig. 3.13. It is clearly visible from the figure how significant the reduction of vagueness is when applying properly (simplified) constrained fuzzy arithmetic instead of (simplified) standard fuzzy arithmetic. Therefore, in order to obtain more reliable results, it is indispensable to take into account all relations between operands when performing arithmetic operations on fuzzy numbers.   △

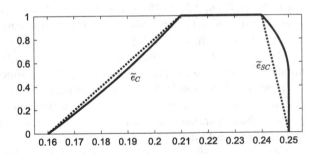

Fig. 3.12 Product of $\widetilde{c}$ and $\widetilde{d}$ obtained by constrained fuzzy arithmetic and by simplified constrained fuzzy arithmetic

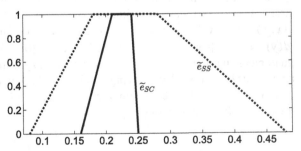

Fig. 3.13 Product of $\widetilde{c}$ and $\widetilde{d}$ obtained by constrained and standard fuzzy arithmetics

Example 34 In Example 32, a simple problem that cannot be solved by using standard fuzzy arithmetic was shown. Let us now apply constrained fuzzy arithmetic to the same problem. As already argued in Example 32, the operands in the subtraction $f(\widetilde{n}, \widetilde{n}) = \widetilde{n} - \widetilde{n}$ are in interaction—they are equal. This interaction can be modeled by the function $g(x, y) = x - y = 0$.

The set $G = \{(x, y) \in \mathbb{R}^2; g(x, y) = x - y = 0\}$ is connected and, further, $\{(x, y) \in \mathbb{R}^2; \widetilde{n}(x) = 1, \widetilde{n}(y) = 1, g(x, y) = x - y = 0\} = \{(x, x) \in \mathbb{R}^2; x \in Core\,\widetilde{n}\} \neq \emptyset$ for $\widetilde{n} \in \mathcal{F}_N(\mathbb{R})$. Applying the extension principle (3.42) to the function $f(\widetilde{n}) = \widetilde{n} - \widetilde{n}$ with the entry $\widetilde{n} = (0.95, 1, 1.05)$, we obtain $\widetilde{m} = (m^L, m^M, m^U) = \widetilde{n} - \widetilde{n}$:

$$
\begin{aligned}
m^L &= \min\{x - y; \ x \in [0.95, 1.05], \ y \in [0.95, 1.05], \ x - y = 0\} = \\
&\quad \min\{x - x; \ x \in [0.95, 1.05]\} = 0, \\
m^M &= \min\{x - y; \ x \in [1, 1], \ y \in [1, 1], \ x - y = 0\} = \\
&\quad \min\{x - x; \ x = 1\} = 0, \\
m^U &= \max\{x - y; \ x \in [0.95, 1.05], \ y \in [0.95, 1.05], \ x - y = 0\} = \\
&\quad \max\{x - x; \ x \in [0.95, 1.05]\} = 0.
\end{aligned}
$$

Thus, we finally obtain the correct solution to our problem. Namely, when there is about 1 l of water ($\widetilde{n} = (0.95, 1, 1.05)$) in the bottle and we drink it all then nothing is left; $\widetilde{n} - \widetilde{n} = (0.95, 1, 1.05) - (0.95, 1, 1.05) = 0$. △

The extension of single arithmetic operations with constraints on operands to functions combining arithmetic operations is straightforward.

Definition 38 Let $\widetilde{n}_i \in \mathcal{F}_N(\mathbb{R})$, $i = 1, \ldots, k$. Further, let $f : \mathbb{R}^k \to \mathbb{R}$ be a function defined by means of any combination of arithmetic operations, $z = f(x_1, \ldots, x_k)$, and let the equality constraints $g_j(x_1, \ldots, x_k) = 0$, $j = 1, \ldots, l$, express interactions between the operands. If $G_j = \{(x_1, \ldots, x_k) \in \mathbb{R}^k; g_j(x_1, \ldots, x_k) = 0\}$, $j = 1, \ldots, l$, are connected sets and if $\{(x_1, \ldots, x_k) \in \mathbb{R}^k; \widetilde{n}_i(x_i) = 1, i = 1, \ldots, k, g_j(x_1, \ldots, x_k) = 0, j = 1, \ldots, l\} \neq \emptyset$, then the extension of the function f to fuzzy numbers, $f : \mathcal{F}_N(\mathbb{R})^k \to \mathcal{F}_N(\mathbb{R})$, is defined as $\widetilde{n} = f(\widetilde{n}_1, \ldots, \widetilde{n}_k)$ with the membership function

$$
\widetilde{n}(z) = \begin{cases} \sup\left\{\min\{\widetilde{n}_1(x_1), \ldots, \widetilde{n}_k(x_k)\}; \begin{array}{l} (x_1, \ldots, x_k) \in \mathbb{R}^k: z = f(x_1, \ldots, x_k), \\ g_j(x_1, \ldots, x_k) = 0, \ j = 1, \ldots, l \end{array}\right\} \\ \qquad \text{if } \left\{\begin{array}{l}(x_1, \ldots, x_k) \in \mathbb{R}^k; \ z = f(x_1, \ldots, x_k), \\ g_j(x_1, \ldots, x_k) = 0, \ j = 1, \ldots, l \end{array}\right\} \neq \emptyset, \\ 0, \qquad \text{otherwise.} \end{cases}
$$

(3.45)

Definition 39 Let $\widetilde{n}_i \in \mathcal{F}_N(\mathbb{R})$ be given by their α-cuts as $\widetilde{n}_i = \bigcup_{\alpha=0}^1 \alpha[n^L_{i(\alpha)}, n^U_{i(\alpha)}]$, $i = 1, \ldots, k$. Further, let $f : \mathbb{R}^k \to \mathbb{R}$ be a function defined by means of any combination of arithmetic operations, $z = f(x_1, \ldots, x_k)$, and let the equality constraints

$g_j(x_1, \ldots, x_k) = 0$, $j = 1, \ldots, l$, express interactions between the operands. If $G_j = \{(x_1, \ldots, x_k) \in \mathbb{R}^k;\ g_j(x_1, \ldots, x_k) = 0\}$, $j = 1, \ldots, l$, are connected sets and if $\{(x_1, \ldots, x_k) \in \mathbb{R}^k;\ \widetilde{n}_i(x_i) = 1,\ i = 1, \ldots, k,\ g_j(x_1, \ldots, x_k) = 0,\ j = 1, \ldots, l\} \neq \emptyset$, then the extension of the function f to fuzzy numbers, $f : \mathcal{F}_N(\mathbb{R})^k \to \mathcal{F}_N(\mathbb{R})$, is defined as $\widetilde{n} = f(\widetilde{n}_1, \ldots, \widetilde{n}_k)$ with the α-cut representation $\widetilde{n} = \bigcup_{\alpha=0}^{1} \alpha[n_{(\alpha)}^L, n_{(\alpha)}^U]$:

$$
\begin{aligned}
n_{(\alpha)}^L &= \min \left\{ f(x_1, \ldots, x_k); \begin{array}{l} x_i \in [n_{i(\alpha)}^L, n_{i(\alpha)}^U],\ i = 1, \ldots, k, \\ g_j(x_1, \ldots, x_k) = 0,\ j = 1, \ldots, l \end{array} \right\}, \\
n_{(\alpha)}^U &= \max \left\{ f(x_1, \ldots, x_k); \begin{array}{l} x_i \in [n_{i(\alpha)}^L, n_{i(\alpha)}^U],\ i = 1, \ldots, k, \\ g_j(x_1, \ldots, x_k) = 0,\ j = 1, \ldots, l \end{array} \right\}.
\end{aligned} \tag{3.46}
$$

The result of applying constrained fuzzy arithmetic in Definitions 38 and 39 to the computations with trapezoidal fuzzy numbers can be approximated according to simplified constrained fuzzy arithmetic by a trapezoidal fuzzy number $\widetilde{n} = (n^\alpha, n^\beta, n^\gamma, n^\delta)$:

$$
\begin{aligned}
n^\alpha &= \min \left\{ f(x_1, \ldots, x_k); \begin{array}{l} x_i \in [n_i^\alpha, n_i^\delta],\ i = 1, \ldots, k, \\ g_j(x_1, \ldots, x_k) = 0,\ j = 1, \ldots, l \end{array} \right\}, \\
n^\beta &= \min \left\{ f(x_1, \ldots, x_k); \begin{array}{l} x_i \in [n_i^\beta, n_i^\gamma],\ i = 1, \ldots, k, \\ g_j(x_1, \ldots, x_k) = 0,\ j = 1, \ldots, l \end{array} \right\}, \\
n^\gamma &= \max \left\{ f(x_1, \ldots, x_k); \begin{array}{l} x_i \in [n_i^\beta, n_i^\gamma],\ i = 1, \ldots, k, \\ g_j(x_1, \ldots, x_k) = 0,\ j = 1, \ldots, l \end{array} \right\}, \\
n^\delta &= \max \left\{ f(x_1, \ldots, x_k); \begin{array}{l} x_i \in [n_i^\alpha, n_i^\delta],\ i = 1, \ldots, k, \\ g_j(x_1, \ldots, x_k) = 0,\ j = 1, \ldots, l \end{array} \right\}.
\end{aligned} \tag{3.47}
$$

The simplified constrained fuzzy arithmetic will be applied in the following chapter in order to preserve the reciprocity of the related PCs of objects in FPCMs. For the simplicity, the terms "standard fuzzy arithmetic" and "constrained fuzzy arithmetic" will be used hereafter. However, the terms will always refer to their simplified versions with results in the form of trapezoidal fuzzy numbers (triangular fuzzy numbers and intervals being special cases) if not specified otherwise.

Chapter 4
Fuzzy Pairwise Comparison Matrices

Abstract This chapter is concerned with the fuzzy extension of the multi-criteria decision making methods based on pairwise comparison and answers the following research question: "Based on a fuzzy pairwise comparison matrix of objects, how should fuzzy priorities of these objects be determined so that they reflect properly all preference information available in the fuzzy pairwise comparison matrices?" Three types of fuzzy pairwise comparison matrices are studied—fuzzy multiplicative pairwise comparison matrices, fuzzy additive pairwise comparison matrices with additive representation, and fuzzy additive pairwise comparison matrices with multiplicative representation. The chapter provides a comprehensive literature review of the fuzzy pairwise comparison methods based on the fuzzy extension of well-known methods originally developed for crisp pairwise comparison matrices. The review includes a detailed study of differences and analogies between the methods and a detailed description of their drawbacks illustrated by numerous numerical examples. Two key properties of the fuzzy pairwise comparison matrices and the related methods—reciprocity of the related pairwise comparisons and the invariance of the methods under permutation of objects—are studied in detail, and it is shown that all reviewed methods violate at least one of these properties. A novel approach to fuzzy extension of the pairwise comparison methods based on constrained fuzzy arithmetic is introduced in order to prevent the drawbacks. Subsequently, suitable methods are introduced for all three types of fuzzy pairwise comparison matrices, showcased on numerous numerical examples, and critically compared with the original methods. Finally, transformations between the approaches for the three types of fuzzy pairwise comparison matrices are studied.

4.1 Introduction

In Chap. 2, crisp PCMs were studied in detail. However, as mentioned in Sect. 1.2.1, crisp PCs are not suitable for every MCDM problem. For example, when linguistic terms are used to provide intensities of preference or when the information about the problem is imprecise, fuzzy numbers seem to be more appropriate for expressing the PCs. In Chap. 3, fuzzy numbers were defined together with all concepts indispensable

© Springer International Publishing AG, part of Springer Nature 2018   85
J. Krejčí, *Pairwise Comparison Matrices and their Fuzzy Extension*, Studies
in Fuzziness and Soft Computing 366, https://doi.org/10.1007/978-3-319-77715-3_4

for properly replacing crisp PCs in PCMs by fuzzy PCs in form of fuzzy numbers and for adapting the related methods accordingly. (The word "crisp" is used here to emphasize the distinction from "fuzzy".)

By a FPCM of n objects $o_1, \ldots, o_n$ we will understand a PCM whose elements are fuzzy numbers, i.e. $\widetilde{C} = \left\{\widetilde{c}_{ij}\right\}_{i,j=1}^n$, $\widetilde{c}_{ij} \in \mathcal{F}_N(\mathbb{R})$, $i, j = 1, \ldots, n$. Note that also a crisp PCM $C = \left\{c_{ij}\right\}_{i,j=1}^n$ is actually a FPCM since crisp numbers are a special case of fuzzy numbers.

In Sect. 2.1, two key properties of PCMs and of the related PC methods were identified—reciprocity of the related PCs and invariance of the PC methods under permutation of objects. When extending crisp PCMs to FPCMs it is necessary to handle properly these two key properties.

As stated in Sect. 2.1, reciprocity is an inherent property of crisp PCMs resulting from the interpretation of the related PCs c_{ij} and c_{ji}. An appropriate extension of the reciprocity relation to FPCMs is of key importance for processing correctly the preference information contained in FPCMs and for deriving conclusions that reliably reflect DM's preferences. For this it is necessary to understand and interpret correctly the information contained in a FPCM. A FPCM $\widetilde{C} = \left\{\widetilde{c}_{ij}\right\}_{i,j=1}^n$ is not just a matrix with entries in form of fuzzy numbers that express uncertain preference information and that are in a reciprocity relation. It is necessary to look at the FPCM $\widetilde{C} = \left\{\widetilde{c}_{ij}\right\}_{i,j=1}^n$ as a set of crisp PCMs $C = \left\{c_{ij}\right\}_{i,j=1}^n$ with different degrees of membership to the FPCM $\widetilde{C}$. Each such PCM $C = \left\{c_{ij}\right\}_{i,j=1}^n$ carries particular preference information and has all the properties of a PCM discussed in Chap. 2. This fact has to be considered when the MCDM methods originally proposed for crisp PCMs are extended to FPCMs.

Similarly as for crisp PCMs, there exists no canonical order for assigning the labels $o_1, \ldots, o_n$ to n objects. Thus, regardless of the permutation of a FPCM $\widetilde{C}$, the methods applied to FPCMs have to lead to the same results, i.e., they have to be invariant under permutation of objects. In particular, the invariance under permutation is required for definitions of consistency and inconsistency indices defined for FPCMs as well as for methods for deriving fuzzy priorities of objects from FPCMs.

Being P a permutation matrix, $\widetilde{C}^\pi = P\widetilde{C}P^T$ is a permutation of $\widetilde{C}$ associated with P. Further, let $\widetilde{\mathcal{C}}$ denote a certain class of PCMs. Then, invariance under permutation for definitions of consistency, inconsistency indices, and methods for deriving priorities of objects from FPCMs can be formally defined as follows.

Definition 40 A definition of consistency for FPCMs in a certain class $\widetilde{\mathcal{C}}$ is invariant under permutation of objects if $\forall \widetilde{C} \in \widetilde{\mathcal{C}}$ the following holds:

$$\widetilde{C} \text{ consistent } \Rightarrow P\widetilde{C}P^T \text{ consistent for every } P,$$

$$\widetilde{C} \text{ not consistent } \Rightarrow P\widetilde{C}P^T \text{ not consistent for any } P,$$

where P is a permutation matrix.

Definition 41 An inconsistency index $I : \widetilde{C} \to \mathcal{F}_N(\mathbb{R})$ defined on a certain class $\widetilde{C}$ of FPCMs is invariant under permutation of objects if

$$I(P\widetilde{C}P^T) = I(\widetilde{C}), \qquad \forall \widetilde{C} \in \widetilde{C} \text{ and for any permutation matrix } P.$$

Definition 42 Let a method for deriving fuzzy priorities $\underline{\widetilde{w}} = (\widetilde{w}_1, \ldots, \widetilde{w}_n)^T$ of objects from FPCMs in a certain class $\widetilde{C}$ be described by a function $f : \widetilde{C} \to \mathcal{F}_N(\mathbb{R})^n$, i.e. $\underline{\widetilde{w}} = f(\widetilde{C})$, $\widetilde{C} \in \widetilde{C}$. Then the method is said to be invariant under permutation of objects if

$$f(P\widetilde{C}P^T) = Pf(\widetilde{C}), \qquad \forall \widetilde{C} \in \widetilde{C} \text{ and for any permutation matrix } P.$$

In Chap. 2, three types of PCMs were studied: MPCMs, APCMs-A, and APCMs-M. In the following sections of this chapter, the fuzzy extension of all three types of PCMs and of the reviewed definitions of consistency, inconsistency idices, and methods for deriving priorities from them is studied in detail. The focus is put on preserving the reciprocity of the related PCs and the invariance under permutation of objects. This is achieved by applying constrained fuzzy arithmetic instead of standard fuzzy arithmetic to the fuzzy extension of the methods reviewed in Chap. 2. As mentioned in Sect. 3.5, constrained fuzzy arithmetic allows for considering constraints on operands when performing arithmetic operations on fuzzy numbers. Thus, it enables us to introduce reciprocity of the related PCs as a constraint on fuzzy arithmetic operations with the entries of a FPCM. After introducing a proper fuzzy extension of the methods reviewed in Chap. 2, it is shown that the three types of FPCMs are equivalent and transformations between the approaches are examined. Note that Gavalec et al. (2015) presented a more general framework for FPCMs. They defined FPCMs as PCMs with fuzzy elements being fuzzy numbers of an abelian linearly ordered group on $\mathbb{R}$.

4.2 Fuzzy Multiplicative Pairwise Comparison Matrices

This section deals with the fuzzy extension of the methods related to MPCMs that were reviewed in Sect. 2.2. In Sect. 4.2.1, a fuzzy MPCM is defined properly and the construction of fuzzy MPCMs is studied. Section 4.2.2 is dedicated to the consistency of fuzzy MPCMs. In particular, the fuzzy extension of the multiplicative-consistency condition (2.4) and of the consistency ratio (2.10) for verifying acceptable level of inconsistency are dealt with, and a detailed study of the fuzzy maximal eigenvalues of fuzzy MPCMs is provided. Finally, Sect. 4.2.3 is focused on methods for obtaining fuzzy priorities from fuzzy MPCMs, in particular on the fuzzy extension of the EVM and the GMM.

4.2.1 Construction of FMPCMs

Definition 43 A fuzzy multiplicative pairwise comparison matrix (FMPCM) of n objects $o_1, \ldots, o_n$ is a square matrix $\tilde{M} = \left\{\tilde{m}_{ij}\right\}_{i,j=1}^{n}$, whose elements $\tilde{m}_{ij}$, $i, j = 1, \ldots, n$, are fuzzy numbers indicating the ratio of preference intensity of object o_i to that of object o_j. That is, element $\tilde{m}_{ij}$ indicates that o_i is $\tilde{m}_{ij}$–times as good as o_j. Further, a FMPCM $\tilde{M} = \left\{\tilde{m}_{ij}\right\}_{i,j=1}^{n}$ has to be multiplicatively reciprocal, i.e.,

$$\tilde{m}_{ij} = \frac{1}{\tilde{m}_{ji}}, \qquad i, j = 1, \ldots, n, \tag{4.1}$$

and

$$\tilde{m}_{ii} = 1, \qquad i = 1, \ldots, n. \tag{4.2}$$

Definition 43 is very general; elements $\tilde{m}_{ij}$ of a FMPCM $\tilde{M}$ are meant to be arbitrary fuzzy numbers satisfying the multiplicative-reciprocity condition (4.1). In practice, these fuzzy numbers can be triangular, trapezoidal, or general fuzzy numbers defined by their α–cuts, intervals, or it can even be a mix of all of them. Note that even a MPCM $M = \left\{m_{ij}\right\}_{ij=1}^{n}$ given by Definition 4 is a FMPCM since crisp numbers are a special case of fuzzy numbers and since a MPCM M satisfies (4.1) as well as (4.2).

In the literature, it is common that FMPCMs are defined by using triangular fuzzy numbers (see, e.g., Laarhoven and Pedrycz 1983; Chang 1996; Enea and Piazza 2004; Ishizaka and Nguyen 2013). However, some papers provide a more general approach by using trapezoidal fuzzy numbers (see, e.g., Buckley 1985a; Csutora and Buckley 2001; Ishizaka 2014).

We already know from Sect. 3.4 that the reciprocal $\frac{1}{\tilde{c}}$ of a trapezoidal fuzzy number $\tilde{c}$ is not a trapezoidal fuzzy number any more; it is a general fuzzy number described uniquely by its α–cuts. This means that if we provide a PC $\tilde{m}_{ij}$ of objects o_i and o_j as a trapezoidal fuzzy number (e.g. $\tilde{m}_{ij} = (2, 3, 4, 5)$) then, according to the multiplicative-reciprocity property (4.1), the entry $\tilde{m}_{ji}$ of the FMPCM $\tilde{M}$ is not a trapezoidal fuzzy number any more. This makes the construction of FMPCMs and the related methods much more complicated. First, there is no simple way to identify uniquely a general fuzzy number $\tilde{m}_{ji}$; it is not possible to use just a quadruple of its representing values as in the case of the trapezoidal fuzzy number $\tilde{m}_{ij}$. Second, having some of the entries in the FMPCM $\tilde{M}$ in the form of general fuzzy numbers, methods related to FMPCMs would become much more complex and not transparent for DMs, and the resulting fuzzy priorities of objects obtained from such FMPCMs would be general fuzzy numbers.

As already mentioned in Chap. 3, this problem is solved in the literature by simply approximating the results of arithmetic operations with trapezoidal fuzzy numbers by trapezoidal fuzzy numbers, i.e. by using the simplified standard fuzzy arithmetic introduced in Sect. 3.4 instead of standard fuzzy arithmetic. This means that the reciprocals of trapezoidal fuzzy PCs in a FMPCM are approximated by trapezoidal fuzzy numbers. Similarly, also the results of fuzzy arithmetic operations with the entries of a trapezoidal FMPCM are approximated by trapezoidal fuzzy numbers. These trapezoidal-fuzzy-number approximations have the same support and the core as the actual results of the arithmetic operations obtainable by applying standard fuzzy arithmetic.

It is obvious that by this approximation some information contained in the original general fuzzy numbers is lost. On the other hand, this approximation allows us to keep the computational process much simpler and transparent. Thus, it is a standard procedure to approximate all results of arithmetic operations with trapezoidal fuzzy numbers by trapezoidal fuzzy numbers. This approach is so deep-rooted in the literature on the fuzzy extension of PC methods (in fact, all papers cited in this book apply this approach) that the authors of the research papers often do not even mention the fact that the simplified standard fuzzy arithmetic is applied to the computations in their papers instead of the standard fuzzy arithmetic. However, as far as I am aware, there are no studies showing how good or bad this approximation really is, and how much the results of fuzzy PC methods with applied simplified standard fuzzy arithmetic vary from the hypothetical actual results obtainable by applying standard fuzzy arithmetic properly. Nevertheless, this simplification consisting in applying simplified standard fuzzy arithmetic to arithmetic operations with trapezoidal fuzzy numbers instead of standard fuzzy arithmetic is used also in this book in order to be in line with the main research stream.

As already mentioned in Sect. 2.2.1, integer numbers from Saaty's scale given in Table 2.1 with assigned linguistic terms and their reciprocals are usually used for expressing PCs in crisp MPCMs. However, since the linguistic terms in Saaty's scale are vague, it is more natural to model their meanings by using fuzzy rather than crisp numbers. Many different approaches to the fuzzy extension of Saaty's scale have been proposed in the literature. Ishizaka and Nguyen (2013) provided a review of various fuzzy extensions of Saaty's scale applied to real-world MCDM problems. Most often, the meaning of the linguistic terms from Saaty's scale is modeled by triangular fuzzy numbers, less often by trapezoidal fuzzy numbers.

However, Saaty's scale is not always fuzzified properly. Most problems are usually related to the modeling of the linguistic term "equal preference". We have to distinguish whether o_i and o_j are the same objects or not. For $i = j$, there is obviously no uncertainty in the comparison as we compare one object with itself. Therefore, $\tilde{m}_{ii}, i = 1, \ldots, n$, has to be set as 1, i.e., it is a crisp number. On the other hand, when two different objects o_i and o_j, $i \neq j$, are assessed to be "equally preferred", then this PC is very likely to contain some uncertainty. In such case, "equal preference" should be modeled by a fuzzy number "about 1", not necessarily by crisp number 1.

In the literature, "equal preference" is usually modeled by triangular fuzzy number $\tilde{1} = (1, 1, c)$, $c = 2$ or $c = 3$; see the literature review provided by Ishizaka and Nguyen (2013). However, Krejčí et al. (2017) showed that $\tilde{1}$ defined in this way is not appropriate for modeling the meaning of the linguistic term "equal preference". If a DM assesses o_i to be equally preferred to o_j, then common sense suggests that also o_j should be equally preferred to o_i. However, if we enter PC $\tilde{m}_{ij} = (1, 1, c)$

into a FMPCM $\widetilde{M}$, then based on the multiplicative-reciprocity property (4.1) it
follows that $\widetilde{m}_{ji} = \frac{1}{(1,1,c)} = \left(\frac{1}{c}, 1, 1\right)$. Obviously, $\left(\frac{1}{c}, 1, 1\right) \neq (1, 1, c)$ for $c > 1$, and
even $\left(\frac{1}{c}, 1, 1\right) < (1, 1, c)$. Thus, $\widetilde{m}_{ji} < \widetilde{m}_{ij}$, which contradicts the statement that o_i is
equally preferred to o_j.

Based on the reasoning in the previous paragraph, if $\widetilde{m}_{ij} = \widetilde{1}$, then it should also
hold that $\widetilde{m}_{ji} = \widetilde{1}$. Therefore, based on the multiplicative-reciprocity property (4.1),
the equality $\widetilde{1} = \dfrac{1}{\widetilde{1}}$ should hold. This means that $\widetilde{1} := (c^L, c^M, c^U)$ has to satisfy

$$(c^L, c^M, c^U) = \frac{1}{(c^L, c^M, c^U)}. \tag{4.3}$$

By solving (4.3) we obtain $\widetilde{1}$ defined as

$$\widetilde{1} = \left(\frac{1}{c}, 1, c\right), \quad c \geq 1. \tag{4.4}$$

Based on the same reasoning, if we wanted to use a trapezoidal fuzzy number instead
of a triangular fuzzy number to model the linguistic term "equal preference", this
would have to be in the form

$$\widetilde{1} = \left(\frac{1}{c}, \frac{1}{b}, b, c\right), \quad c \geq b \geq 1. \tag{4.5}$$

Interestingly, the appropriate representation of "equal preference" in the form (4.4)
(or 4.5 for trapezoidal representation) has been found only in two papers (Enea and
Piazza 2004 and Javanbarg et al. 2012).

Note 7 *From now on, the term "FMPCM" will be used exclusively for a FMPCM
that is given by Definition 43 and that satisfies the indispensable condition $\widetilde{1} = \frac{1}{\widetilde{1}}$ for
the fuzzy number modeling the meaning of the linguistic term "equal preference", i.e.,
(4.4) and (4.5) in the case of triangular and trapezoidal fuzzy numbers, respectively.*

As for the representation of other linguistic terms from Saaty's scale, it is reasonable to define the respective fuzzy numbers and their reciprocals in such a way
that they form Ruspini's fuzzy partition of interval $\left[\frac{1}{9}, 9\right]$. In this way any element
in the interval $[\frac{1}{9}, 9]$ has linguistic interpretation (Stoklasa 2014). Using triangular
fuzzy numbers, such fuzzy extension of Saaty's scale with the main 5 linguistic terms
is given in Table 4.1, while the fuzzy extension of Saaty's scale with intermediate
linguistic terms included is given in Table 4.2 (Krejčí and Talašová 2013).

The triangular fuzzy numbers in the fuzzy scales given in Table 4.1 and in
Table 4.2 model the meanings of the corresponding linguistic terms more naturally
than the crisp numbers in original Saaty's scale. However, as already emphasized in
Sect. 2.2.1, the uniform distribution of the numerical values assigned to the linguistic
terms in Saaty's scale on interval [1, 9] does not seem to be appropriate (e.g., "weakly
preferred" does not really correspond to "3-times preferred"). Thus, before actually

Table 4.1 Fuzzy extension of Saaty's 5-point scale

Intensity of preference	Linguistic term
$\left(\frac{1}{3}, 1, 3\right)$	Equal preference
$(1, 3, 5)$	Weak preference
$(3, 5, 7)$	Strong preference
$(5, 7, 9)$	Demonstrated preference
$(7, 9, 9)$	Absolute preference

Table 4.2 Fuzzy extension of Saaty's 9-point scale

Intensity of preference	Linguistic term
$\left(\frac{1}{2}, 1, 2\right)$	Equal preference
$(1, 2, 3)$	Between equal and weak preference
$(2, 3, 4)$	Weak preference
$(3, 4, 5)$	Between weak and strong preference
$(4, 5, 6)$	Strong preference
$(5, 6, 7)$	Between strong and demonstrated preference
$(6, 7, 8)$	Demonstrated preference
$(7, 8, 9)$	Between demonstrated and absolute preference
$(8, 9, 9)$	Absolute preference

modeling the meanings of the linguistic terms by fuzzy numbers, the original scale should be revised and more intuitive meanings should be found for each linguistic term. Moreover, since every DM perceives the linguistic terms from Saaty's scale differently, the scale should be calibrated for each particular decision-making problem in cooperation with the DM. Such idea was applied, e.g., by Ishizaka and Nguyen (2013) to a current bank account selection.

The idea of customizing Saaty's scale presented by Ishizaka and Nguyen (2013) is an innovative step in the direction of taking into account subjectivity of every single DM. However, the proposed process of calibration suffers from some severe drawbacks. For instance, an inappropriate form of the triangular fuzzy number modeling the meaning of the linguistic term "equal preference" is used for the calibration. In particular, Ishizaka and Nguyen (2013) suggest to model the meaning of the linguistic term "equal preference" by the triangular fuzzy number $(1, 1.1, 1.2)$. Thus, if $\tilde{m}_{ij} = (1, 1.1, 1.2) > 1$, then $\tilde{m}_{ji} = \frac{1}{\tilde{m}_{ij}} = (\frac{1}{1.2}, \frac{1}{1.1}, 1) < 1$. This clearly does not model the desired preference information "objects o_i and o_j are equally preferred" but rather "o_i is about 1.1-times preferred to o_j". Thus, using the triangular fuzzy number $(1, 1.1, 1.2)$ to model "equal preference" would actually have more severe negative impact on the result of a MCDM problem than using the crisp number 1 as suggested in original Saaty's scale or the fuzzy number $\tilde{1} = (1, 1, c), c > 1$.

Besides providing preference information linguistically by using linguistic terms from Saaty's scale, it is possible to enter expert numerical judgments into a MPCM. Especially when the information about the given problem is uncertain or incomplete, it is more appropriate to provide numerical judgments in form of fuzzy numbers rather than crisp numbers as illustrated by the following example.

Example 35 Let us assume we are searching for a new job. We have two interesting job offers, J_1 and J_2, and we evaluate them based on the expected income. Let us express our preferences by using the interval $[\frac{1}{9}, 9]$, where 1 stands for equal preference and 9 stands for absolute preference.

First, let us assume the expected income from J_1 is 2 500 € per month, and the expected income from J_2 is 1 000 € per month. For me, for example, the income 2 500 € may be 5-times preferred to 1 000 €, i.e. $m_{12} = 5$.

Now, let us assume, the income from J_1 is not known precisely, but it may be between 2 000 € and 3 000 € with the most possible income 2 500 €. In this case it would be quite difficult to express our preferences by using one crisp number, would not it? The preference information obtained from a DM in this case may have the following form. "If the income from J_1 is 2 500 €, then I prefer J_1 5-times over J_2. If the income from J_1 is 3 000 €, then J_1 is 7-times preferred to J_2. However, if the income from J_1 is only 2 000 €, then J_1 is only 4-times preferred to J_2." It is very natural to model this preference information by a triangular fuzzy number (i.e. $\widetilde{m}_{12} = (4, 5, 7)$ in this case) rather than by just one crisp number.   △

Saaty (2006) argues that "what fuzziness does by wholesale change of judgments numerically without obtaining the consent of the DM for each judgment goes against the grain of what decision making is all about, namely using experts to input valid judgments to obtain valid decisions" (Saaty 2006, p. 462). I fully agree with this argument—replacing crisp numerical judgments in a PCM by fuzzy numbers blindly without the consent of the DM does not lead to valid results. When linguistic terms are used for providing PCs, the meaning of the linguistic terms should be modeled for every particular MCDM problem in cooperation with the DM, i.e., calibration of the scale should be done. Nevertheless, using linguistic scales given in Table 4.1 and Table 4.2 is sufficient for the scope of this book since only theoretical results are presented here without any particular MCDM application and a particular DM.

There is no doubt that by means of fuzzy numbers we can describe possible uncertainty involved in DM's judgments or incompleteness of information in a decision-making problem. For instance, by the triangular fuzzy number $\widetilde{m}_{ij} = (3, 4, 6)$ we can model the case when a DM says that o_i is about 4−times preferred to o_j, definitely not less than 3−times preferred and not more than 6−times preferred to o_j. On the other hand, if the DM is sure about his/her judgment, e.g., if he/she is sure that o_i is 4−times preferred to o_j, we can appropriately describe this information by the triangular fuzzy number $\widetilde{m}_{ij} = (4, 4, 4)$ that corresponds to the crisp number 4.

4.2.2 Multiplicative Consistency of FMPCMs

As stated in Sect. 2.2.2, examining consistency or acceptable inconsistency of MPCMs is crucial in order to derive reliable priorities of objects. In the case of

FMPCMs, this task is of the same importance. That is the reason why the fuzzy extension of definitions of consistency for FMPCMs has been studied extensively (see, e.g., Buckley 1985a; Wang et al. 2005b; Liu 2009; Liu et al. 2014; Zheng et al. 2012; Gavalec et al. 2015; Wang 2015a, b; Li et al. 2016; Krejčí 2017b). Multiplicative consistency (2.4) is the basic and the most often applied consistency condition for MPCMs. Therefore, in this section, the fuzzy extension of this consistency condition is of interest.

In Sect. 4.2.2.1, a review of definitions of multiplicatively consistent FMPCMs proposed in the literature is given. In Sect. 4.2.2.2, a new fuzzy extension of the definition of multiplicative consistency based on the constrained fuzzy arithmetic is proposed. In Sect. 4.2.2.3, a fuzzy extension of Consistency Index (2.9) is proposed. Finally, in Sect. 4.2.2.4, the fuzzy maximal eigenvalue of a FMPCM is studied in detail.

4.2.2.1 Review of Fuzzy Extensions of Multiplicative Consistency

Many definitions of consistency based on the extension of multiplicative-transitivity property (2.4) have been proposed in the literature. Buckley (1985a) proposed a fuzzy extension of multiplicative consistency to FMPCMs based on a parameter. Wang et al. (2005a) proposed a definition of multiplicative consistency for interval FMPCMs, and they proposed an algorithm for deriving interval priorities from interval FMPCMs based on minimizing the inconsistency. Liu (2009) defined multiplicative consistency and acceptable inconsistency for interval FMPCMs. Liu et al. (2014) extended this definition to triangular FMPCMs, and they proposed a procedure for obtaining a consistent triangular FMPCM from n-1 fuzzy PCs. Wang (2015a) showed that the definition of consistency proposed by Liu (2009) is not invariant under permutation of objects, and he introduced a new definition of multiplicative consistency for interval FMPCMs. Similarly, (Wang 2015b) showed that the definition of consistency proposed by Liu et al. (2014) is not invariant under permutation of objects. Afterwards, he introduced a new definition of multiplicative consistency for triangular FMPCMs and proposed formulas for converting normalized fuzzy priorities into a consistent triangular FMPCM. Gavalec et al. (2015) introduced a more general definition of multiplicative consistency for FMPCMs, so called $\alpha-$consistency, $\alpha \in [0, 1]$, which can be looked at as an acceptable-consistency definition.

In this section, the definitions of multiplicative consistency for interval, triangular, and trapezoidal FMPCMs introduced by Buckley (1985a), Wang et al. (2005a), Liu (2009), Liu et al. (2014), and Wang (2015a, b) are reviewed. Furthermore, the drawbacks in the definitions proposed by Liu (2009), Liu et al. (2014), and Wang (2015a, b) are pointed out, based on the critical review originally conducted by Krejčí (2017b). In particular, violation of the invariance under permutation of objects and violation of the reciprocity of the related PCs, which is the key property of MPCMs, are emphasized.

Buckley (1985a) defined multiplicative consistency for FMPCMs with trapezoidal fuzzy numbers.

Definition 44 (Buckley 1985a) Let $\tilde{M} = \{\tilde{m}_{ij}\}_{i,j=1}^{n}$, $\tilde{m}_{ij} = \left(m_{ij}^{\alpha}, m_{ij}^{\beta}, m_{ij}^{\gamma}, m_{ij}^{\delta}\right)$, be a trapezoidal FMPCM. Then $\tilde{M}$ is said to be *multiplicatively consistent* if $\tilde{m}_{ik}\tilde{m}_{kj} \approx \tilde{m}_{ij}$. Otherwise, $\tilde{M}$ is said to be *multiplicatively inconsistent*.

In Definition 44, Buckley defined the approximate equality $\approx$ as follows. Having

$$\nu(\tilde{c} \geq \tilde{d}) = \sup_{x \geq y}\left\{\min\left\{\tilde{c}(x), \tilde{d}(y)\right\}\right\},$$

the fuzzy number $\tilde{c}$ is greater than the fuzzy number $\tilde{d}$, $\tilde{c} > \tilde{d}$, if $\nu(\tilde{c} \geq \tilde{d}) = 1$ and $\nu(\tilde{d} \geq \tilde{c}) < \theta$, where θ is a fixed value from interval $]0, 1]$. If $\tilde{c}$ is not greater than $\tilde{d}$ and $\tilde{d}$ is not greater than $\tilde{c}$, i.e. $\min\left\{\nu(\tilde{c} \geq \tilde{d}), \nu(\tilde{d} \geq \tilde{c})\right\} \geq \theta$, then $\tilde{c}$ and $\tilde{d}$ are said to be approximately equal, $\tilde{c} \approx \tilde{d}$.

Further, Buckley (1985a) derived the following theorem for trapezoidal FMPCMs.

Theorem 17 (Buckley 1985a) *Let* $\tilde{M} = \{\tilde{m}_{ij}\}_{i,j=1}^{n}$, $\tilde{m}_{ij} = \left(m_{ij}^{\alpha}, m_{ij}^{\beta}, m_{ij}^{\gamma}, m_{ij}^{\delta}\right)$, *be a trapezoidal FMPCM, and let* $m_{ij} \in \left[m_{ij}^{\beta}, m_{ij}^{\gamma}\right], i, j = 1, \ldots, n$. *If* $M = \{m_{ij}\}_{i,j=1}^{n}$ *is multiplicatively consistent according to (2.4), then* $\tilde{M}$ *is multiplicatively consistent according to Definition 44.*

Definition 44 is invariant under permutation of objects in trapezoidal FMPCM $\tilde{M}$. However, it is dependent on the value of the parameter θ, and there are no studies regarding an appropriate choice of the value of θ. Further, Theorem 17 is not very helpful in verifying multiplicative consistency of trapezoidal FMPCMs. First of all, the theorem helps to identify only a small part of multiplicatively consistent trapezoidal FMPCMs; trapezoidal FMPCMs multiplicatively consistent according to Definition 44 for which there do not exist $m_{ij} \in \left[m_{ij}^{\beta}, m_{ij}^{\gamma}\right], i, j = 1, \ldots, n$, such that $M = \{m_{ij}\}_{i,j=1}^{n}$ would be multiplicatively consistent according to (2.4) cannot be identified by using this theorem. Furthermore, the theorem does not even tell how to identify the given multiplicative consistent MPCM $M = \{m_{ij}\}_{i,j=1}^{n}$, $m_{ij} \in \left[m_{ij}^{\beta}, m_{ij}^{\gamma}\right], i, j = 1, \ldots, n$.

Another definition of consistency was proposed by Wang et al. (2005a). This definition of consistency will be studied in more detail in the following section. To distinguish easily this definition of consistency from others, consistency according to this definition will be called multiplicative weak consistency. The reason for the use of the word "weak" in the name will be clarified right after providing the definition.

Definition 45 (Wang et al. 2005a) Let $\overline{M} = \{\overline{m}_{ij}\}_{i,j=1}^{n}$, $\overline{m}_{ij} = \left[m_{ij}^{L}, m_{ij}^{U}\right]$, be an interval FMPCM. If $S = \left\{(w_1, \ldots, w_n) \in \mathbb{R}^n; m_{ij}^{L} \leq \frac{w_i}{w_j} \leq m_{ij}^{U}, \sum_{i=1}^{n} w_i = 1,\right.$ $\left. w_i > 0, i = 1, \ldots, n\right\} \neq \emptyset$, then $\overline{M}$ is said to be *multiplicatively weakly consistent*. Otherwise, $\overline{M}$ is said to be *multiplicatively weakly inconsistent*.

Definition 45 of multiplicative weak consistency for interval FMPCMs is clearly based on Proposition 1 for MPCMs. According to the definition, an interval FMPCM $\overline{M}$ is multiplicatively weakly consistent if there exists a vector $\underline{w} = (w_1, \ldots, w_n)^T$ which we can use to construct a multiplicatively consistent MPCM $M^* = \left\{ m_{ij}^* \right\}_{i,j=1}^n$ such that $m_{ij}^* \in [m_{ij}^L, m_{ij}^U], i, j = 1, \ldots, n$.

The requirement of at least one multiplicatively consistent MPCM obtainable from the interval FMPCM is very weak (that is why the name multiplicative "weak" consistency). Therefore, it is quite easy to satisfy the multiplicative-consistency condition in Definition 45 when constructing interval FMPCMs. The multiplicative-consistency conditions reviewed in the rest of this section are significantly stronger and thus much more difficult to fulfill.

Wang et al. (2005b) derived the following theorem which provides a useful tool for verifying the multiplicative weak consistency according to Definition 45.

Theorem 18 (Wang et al. 2005b) *An interval FMPCM* $\overline{M} = \left\{ \overline{m}_{ij} \right\}_{i,j=1}^n, \overline{m}_{ij} = \left[m_{ij}^L, m_{ij}^U \right]$, *is multiplicatively weakly consistent if and only if it satisfies the inequalities*

$$\max_{k=1,\ldots,n} \left\{ m_{ik}^L m_{kj}^L \right\} \leq \min_{k=1,\ldots,n} \left\{ m_{ik}^U m_{kj}^U \right\}, \qquad i, j = 1, \ldots, n. \tag{4.6}$$

The following theorem shows the relation between Definitions 44 and 45.

Theorem 19 *An interval FMPCM* $\overline{M} = \left\{ \overline{m}_{ij} \right\}_{i,j=1}^n, \overline{m}_{ij} = \left[m_{ij}^L, m_{ij}^U \right]$, *multiplicatively weakly consistent according to Definition 45 is also multiplicatively consistent according to Definition 44.*

Proof Let $\overline{M} = \left\{ \overline{m}_{ij} \right\}_{i,j=1}^n, \overline{m}_{ij} = \left[m_{ij}^L, m_{ij}^U \right]$, be multiplicatively consistent according to Definition 45. Then, from Proposition 1 it follows that there exists a MPCM $M = \left\{ m_{ij} \right\}_{i,j=1}^n, m_{ij} = \frac{w_i}{w_j} \in [m_{ij}^L, m_{ij}^U], i, j = 1, \ldots, n$, that is multiplicatively consistent according to (2.4). Thus, based on Theorem 17, intervals being a particular case of trapezoidal fuzzy numbers, it follows that $\overline{M}$ is multiplicatively consistent according to Definition 44. □

Another definition of multiplicative consistency for interval FMPCMs was given by Liu (2009).

Definition 46 (Liu 2009) Let $\overline{M} = \left\{ \overline{m}_{ij} \right\}_{i,j=1}^n, \overline{m}_{ij} = \left[m_{ij}^L, m_{ij}^U \right]$, be an interval FMPCM. Further, let MPCMs $C = \left\{ c_{ij} \right\}_{i,j=1}^n, D = \left\{ d_{ij} \right\}_{i,j=1}^n$ be constructed from the interval FMPCM $\overline{M}$ as

$$c_{ij} = \begin{cases} m_{ij}^L, & i < j \\ 1, & i = j \\ m_{ij}^U, & i > j \end{cases}, \qquad d_{ij} = \begin{cases} m_{ij}^U, & i < j \\ 1, & i = j \\ m_{ij}^L, & i > j \end{cases}. \tag{4.7}$$

If the matrices C, D are multiplicatively consistent according to (2.4), then $\overline{M}$ is said to be *multiplicatively consistent*. Otherwise, $\overline{M}$ is said to be *multiplicatively inconsistent*.

Later, Liu et al. (2014) extended Definition 46 of multiplicative consistency for interval FMPCMs to triangular and trapezoidal FMPCMs.

Definition 47 (Liu et al. 2014) Let $\tilde{M} = \{\tilde{m}_{ij}\}_{i,j=1}^{n}$, $\tilde{m}_{ij} = \left(m_{ij}^{\alpha}, m_{ij}^{\beta}, m_{ij}^{\gamma}, m_{ij}^{\delta}\right)$, be a trapezoidal FMPCM. Further, let MPCMs $A = \{a_{ij}\}_{i,j=1}^{n}$, $B = \{b_{ij}\}_{i,j=1}^{n}$, $C = \{c_{ij}\}_{i,j=1}^{n}$, and $D = \{d_{ij}\}_{i,j=1}^{n}$ be constructed from the trapezoidal FMPCM $\tilde{M}$ as

$$a_{ij} = \begin{cases} m_{ij}^{\alpha}, & i<j \\ 1, & i=j, \\ m_{ij}^{\delta}, & i>j \end{cases} \quad b_{ij} = \begin{cases} m_{ij}^{\beta}, & i<j \\ 1, & i=j, \\ m_{ij}^{\gamma}, & i>j \end{cases} \quad c_{ij} = \begin{cases} m_{ij}^{\gamma}, & i<j \\ 1, & i=j, \\ m_{ij}^{\beta}, & i>j \end{cases} \quad d_{ij} = \begin{cases} m_{ij}^{\delta}, & i<j \\ 1, & i=j \\ m_{ij}^{\alpha}, & i>j \end{cases}. \quad (4.8)$$

If the matrices A, B, C, and D are multiplicatively consistent according to (2.4), then $\tilde{M}$ is said to be multiplicatively consistent. Otherwise, $\tilde{M}$ is said to be multiplicatively inconsistent.

The definition of multiplicative consistency for triangular FMPCMs proposed by Liu et al. (2014) is actually obtainable by adjusting Definition 47 to triangular fuzzy numbers. Note that also Definition 46 of multiplicative consistency for interval FMPCMs is just a particular case of Definition 47.

Liu et al. (2014) stated that Definition 47 of multiplicative consistency naturally reflects the multiplicative-reciprocity property of trapezoidal FMPCMs since the matrices A, B, C, and D are clearly reciprocal. Furthermore, the authors claim that the definition of multiplicative consistency is closely related to Definition 5 of multiplicative consistency for MPCMs given by Saaty (1980). However, as demonstrated by Wang (2015b), Definition 47 highly depends on the ordering of objects compared in the trapezoidal FMPCM, i.e., it is not invariant under permutation of objects. Analogously, Wang (2015a) demonstrated the invalidity of Definition 46 for interval FMPCMs.

The drawback in Definition 47 is caused by the fact that the MPCMs A, B, C, and D given as (4.8) are not invariant under permutation of objects. There is no logical justification for choosing the lower boundary values of trapezoidal fuzzy numbers above the main diagonal and the upper boundary values of the trapezoidal fuzzy numbers below the main diagonal to construct the MPCM A. A similar problem appears also for the MPCMs B, C, and D. This way of constructing MPCMs from a trapezoidal FMPCM does not reflect naturally the multiplicative-reciprocity property of trapezoidal FMPCMs; the MPCMs A, B, C, and D change completely with a permutation of compared objects. Definition 46 of multiplicative consistency for interval FMPCMs suffers from the same drawbacks, and the same is valid also for the multiplicative consistency of triangular FMPCMs defined by Liu et al. (2014).

Clearly, Definitions 46 and 47 are not proper fuzzy extensions of Definition 5 of multiplicative consistency proposed by Saaty (1980). Properly defined multiplicative consistency for FMPCMs has to be invariant under permutation of objects and, naturally, as already pointed out by Liu et al. (2014), it has to preserve multiplicative reciprocity.

Wang (2015a, b) proposed definitions of multiplicative consistency for interval and triangular FMPCMs invariant under permutation of objects as follows.

Definition 48 (Wang 2015a) Let $\overline{M} = \left\{ \overline{m}_{ij} \right\}_{i,j=1}^{n}$, $\overline{m}_{ij} = \left[m_{ij}^{L}, m_{ij}^{U} \right]$, be an interval FMPCM. $\overline{M} = \left\{ \overline{m}_{ij} \right\}_{i,j=1}^{n}$ is said to be *multiplicatively consistent* if

$$m_{ij}^{L} m_{ij}^{U} = m_{ik}^{L} m_{ik}^{U} m_{kj}^{L} m_{kj}^{U}, \qquad i, j, k = 1, \ldots, n. \tag{4.9}$$

Definition 49 (Wang 2015b) Let $\tilde{M} = \left\{ \tilde{m}_{ij} \right\}_{i,j=1}^{n}$, $\tilde{m}_{ij} = \left(m_{ij}^{L}, m_{ij}^{M}, m_{ij}^{U} \right)$, be a triangular FMPCM. $\tilde{M} = \left\{ \tilde{m}_{ij} \right\}_{i,j=1}^{n}$ is said to be *multiplicatively consistent* if

$$\tilde{m}_{ij} \tilde{m}_{jk} \tilde{m}_{ki} = \tilde{m}_{ik} \tilde{m}_{kj} \tilde{m}_{ji}, \qquad i, j, k = 1, \ldots, n. \tag{4.10}$$

Furthermore, Wang (2015b) formulated the following theorem.

Theorem 20 *(Wang 2015b) The following statements are equivalent for a triangular FMPCM $\tilde{M} = \left\{ \tilde{m}_{ij} \right\}_{i,j=1}^{n}$, $\tilde{m}_{ij} = \left(m_{ij}^{L}, m_{ij}^{M}, m_{ij}^{U} \right)$:*

(i) $\tilde{M}$ is multiplicatively consistent according to Definition 49,
(ii) $m_{ij}^{M} = m_{ik}^{M} m_{kj}^{M}$, $m_{ij}^{L} m_{ij}^{U} = m_{ik}^{L} m_{ik}^{U} m_{kj}^{L} m_{kj}^{U}$, $i, j, k = 1, \ldots, n$,
(iii) $m_{ij}^{M} m_{jk}^{M} m_{ki}^{M} = m_{ik}^{M} m_{kj}^{M} m_{ji}^{M}$, $m_{ij}^{L} m_{jk}^{L} m_{ki}^{L} = m_{ik}^{L} m_{kj}^{L} m_{ji}^{L}$, $i, j, k = 1, \ldots, n$,
(iv) $m_{ij}^{M} m_{jk}^{M} m_{ki}^{M} = m_{ik}^{M} m_{kj}^{M} m_{ji}^{M}$, $m_{ij}^{U} m_{jk}^{U} m_{ki}^{U} = m_{ik}^{U} m_{kj}^{U} m_{ji}^{U}$, $i, j, k = 1, \ldots, n$.

A theorem similar to Theorem 20 could be formulated for interval FMPCMs $\overline{M} = \left\{ \overline{m}_{ij} \right\}_{i,j=1}^{n}$ by just removing the middle values m_{ij}^{M}, $i, j = 1, \ldots, n$, and the associated equations.

Remark 10 Multiplication in the formula (4.10) is done according to the simplified standard fuzzy arithmetic as defined in Sect. 3.4, in particular by the formula (3.28). In fact, using the simplified standard fuzzy arithmetic, the formula (4.10) can be written as (iii) and (iv) of Theorem 20. Therefore, based on Theorem 20 it follows that Definitions 48 and 49 for interval and triangular FMPCMs, respectively, are practically the same (keeping in mind that one is given for interval FMPCMs and one for triangular FMPCMs), although the multiplicative-consistency conditions (4.9) and (4.10) seem to have different forms.

In order to demonstrate the inappropriateness of Definitions 48 and 49, let us analyze in more detail the requirement of multiplicative reciprocity of the related PCs in FMPCMs and its impact on the multiplicative-consistency condition. As stated in Sect. 2.2.1, multiplicative reciprocity of the related PCs is an inherent property of every MPCM $M = \{m_{ij}\}_{i,j=1}^{n}$. When, for example, the intensity of preference of object o_i over object o_j is $m_{ij} = 3$ (i.e. o_i is 3-times preferred to o_j), then the intensity of preference of object o_j over object o_i has to be $m_{ji} = \frac{1}{3}$ (i.e. o_j has to be 3-times less preferred to o_i). Therefore, multiplicative reciprocity is a reasonable property that results from the interpretation of PCs of objects in MPCMs. Because of the multiplicative reciprocity of the related PCs, the multiplicative-consistency condition (2.4) for MPCMs is equivalent to the statements (ii) and (iii) in Theorem 1.

Conception of the multiplicative reciprocity becomes more complicated when extended to fuzzy numbers. For a triangular FMPCM $\widetilde{M} = \{\widetilde{m}_{ij}\}_{i,j=1}^{n}$, $\widetilde{m}_{ij} = \left(m_{ij}^{L}, m_{ij}^{M}, m_{ij}^{U}\right)$, multiplicative reciprocity is defined as $\widetilde{m}_{ji} = \frac{1}{\widetilde{m}_{ij}} = \left(\frac{1}{m_{ij}^{U}}, \frac{1}{m_{ij}^{M}}, \frac{1}{m_{ij}^{L}}\right)$. According to this property, when, e.g., the highest possible intensity of preference m_{ij}^{U} of object o_i over object o_j is $m_{ij}^{U} = 5$ (i.e. o_i is at most 5-times preferred to o_j), this means that the lowest possible intensity of preference m_{ji}^{L} of object o_j over object o_i is automatically $m_{ji}^{L} = \frac{1}{5}$ (i.e. o_j is at least 5-times less preferred to o_i). However, this is not all.

FMPCMs carry much more information about the preference intensities. For example, when we consider any particular value m_{ij}^{*} of $\widetilde{m}_{ij} = \left(m_{ij}^{L}, m_{ij}^{M}, m_{ij}^{U}\right)$, i.e. $m_{ij}^{*} \in [m_{ij}^{L}, m_{ij}^{U}]$, as a possible intensity of preference of object o_i over object o_j, this intensity of preference is associated inseparably with the corresponding intensity of preference m_{ji}^{*} of $\widetilde{m}_{ji} = \left(m_{ji}^{L}, m_{ji}^{M}, m_{ji}^{U}\right)$ such that $m_{ji}^{*} = \frac{1}{m_{ij}^{*}}$; remember that m_{ij}^{*} and m_{ji}^{*} have to express the same preference information about o_i and o_j. This property results naturally from the meaning of PCs.

Wang (2015b) defined multiplicative consistency of triangular FMPCMs by fuzzifying the expression (iii) in Theorem 1 using the simplified standard fuzzy arithmetic; see Definition 49. As emphasized by Wang (2015b), the definition is invariant under permutation of objects in the triangular FMPCM. This is definitely an advantage over Definition 47 proposed by Liu et al. (2014).

Since Wang (2015b) applied simplified standard fuzzy arithmetic to the computations with triangular fuzzy numbers, the expression (4.10) is nothing else but the statements (iii) and (iv) in Theorem 20. However, the expressions $m_{ij}^{L} m_{jk}^{L} m_{ki}^{L} = m_{ik}^{L} m_{kj}^{L} m_{ji}^{L}$ and $m_{ij}^{U} m_{jk}^{U} m_{ki}^{U} = m_{ik}^{U} m_{kj}^{U} m_{ji}^{U}$ in the statements (iii) and (iv) violate the multiplicative reciprocity of the related PCs. For example, in the first expression, the intensity of preference m_{ij}^{L} of o_i over o_j and the intensity of preference m_{ji}^{L} of o_j over o_i are used at the same time. This clearly violates the multiplicative reciprocity of the related PCs since $m_{ji}^{L} = \frac{1}{m_{ij}^{U}} \neq \frac{1}{m_{ij}^{L}}$ (unless $m_{ij}^{L} = m_{ij}^{U} = \frac{1}{m_{ji}^{L}} = \frac{1}{m_{ji}^{U}}$).

Example 36 Let us consider the interval FMPCM

$$\overline{M} = \begin{pmatrix} 1 & \left[\frac{3}{2}, 2\right] & [x, y] \\ \left[\frac{1}{2}, \frac{2}{3}\right] & 1 & \left[\frac{3}{2}, 2\right] \\ \left[\frac{1}{y}, \frac{1}{x}\right] & \left[\frac{1}{2}, \frac{2}{3}\right] & 1 \end{pmatrix} \tag{4.11}$$

where the PCs $\overline{m}_{13}$ and $\overline{m}_{31} = \frac{1}{\overline{m}_{13}}$ are unknown. First, let us apply Definition 48 proposed by Wang (2015a) in order to find out which values of $\overline{m}_{13}$ are allowed to preserve the multiplicative consistency.

By applying Definition 48, for $i = 1, k = 2, j = 3$, we obtain

$$xy = \frac{3}{2} \cdot 2 \cdot \frac{3}{2} \cdot 2 = 9.$$

Therefore, the interval PC $\overline{m}_{13} = [x, y]$, $x \leq y$, can be in the form $\left[\frac{9}{y}, y\right]$, $y \in [3, 9]$, in order to keep multiplicative consistency of $\overline{M}$. This means that even the whole interval $[1, 9]$ can be used to model the intensity of preference of o_1 over o_3, i.e. $\overline{m}_{13} = [1, 9]$. Let us have a closer look on the intensities of preferences in such interval FMPCM. Clearly, object o_1 is preferred to object o_2 and object o_2 is preferred to object o_3 since $\overline{m}_{12} = \overline{m}_{23} = \left[\frac{3}{2}, 2\right]$, $\frac{3}{2} > 1$. Therefore, object o_1 should be also preferred to object o_3. However, according to $\overline{m}_{13} = [1, 9]$, equal preference ($m_{13}^L = 1$) of objects o_1 and o_3 is admitted.

Moreover, the intensity of preference of object o_1 over object o_2 is at most 2 (between equal and weak preference) and also the intensity of preference of object o_2 over object o_3 is at most 2. However, the intensity of preference of object o_1 over object o_3 can be up to 9 (absolute preference), which is much higher than $2 \cdot 2 = 4$. In fact, there are no intensities of preference $m_{12} \in \left[m_{12}^L, m_{12}^U\right]$ and $m_{23} \in \left[m_{23}^L, m_{23}^U\right]$ such that $m_{12}m_{23} = 1$ or $m_{12}m_{23} = 9$.

According to Theorem 20 (properly adjusted for interval FMPCMs) the multiplicative consistency can be checked by using the equivalent multiplicative-consistency condition

$$m_{ij}^L m_{jk}^L m_{ki}^L = m_{ik}^L m_{kj}^L m_{ji}^L, \qquad i, j, k = 1, \ldots, n.$$

Assuming $\overline{m}_{13} = [1, 9]$ in the interval FMPCM $\overline{M}$ given by (4.11), this basically means to verify the multiplicative consistency of the matrix

$$M^L = \begin{pmatrix} 1 & \dfrac{3}{2} & 1 \\ \dfrac{1}{2} & 1 & \dfrac{3}{2} \\ \dfrac{1}{9} & \dfrac{1}{2} & 1 \end{pmatrix} \tag{4.12}$$

by using the property (iii) of Theorem 1. However, the matrix (4.12) is not multiplicatively reciprocal, i.e., it is not even a MPCM. Therefore, verifying its multiplicative consistency is meaningless.   △

Even though Definitions 48 and 49 of multiplicatively consistent interval and triangular FMPCMs are invariant under permutation of objects in the FMPCMs, their meaning is questionable since they violate the multiplicative reciprocity of the related PCs. Furthermore, when applying the simplified standard fuzzy arithmetic, as Wang (2015b) did, the condition (4.10) is not equivalent neither to $\tilde{m}_{ij} = \tilde{m}_{ik}\tilde{m}_{kj}$, $i, j, k = 1, \ldots, n$, nor to $\tilde{m}_{ij}\tilde{m}_{jk}\tilde{m}_{ki} = 1$, $i, j, k = 1, \ldots, n$. The same is true for Definition 48 of multiplicative consistency for interval FMPCMs. This means that for Definitions 48 and 49 of multiplicative consistency proposed by Wang (2015a, b) Theorem 1 cannot be extended to interval and triangular FMPCMs, respectively. Definition 5 of multiplicative consistency for MPCMs should be extended to FMPCMs in such a way that Theorem 1 can be extended to FMPCMs accordingly. This is possible to do by employing properly the multiplicative-reciprocity property.

Another serious drawback of applying the simplified standard fuzzy arithmetic to computation with PCs in triangular FMPCMs (and, of course, FMPCMs in general) is the fact that

$$\tilde{m}_{ij}\tilde{m}_{ji} = \left(m_{ij}^L m_{ji}^L, m_{ij}^M m_{ji}^M, m_{ij}^U m_{ji}^U \right) = \left(\dfrac{m_{ij}^L}{m_{ij}^U}, 1, \dfrac{m_{ij}^U}{m_{ij}^L} \right) \neq 1, \quad i, j = 1, \ldots, n. \tag{4.13}$$

From the multiplicative-reciprocity property $\tilde{m}_{ij} = \frac{1}{\tilde{m}_{ji}}$ of FMPCMs, it should follow that $\tilde{m}_{ij}\tilde{m}_{ji} = 1$. Obviously, the expression (4.13) violates the multiplicative-reciprocity property of PCs.

For reader's convenience the properties regarding the reciprocity of the related PCs and invariance under permutations of all definitions of multiplicative consistency reviewed in this section are listed in Table 4.3. Only Definition 45 of multiplicative weak consistency introduced by Wang et al. (2005a) for interval FMPCMs keeps both properties. Definition 44 is based on similarity of fuzzy numbers, and the reciprocity of the related PCs does not play a role in the definition; thus this property is left out in Table 4.3. All other reviewed definitions of multiplicative consistency violate either the reciprocity of the related PCs or invariance under permutation.

We know that constrained fuzzy arithmetic has to be applied whenever there are any interactions among operands. Clearly, there is an interaction between fuzzy PCs $\tilde{m}_{ij}$ and $\tilde{m}_{ji}$; from the multiplicative reciprocity of the related PCs it follows that any intensity of preference $m_{ij}^* \in \tilde{m}_{ij}$ of object o_i over object o_j is associated

Table 4.3 Properties of the definitions of multiplicative consistency for FMPCMs

Definition	Authors	Invariance under permutation	Multiplicative reciprocity
Definition 44	Buckley (1985a)	✓	–
Definition 45	Wang et al. (2005a)	✓	✓
Definition 46	Liu (2009)	✗	✓
Definition 47	Liu et al. (2014)	✗	✓
Definition 48	Wang (2015a)	✓	✗
Definition 49	Wang (2015b)	✓	✗

with the corresponding intensity of preference $m_{ji}^* = \frac{1}{m_{ij}^*}$ of object o_j over object o_i. Thus, $\tilde{c} = \tilde{m}_{ij}\tilde{m}_{ji}$ should be correctly computed according to the constrained fuzzy arithmetic (3.43) as $\tilde{c} = (c^L, c^M, c^U)$:

$$
c^L = \min\left\{ m_{ij}m_{ji};\ m_{ij} \in [m_{ij}^L, m_{ij}^U], m_{ji} \in [m_{ji}^L, m_{ji}^U], m_{ji} = \frac{1}{m_{ij}} \right\} =
$$
$$
= \min\left\{ m_{ij}\frac{1}{m_{ij}};\ m_{ij} \in [m_{ij}^L, m_{ij}^U] \right\} = 1,
$$
$$
c^M = m_{ij}^M m_{ji}^M = m_{ij}^M \frac{1}{m_{ij}^M} = 1,
$$
$$
c^U = \max\left\{ m_{ij}m_{ji};\ m_{ij} \in [m_{ij}^L, m_{ij}^U], m_{ji} \in [m_{ji}^L, m_{ji}^U], m_{ji} = \frac{1}{m_{ij}} \right\} =
$$
$$
= \max\left\{ m_{ij}\frac{1}{m_{ij}};\ m_{ij} \in [m_{ij}^L, m_{ij}^U] \right\} = 1.
$$

Keeping in mind the importance of the multiplicative-reciprocity property of PCs in FMPCMs, multiplicative consistency needs to be defined accordingly so that it does not violate the multiplicative reciprocity. In the following section, two definitions of multiplicatively consistent trapezoidal FMPCMs respecting the multiplicative reciprocity of the related PCs and invariant under permutation of objects are proposed.

4.2.2.2 New Fuzzy Extension of Multiplicative Consistency

In this section, Definition 45 of multiplicative weak consistency given by Wang et al. (2005a) is extended to trapezoidal FMPCMs and another definition of multiplicative consistency much stronger than Definition 45 is proposed by extending the definition of Krejčí (2017b) for triangular FMPCMs. Tools for verifying both the multiplicative weak consistency and the multiplicative consistency are provided and some properties of multiplicatively weakly consistent and multiplicatively consistent trapezoidal FMPCMs are derived. Both definitions preserve two desired properties—invariance under permutation of objects and multiplicative reciprocity of the related PCs.

Definition 50 Let $\widetilde{M} = \left\{\widetilde{m}_{ij}\right\}_{i,j=1}^{n}$, $\widetilde{m}_{ij} = \left(m_{ij}^{\alpha}, m_{ij}^{\beta}, m_{ij}^{\gamma}, m_{ij}^{\delta}\right)$, be a trapezoidal FMPCM. $\widetilde{M}$ is said to be multiplicatively weakly consistent if there exists a positive vector $\underline{w} = (w_1, \ldots, w_n)^T$ such that

$$m_{ij}^{\alpha} \le \frac{w_i}{w_j} \le m_{ij}^{\delta}, \qquad i,j = 1, \ldots, n. \qquad (4.14)$$

Notice that when Definition 50 is applied to interval FMPCMs, it is identical to Definition 45 proposed by Wang et al. (2005a).

Proposition 10 *Let* $\widetilde{M} = \left\{\widetilde{m}_{ij}\right\}_{i,j=1}^{n}$, $\widetilde{m}_{ij} = \left(m_{ij}^{\alpha}, m_{ij}^{\beta}, m_{ij}^{\gamma}, m_{ij}^{\delta}\right)$, *be a trapezoidal FMPCM.* $\widetilde{M} = \left\{\widetilde{m}_{ij}\right\}_{i,j=1}^{n}$ *is multiplicatively weakly consistent according to Definition 50 if and only if there exist elements* $m_{ij}^{*} \in \left[m_{ij}^{\alpha}, m_{ij}^{\delta}\right]$, $i,j = 1, \ldots, n$, *such that* $M^{*} = \left\{m_{ij}^{*}\right\}_{i,j=1}^{n}$ *is a MPCM multiplicatively consistent according to Definition 5.*

Proof First, let $\widetilde{M} = \left\{\widetilde{m}_{ij}\right\}_{i,j=1}^{n}$, $\widetilde{m}_{ij} = \left(m_{ij}^{\alpha}, m_{ij}^{\beta}, m_{ij}^{\gamma}, m_{ij}^{\delta}\right)$, be a trapezoidal FMPCM multiplicatively weakly consistent according to Definition 50. Let us denote $m_{ij}^{*} := \frac{w_i}{w_j}$. From (4.14) it follows that $m_{ij}^{*} \in \left[m_{ij}^{\alpha}, m_{ij}^{\delta}\right]$, $i,j = 1, \ldots, n$. Further, we have $m_{ii}^{*} = \frac{w_i}{w_i} = 1$, and $m_{ji}^{*} = \frac{w_j}{w_i} = \frac{1}{\frac{w_i}{w_j}} = \frac{1}{m_{ij}^{*}}$, $i,j = 1, \ldots, n$. From $[m_{ij}^{\alpha}, m_{ij}^{\delta}] \subseteq [\frac{1}{9}, 9]$, $i,j = 1, \ldots, n$, it follows that also $m_{ij}^{*} \subseteq [\frac{1}{9}, 9]$, $i,j = 1, \ldots, n$. Therefore, $M^{*} = \left\{m_{ij}^{*}\right\}_{i,j=1}^{n}$ is a MPCM. Finally, $m_{ik}^{*} m_{kj}^{*} = \frac{w_i}{w_k} \frac{w_k}{w_j} = \frac{w_i}{w_j} = m_{ij}^{*}$, $i,j,k = 1, \ldots, n$, which means that $M^{*} = \left\{m_{ij}^{*}\right\}_{i,j=1}^{n}$ is multiplicatively consistent according to (2.4).

In the opposite direction, let $\widetilde{M} = \left\{\widetilde{m}_{ij}\right\}_{i,j=1}^{n}$, $\widetilde{m}_{ij} = \left(m_{ij}^{\alpha}, m_{ij}^{\beta}, m_{ij}^{\gamma}, m_{ij}^{\delta}\right)$, be a trapezoidal FMPCM and let $M^{*} = \left\{m_{ij}^{*}\right\}_{i,j=1}^{n}$, $m_{ij}^{*} \in \left[m_{ij}^{\alpha}, m_{ij}^{\delta}\right]$, $i,j = 1, \ldots, n$, be a MPCM multiplicatively consistent according to Definition 5. Then, from Proposition 1 it follows that there exists a vector $\underline{w} = (w_1, \ldots, w_n)^T$ such that $m_{ij}^{*} = \frac{w_i}{w_j}$, $i,j = 1, \ldots, n$. Because, $m_{ij}^{*} \in \left[m_{ij}^{\alpha}, m_{ij}^{\delta}\right]$, $i,j = 1, \ldots, n$, then clearly (4.14) holds. $\qquad\square$

Remark 11 According to Proposition 10 and its proof, a trapezoidal FMPCM $\widetilde{M} = \left\{\widetilde{m}_{ij}\right\}_{i,j=1}^{n}$, $\widetilde{m}_{ij} = \left(m_{ij}^{\alpha}, m_{ij}^{\beta}, m_{ij}^{\gamma}, m_{ij}^{\delta}\right)$, is multiplicatively weakly consistent if and only if there exists a multiplicatively consistent MPCM $M^{*} = \left\{m_{ij}^{*}\right\}_{i,j=1}^{n}$ such that $m_{ij}^{*} \in \left[m_{ij}^{\alpha}, m_{ij}^{\delta}\right]$. This consistency condition is quite easy to reach. That is why the consistency according to Definition 50 is called weak. Later in this section, also a much stronger definition of multiplicative consistency for trapezoidal FMPCMs will be given.

Definition 50 of multiplicative weak consistency satisfies two desirable properties—it is invariant under permutation of objects and it preserves the multiplicative reciprocity of the related PCs in trapezoidal FMPCMs.

Theorem 21 *Definition 50 of multiplicative weak consistency is invariant under permutation of objects in trapezoidal FMPCMs.*

Proof For a multiplicatively consistent trapezoidal FMPCM $\widetilde{M} = \left\{ \widetilde{m}_{ij} \right\}_{i,j=1}^{n}$, $\widetilde{m}_{ij} = \left(m_{ij}^{\alpha}, m_{ij}^{\beta}, m_{ij}^{\gamma}, m_{ij}^{\delta} \right)$, there exists a positive priority vector $\underline{w} = (w_1, \ldots, w_n)^T$, such that the inequality $m_{ij}^{\alpha} \leq \frac{w_i}{w_j} \leq m_{ij}^{\delta}$ holds for every single PC $\widetilde{m}_{ij}$. By permuting the FMPCM $\widetilde{M}$ to $\widetilde{M}^{\pi} = P\widetilde{M}P^T$, the original PC $\widetilde{m}_{ij}$ in the i−th row and in the j−th column of $\widetilde{M}$ is moved to the $\pi(i)$−th row and the $\pi(j)$−th column of the permuted trapezoidal FMPCM $\widetilde{M}^{\pi}$ as $\widetilde{m}_{\pi(i)\pi(j)}^{\pi}$, but still keeping $\widetilde{m}_{ij} = \widetilde{m}_{\pi(i)\pi(j)}^{\pi}$, $i, j = 1, \ldots, n$. Thus, there exists a vector $\underline{w}^{\pi} = (w_1^{\pi}, \ldots, w_n^{\pi})^T$ obtained by permuting the vector $\underline{w}$, i.e. $\underline{w}^{\pi} = P\underline{w}$, with the components satisfying the inequalities $m_{ij}^{\pi\alpha} \leq \frac{w_i^{\pi}}{w_j^{\pi}} \leq m_{ij}^{\pi\delta}$ for every $i, j = 1, \ldots, n$. □

Theorem 22 *Definition 50 of multiplicative weak consistency preserves the multiplicative reciprocity of the related PCs in trapezoidal FMPCMs in the sense that any fixed value $m_{ij} \in [m_{ij}^{\alpha}, m_{ij}^{\delta}]$, $i, j \in \{1, \ldots, n\}$, representing the intensity of preference of object o_i over object o_j is associated with the corresponding value $m_{ji} \in [m_{ji}^{\alpha}, m_{ji}^{\delta}]$, representing the intensity of preference of object o_j over object o_i such that $m_{ji} = \frac{1}{m_{ij}}$.*

Proof The existence of a priority vector $\underline{w} = (w_1, \ldots, w_n)^T$ satisfying the inequalities (4.14) means that there exists a MPCM $M = \left\{ m_{ij} \right\}_{i,j=1}^{n}$, $m_{ij} \in [m_{ij}^{\alpha}, m_{ij}^{\delta}]$, such that $m_{ij} = \frac{w_i}{w_j}$, $i, j = 1, \ldots, n$. M is multiplicatively reciprocal from the definition, i.e. every PC m_{ij} is associated with the PC m_{ji} such that $m_{ji} = \frac{1}{m_{ij}}$. □

Remark 12 Note that Theorem 22 does not simply state that a FMPCM $\widetilde{M} = \left\{ \widetilde{m}_{ij} \right\}_{i,j=1}^{n}$ multiplicatively weakly consistent according to Definition 50 is multiplicatively reciprocal, i.e. $\widetilde{m}_{ji} = \frac{1}{\widetilde{m}_{ij}}$, $i, j = 1, \ldots, n$. The validity of this property automatically follows from Definition 43 of a FMPCM; every FMPCM is multiplicatively reciprocal, and thus also a FMPCM that is multiplicatively weakly consistent according to Definition 50 is multiplicatively reciprocal.

As explained on p. 98, the extension of the multiplicative-reciprocity property from MPCMs to FMPCMs does not concern only the "simple" multiplicative reciprocity of the related fuzzy PCs $\widetilde{m}_{ij}$ and $\widetilde{m}_{ji}$ in the sense that $\widetilde{m}_{ji} = \frac{1}{\widetilde{m}_{ij}}$, $i, j = 1, \ldots, n$. The conception of the multiplicative reciprocity becomes more complex for FMPCMs. In particular, every intensity of preference $m_{ij}^{*} \in \widetilde{m}_{ij}$ of object o_i over object o_j is associated inseparably with the corresponding intensity of preference $m_{ji}^{*} \in \widetilde{m}_{ji}$ such that $m_{ji}^{*} = \frac{1}{m_{ij}^{*}}$ since both m_{ij}^{*} and m_{ji}^{*} have to express the same preference information about the objects o_i and o_j. (If the reader has difficulties to understand the general ideal behind this argument, he/she should again have a look on Examples 1,

32 and 34 on p. 11, 78, and 83, respectively, where the problem is explained in detail on very simple and easily understandable problems.) Theorem 22 states that Definition 50 is in accordance with this conception of multiplicative reciprocity, i.e., only multiplicatively reciprocal PCs are involved in Definition 50 of multiplicative weak consistency.

Theorem 18 for verifying multiplicative weak consistency of interval FMPCMs can be extended to trapezoidal FMPCMs as follows.

Theorem 23 *A trapezoidal FMPCM* $\widetilde{M} = \{\widetilde{m}_{ij}\}_{i,j=1}^{n}$, $\widetilde{m}_{ij} = \left(m_{ij}^{\alpha}, m_{ij}^{\beta}, m_{ij}^{\gamma}, m_{ij}^{\delta}\right)$, *is multiplicatively weakly consistent according to Definition 50 if and only if*

$$\max_{k=1,\ldots,n} \left\{m_{ik}^{\alpha} m_{kj}^{\alpha}\right\} \leq \min_{k=1,\ldots,n} \left\{m_{ik}^{\delta} m_{kj}^{\delta}\right\}, \qquad i,j = 1,\ldots,n. \tag{4.15}$$

Proof The proof is the same as the proof of Theorem 18; it is sufficient to substitute $m_{ik}^{L}, m_{kj}^{L}, m_{ik}^{U}, m_{kj}^{U}$ with $m_{ik}^{\alpha}, m_{kj}^{\alpha}, m_{ik}^{\delta}, m_{kj}^{\delta}$, respectively.   □

The following theorem shows that it is sufficient to verify the inequality (4.15) only for $i,j = 1,\ldots,n$, $i < j$, thus saving half of the computations.

Theorem 24 *A trapezoidal FMPCM* $\widetilde{M} = \{\widetilde{m}_{ij}\}_{i,j=1}^{n}$, $\widetilde{m}_{ij} = \left(m_{ij}^{\alpha}, m_{ij}^{\beta}, m_{ij}^{\gamma}, m_{ij}^{\delta}\right)$, *is multiplicatively weakly consistent according to Definition 50 if and only if*

$$\max_{k=1,\ldots,n} \left\{m_{ik}^{\alpha} m_{kj}^{\alpha}\right\} \leq \min_{k=1,\ldots,n} \left\{m_{ik}^{\delta} m_{kj}^{\delta}\right\}, \qquad i,j = 1,\ldots,n, \ i < j. \tag{4.16}$$

Proof It is sufficient to show that the validity of inequalities (4.16) for $i,j = 1,\ldots,n$, $i < j$ implies automatically their validity for all $i,j = 1,\ldots,n$, i.e. the validity of (4.15). The validity of inequalities (4.15) for $i = j$ is trivial from the definition of trapezoidal FMPCMs since

$$\max_{k=1,\ldots,n} \left\{m_{ik}^{\alpha} m_{ki}^{\alpha}\right\} = \max_{k=1,\ldots,n} \left\{\frac{m_{ik}^{\alpha}}{m_{ik}^{\delta}}\right\} \leq 1 \leq \min_{k=1,\ldots,n} \left\{\frac{m_{ik}^{\delta}}{m_{ik}^{\alpha}}\right\} = \min_{k=1,\ldots,n} \left\{m_{ik}^{\delta} m_{ki}^{\delta}\right\}.$$

Further, for $i > j$, by using (4.16) and the multiplicative-reciprocity properties, we get

$$\max_{k=1,\ldots,n} \left\{m_{ik}^{\alpha} m_{kj}^{\alpha}\right\} = \max_{k=1,\ldots,n} \left\{\frac{1}{m_{ki}^{\delta}} \frac{1}{m_{jk}^{\delta}}\right\} = \frac{1}{\min\limits_{k=1,\ldots,n} \left\{m_{jk}^{\delta} m_{ki}^{\delta}\right\}} \leq$$

$$\frac{1}{\max\limits_{k=1,\ldots,n} \left\{m_{jk}^{\alpha} m_{ki}^{\alpha}\right\}} = \min_{k=1,\ldots,n} \left\{\frac{1}{m_{jk}^{\alpha}} \frac{1}{m_{ki}^{\alpha}}\right\} = \min_{k=1,\ldots,n} \left\{m_{ik}^{\delta} m_{kj}^{\delta}\right\}.$$

□

Remark 13 An alternative definition of multiplicative weak consistency to Definition 50 might be formulated as follows.

Let $\widetilde{M} = \{\widetilde{m}_{ij}\}_{i,j=1}^{n}$, $\widetilde{m}_{ij} = \left(m_{ij}^{\alpha}, m_{ij}^{\beta}, m_{ij}^{\gamma}, m_{ij}^{\delta} \right)$, be a trapezoidal FMPCM. $\widetilde{M}$ is said to be multiplicatively weakly consistent if there exists a positive vector $\underline{w} = (w_1, \ldots, w_n)^T$ such that

$$m_{ij}^{\beta} \leq \frac{w_i}{w_j} \leq m_{ij}^{\gamma}, \qquad i, j = 1, \ldots, n. \tag{4.17}$$

Notice that this definition is stronger than Definition 50. In fact, every trapezoidal FMPCM multiplicatively weakly consistent according to this definition is also multiplicatively weakly consistent according to Definition 50 since (4.17) implies (4.14). Further, when this definition is applied to interval FMPCMs, it is again identical to Definition 45 proposed by Wang et al. (2005a).

All theorems regarding FMPCMs multiplicatively weakly consistent according to Definition 50 formulated above can be easily reformulated for FMPCMs multiplicatively weakly consistent according to this definition; it is sufficient to consider m_{ij}^{β} and m_{ij}^{γ} instead of m_{ij}^{α} and m_{ij}^{δ}, respectively, where appropriate.

In the following definition, a stronger version of multiplicative consistency for trapezoidal FMPCMs is formulated.

Definition 51 Let $\widetilde{M} = \{\widetilde{m}_{ij}\}_{i,j=1}^{n}$, $\widetilde{m}_{ij} = \left(m_{ij}^{\alpha}, m_{ij}^{\beta}, m_{ij}^{\gamma}, m_{ij}^{\delta} \right)$, be a trapezoidal FMPCM. $\widetilde{M}$ is said to be multiplicatively consistent if for each triplet $(i, j, k) \subseteq \{1, \ldots, n\}$ the following holds:

$$\forall m_{ij} \in \left[m_{ij}^{\alpha}, m_{ij}^{\delta} \right] \exists m_{ik} \in \left[m_{ik}^{\alpha}, m_{ik}^{\delta} \right] \wedge \exists m_{kj} \in \left[m_{kj}^{\alpha}, m_{kj}^{\delta} \right] : m_{ij} = m_{ik} m_{kj}, \tag{4.18}$$

$$\forall m_{ij} \in \left[m_{ij}^{\beta}, m_{ij}^{\gamma} \right] \exists m_{ik} \in \left[m_{ik}^{\beta}, m_{ik}^{\gamma} \right] \wedge \exists m_{kj} \in \left[m_{kj}^{\beta}, m_{kj}^{\gamma} \right] : m_{ij} = m_{ik} m_{kj}. \tag{4.19}$$

Remark 14 Definition 51 is a natural fuzzy extension of Definition 5 of multiplicative consistency proposed by Saaty (1980). According to this definition, for any value $m_{ij} \in \widetilde{m}_{ij}, i, j \in \{1, \ldots, n\}$, there exist values $m_{ik} \in \widetilde{m}_{ik}$ and $m_{kj} \in \widetilde{m}_{kj}, k \in \{1, \ldots, n\}$, such that they satisfy the multiplicative-consistency property (2.4). Analogously, for any value $m_{ij} \in Core\,\widetilde{m}_{ij}, i, j \in \{1, \ldots, n\}$, there exist possible values $m_{ik} \in Core\,\widetilde{m}_{ik}$ and $m_{kj} \in Core\,\widetilde{m}_{kj}, k \in \{1, \ldots, n\}$, such that they satisfy (2.4). Clearly, in comparison to the multiplicative weak consistency given by Definition 50, the multiplicative consistency given by Definition 51 is very strong.

Unlike Definitions 46 and 47 of multiplicative consistency for FMPCMs proposed by Liu (2009) and Liu et al. (2014), respectively, Definition 51 is invariant under permutation of objects compared in trapezoidal FMPCMs.

Theorem 25 *Definition 51 of multiplicative consistency is invariant under permutation of objects in trapezoidal FMPCMs.*

Proof For a multiplicatively consistent trapezoidal FMPCM $\widetilde{M} = \{\widetilde{m}_{ij}\}_{i,j=1}^{n}$, $\widetilde{m}_{ij} = (m_{ij}^{\alpha}, m_{ij}^{\beta}, m_{ij}^{\gamma}, m_{ij}^{\delta})$, the conditions (4.18) and (4.19) are satisfied for every triplet $(i, j, k) \subseteq \{1, \ldots, n\}$. By permuting the FMPCM $\widetilde{M}$ to $\widetilde{M}^{\pi} = P\widetilde{M}P^{T}$, the original PC $\widetilde{m}_{ij}$ in the i-th row and in the j-th column of $\widetilde{M}$ moves to the $\pi(i)$-th row and the $\pi(j)$-th column of $\widetilde{M}^{\pi}$ preserving $\widetilde{m}_{\pi(i)\pi(j)}^{\pi} = \widetilde{m}_{ij}$. Thus, by permuting $\widetilde{M}$, also the validity of the conditions (4.18) and (4.19) is preserved, i.e.,

$$\forall m_{ij}^{\pi} \in \left[m_{ij}^{\pi\alpha}, m_{ij}^{\pi\delta}\right] \exists m_{ik}^{\pi} \in \left[m_{ik}^{\pi\alpha}, m_{ik}^{\pi\delta}\right] \wedge \exists m_{kj}^{\pi} \in \left[m_{kj}^{\pi\alpha}, m_{kj}^{\pi\delta}\right]: m_{ij}^{\pi} = m_{ik}^{\pi} m_{kj}^{\pi},$$

$$\forall m_{ij}^{\pi} \in \left[m_{ij}^{\pi\beta}, m_{ij}^{\pi\gamma}\right] \exists m_{ik}^{\pi} \in \left[m_{ik}^{\pi\beta}, m_{ik}^{\pi\gamma}\right] \wedge \exists m_{kj}^{\pi} \in \left[m_{kj}^{\pi\beta}, m_{kj}^{\pi\gamma}\right]: m_{ij}^{\pi} = m_{ik}^{\pi} m_{kj}^{\pi},$$

for every triplet $(i, j, k) \subseteq \{1, \ldots, n\}$. Thus, $\widetilde{M}^{\pi}$ is multiplicatively consistent according to Definition 51. □

Further, unlike Definitions 48 and 49 of multiplicatively consistent interval and triangular FMPCMs proposed by Wang (2015a) and Wang (2015b), respectively, new Definition 51 does not violate the multiplicative reciprocity of the related PCs.

Theorem 26 *Definition 51 of multiplicative consistency preserves the multiplicative reciprocity of the related PCs in trapezoidal FMPCMs in the sense that any fixed value $m_{ij} \in \left[m_{ij}^{\alpha}, m_{ij}^{\delta}\right]$, $i, j \in \{1, \ldots, n\}$, representing the intensity of preference of object o_i over object o_j is associated with the corresponding value $m_{ji} \in \left[m_{ji}^{\alpha}, m_{ji}^{\delta}\right]$ representing the intensity of preference of object o_j over object o_i such that $m_{ji} = \frac{1}{m_{ij}}$.*

Proof It is sufficient to show that expressions (4.18) and (4.19) do not violate the multiplicative-reciprocity property in the sense that when two particular intensities of preference $m_{ij} \in \widetilde{m}_{ij}$ and $m_{ji} \in \widetilde{m}_{ji}$ on the pair of objects o_i and o_j are considered at the same time in the expressions (4.18) and (4.19), then they are such that $m_{ji} = \frac{1}{m_{ij}}$.

For a triplet $(i, j, k) \subseteq \{1, \ldots, n\}$, $i \neq j \neq k$, no reciprocals appear in expression $m_{ij} = m_{ik}m_{kj}$ for any $m_{ij} \in \left[m_{ij}^{\alpha}, m_{ij}^{\delta}\right]$. For $i = j = k$, expression (4.18) reduces to: $\forall m_{ii} = 1 \exists m_{ii}^{*} = 1 \wedge \exists m_{ii}^{**} = 1: 1 = 1 \cdot 1$, which again does not violate the multiplicative reciprocity. Further, for $i \neq j = k$, expression (4.18) is as: $\forall m_{ij} \in \left[m_{ij}^{\alpha}, m_{ij}^{\delta}\right]$ $\exists m_{ij}^{*} \in \left[m_{ij}^{\alpha}, m_{ij}^{\delta}\right] \wedge \exists m_{jj} = 1: m_{ij} = m_{ij}^{*} \cdot 1$. This means that $m_{ij}^{*} = m_{ij}$, and therefore, the multiplicative reciprocity is not violated. For $i = k \neq j$ the proof is analogous. Finally, for $i = j \neq k$, expression (4.18) is as

$$\forall m_{ii} = 1 \exists m_{ik} \in \left[m_{ik}^{\alpha}, m_{ik}^{\delta}\right] \wedge \exists m_{ki}^{*} \in \left[m_{ki}^{\alpha}, m_{ki}^{\delta}\right] + : 1 = m_{ik}m_{ki}^{*}.$$

This means that $m_{ki}^{*} = \frac{1}{m_{ik}}$, and therefore, the multiplicative reciprocity is preserved. The proof for the expression (4.19) is analogous. □

Remark 15 Similarly to Theorem 22, also Theorem 26 does not simply state that a FMPCM $\tilde{M} = \{\tilde{m}_{ij}\}_{i,j=1}^n$ multiplicatively consistent according to Definition 51 is multiplicatively reciprocal since this property automatically follows from Definition 43 of a FMPCM. Theorem 26 states that Definition 51 is in accordance with the conception of multiplicative reciprocity discussed on p. 98, i.e., only multiplicatively reciprocal PCs are involved in Definition 51 of multiplicative consistency. For more details see Remark 12.

By handling properly the multiplicative reciprocity of the related PCs, Theorem 1 can be extended to trapezoidal FMPCMs as follows.

Theorem 27 *For a trapezoidal FMPCM* $\tilde{M} = \{\tilde{m}_{ij}\}_{i,j=1}^n$, $\tilde{m}_{ij} = \left(m_{ij}^\alpha, m_{ij}^\beta, m_{ij}^\gamma, m_{ij}^\delta\right)$, *the following statements are equivalent:*

(i) $\tilde{M}$ *is multiplicatively consistent according to Definition 51,*
(ii) *For every* $i, j, k = 1, \ldots, n$:

$$\forall m_{ij} \in \left[m_{ij}^\alpha, m_{ij}^\delta\right] \exists m_{jk} \in \left[m_{jk}^\alpha, m_{jk}^\delta\right] \land \exists m_{ki} \in \left[m_{ki}^\alpha, m_{ki}^\delta\right] : m_{ij}m_{jk}m_{ki} = 1,$$
(4.20)
$$\forall m_{ij} \in \left[m_{ij}^\beta, m_{ij}^\gamma\right] \exists m_{jk} \in \left[m_{jk}^\beta, m_{jk}^\gamma\right] \land \exists m_{ki} \in \left[m_{ki}^\beta, m_{ki}^\gamma\right] : m_{ij}m_{jk}m_{ki} = 1,$$
(4.21)

(iii) *For every* $i, j, k = 1, \ldots, n$:

$$\forall m_{ij} \in \left[m_{ij}^\alpha, m_{ij}^\delta\right] \exists m_{jk} \in \left[m_{jk}^\alpha, m_{jk}^\delta\right] \land \exists m_{ki} \in \left[m_{ki}^\alpha, m_{ki}^\delta\right] :$$

$$m_{ij}m_{jk}m_{ki} = m_{ik}m_{kj}m_{ji}, \; m_{ji} = \frac{1}{m_{ij}}, m_{ki} = \frac{1}{m_{ik}}, m_{jk} = \frac{1}{m_{kj}}, \quad (4.22)$$

$$\forall m_{ij} \in \left[m_{ij}^\beta, m_{ij}^\gamma\right] \exists m_{jk} \in \left[m_{jk}^\beta, m_{jk}^\gamma\right] \land \exists m_{ki} \in \left[m_{ki}^\beta, m_{ki}^\gamma\right] :$$

$$m_{ij}m_{jk}m_{ki} = m_{ik}m_{kj}m_{ji}, \; m_{ji} = \frac{1}{m_{ij}}, m_{ki} = \frac{1}{m_{ik}}, m_{jk} = \frac{1}{m_{kj}}. \quad (4.23)$$

Proof From the multiplicative-reciprocity property $\tilde{m}_{ij} = \frac{1}{\tilde{m}_{ji}}$, $i, j = 1, \ldots, n$, it follows that $\forall m_{ij} \in \left[m_{ij}^\alpha, m_{ij}^\delta\right] \exists m_{ji} \in \left[m_{ji}^\alpha, m_{ji}^\delta\right] : m_{ji} = \frac{1}{m_{ij}}$, and $\forall m_{ij} \in \left[m_{ij}^\beta, m_{ij}^\gamma\right] \exists m_{ji} \in \left[m_{ji}^\beta, m_{ji}^\gamma\right] : m_{ji} = \frac{1}{m_{ij}}$.

(a) First, let us show that the statements (i) and (ii) are equivalent. Because of the multiplicative-reciprocity property, (4.18) can be equivalently written as

$$\forall m_{ij} \in \left[m_{ij}^\alpha, m_{ij}^\delta\right] \exists m_{ki} \in \left[m_{ki}^\alpha, m_{ki}^\delta\right] \land \exists m_{jk} \in \left[m_{jk}^\alpha, m_{jk}^\delta\right] : m_{ij} = \frac{1}{m_{jk}} \frac{1}{m_{ki}},$$

which is equivalent to (4.20). Analogously, the equivalence of (4.19) and (4.21)
is proved.

(b) Now, let us show that the statements (ii) and (iii) are equivalent. The expression
(4.20) can be equivalently written as

$$\forall m_{ij} \in \left[m_{ij}^{\alpha}, m_{ij}^{\delta}\right] \exists m_{jk} \in \left[m_{jk}^{\alpha}, m_{jk}^{\delta}\right] \wedge \exists m_{ki} \in \left[m_{ki}^{\alpha}, m_{ki}^{\delta}\right] : m_{ij}^2 m_{jk}^2 m_{ki}^2 = 1.$$
(4.24)

Because all fuzzy numbers in a trapezoidal FMPCM are positive, the second
power can be removed from expression $m_{ij}^2 m_{jk}^2 m_{ki}^2 = 1$, which means that (4.20)
is equivalent to (4.22). Analogously, the equivalence of (4.21) and (4.23) is
proved. $\qquad\square$

The following theorems give us useful tools for verifying multiplicative consistency of trapezoidal FMPCMs.

Theorem 28 *A trapezoidal FMPCM* $\tilde{M} = \left\{\tilde{m}_{ij}\right\}_{i,j=1}^{n}$, $\tilde{m}_{ij} = \left(m_{ij}^{\alpha}, m_{ij}^{\beta}, m_{ij}^{\gamma}, m_{ij}^{\delta}\right)$, *is multiplicatively consistent according to Definition 51 if and only if the inequalities*

$$m_{ij}^{\alpha} \geq m_{ik}^{\alpha} m_{kj}^{\alpha}, \qquad m_{ij}^{\delta} \leq m_{ik}^{\delta} m_{kj}^{\delta},$$
(4.25)

$$m_{ij}^{\beta} \geq m_{ik}^{\beta} m_{kj}^{\beta}, \qquad m_{ij}^{\gamma} \leq m_{ik}^{\gamma} m_{kj}^{\gamma},$$
(4.26)

hold for every $i, j, k = 1, \ldots, n$, $i < j$, $k \neq i, j$.

Proof It is sufficient to demonstrate the equivalence of the expressions (4.25) and
(4.18). The demonstration of the equivalence of (4.26) and (4.19) is analogous.

First, let us demonstrate that when the inequalities (4.25) hold for every $i, j, k =
1, \ldots, n$, $i < j$, $k \neq i, j$, then they hold for every $i, j, k = 1, \ldots, n$. The inequalities
(4.25) are always satisfied for $i, j, k = 1, \ldots, n$, such that $i = j \neq k$, or $i \neq j = k$,
or $j \neq k = i$, or $i = j = k$:

$$m_{ik}^{\alpha} m_{ki}^{\alpha} = \frac{m_{ik}^{\alpha}}{m_{ik}^{\delta}} \leq 1 = m_{ii}^{\alpha}, \qquad m_{ik}^{\delta} m_{ki}^{\delta} = \frac{m_{ik}^{\delta}}{m_{ik}^{\alpha}} \geq 1 = m_{ii}^{\delta},$$

$$m_{ij}^{\alpha} m_{jj}^{\alpha} = m_{ij}^{\alpha}, \qquad m_{ij}^{\delta} m_{jj}^{\delta} = m_{ij}^{\delta},$$

$$m_{ii}^{\alpha} m_{ij}^{\alpha} = m_{ij}^{\alpha}, \qquad m_{ii}^{\delta} m_{ij}^{\delta} = m_{ij}^{\delta},$$

$$m_{ii}^{\alpha} m_{ii}^{\alpha} = 1 = m_{ii}^{\alpha}, \qquad m_{ii}^{\delta} m_{ii}^{\delta} = 1 = m_{ij}^{\delta}.$$

Further, when the inequalities (4.25) are satisfied for $i, j, k = 1, \ldots, n$, $i < j$, $k \neq i, j$, then they are satisfied also for $j, i, k = 1, \ldots, n$, $j > i$, $k \neq i, j$:

$$m_{jk}^\alpha m_{ki}^\alpha = \frac{1}{m_{kj}^\delta} \frac{1}{m_{ik}^\delta} = \frac{1}{m_{ik}^\delta m_{kj}^\delta} \leq \frac{1}{m_{ij}^\delta} = m_{ji}^\alpha,$$

$$m_{jk}^\delta m_{ki}^\delta = \frac{1}{m_{kj}^\alpha} \frac{1}{m_{ik}^\alpha} = \frac{1}{m_{ik}^\alpha m_{kj}^\alpha} \geq \frac{1}{m_{ij}^\alpha} = m_{ji}^\delta.$$

To finalize the proof, it is sufficient to show that the inequalities (4.25) are equivalent to the condition (4.18) for every $i, j, k = 1, \ldots, n$. First, let $\tilde{M}$ be a trapezoidal FMPCM multiplicatively consistent according to Definition 51. Then, for $m_{ij} := m_{ij}^\alpha \; \exists m_{ik} \in \left[m_{ik}^\alpha, m_{ik}^\delta \right] \wedge \exists m_{kj} \in \left[m_{kj}^\alpha, m_{kj}^\delta \right] : m_{ij}^\alpha = m_{ik} m_{kj}$. Since $m_{ik} \geq m_{ik}^\alpha$, $m_{kj} \geq m_{kj}^\alpha$, then clearly $m_{ij}^\alpha \geq m_{ik}^\alpha m_{kj}^\alpha$. Analogously, for $m_{ij} := m_{ij}^\delta \; \exists m_{ik} \in \left[m_{ik}^\alpha, m_{ik}^\delta \right] \wedge \exists m_{kj} \in \left[m_{kj}^\alpha, m_{kj}^\delta \right] : m_{ij}^\delta = m_{ik} m_{kj}$. Since $m_{ik} \leq m_{ik}^\delta$, $m_{kj} \leq m_{kj}^\delta$, then $m_{ij}^\delta \leq m_{ik}^\delta m_{kj}^\delta$.

Second, let (4.25) be valid for a trapezoidal FMPCM $\tilde{M}$. Then, from inequalities (4.25) we get $\forall m_{ij} \in \left[m_{ij}^\alpha, m_{ij}^\delta \right] : m_{ik}^\alpha m_{kj}^\alpha \leq m_{ij} \leq m_{ik}^\delta m_{kj}^\delta$, and therefore, (4.18) is satisfied. $\qquad \square$

Theorem 29 *A trapezoidal FMPCM* $\tilde{M} = \{\tilde{m}_{ij}\}_{i,j=1}^n$, $\tilde{m}_{ij} = \left(m_{ij}^\alpha, m_{ij}^\beta, m_{ij}^\gamma, m_{ij}^\delta \right)$, *is multiplicatively consistent according to Definition 51 if and only if the inequalities*

$$m_{ij}^\alpha \geq \max_{\substack{k=1,\ldots,n \\ k \neq i,j}} \left\{ m_{ik}^\alpha m_{kj}^\alpha \right\}, \qquad m_{ij}^\delta \leq \min_{\substack{k=1,\ldots,n \\ k \neq i,j}} \left\{ m_{ik}^\delta m_{kj}^\delta \right\}, \qquad (4.27)$$

$$m_{ij}^\beta \geq \min_{\substack{k=1,\ldots,n \\ k \neq i,j}} \left\{ m_{ik}^\beta m_{kj}^\beta \right\}, \qquad m_{ij}^\gamma \leq \min_{\substack{k=1,\ldots,n \\ k \neq i,j}} \left\{ m_{ik}^\gamma m_{kj}^\gamma \right\}, \qquad (4.28)$$

hold for every $i, j = 1, \ldots, n$, $i < j$.

Proof The inequalities (4.27) and (4.28) follow immediately from Theorem 28. $\quad \square$

In the following example, Definition 51 of multiplicative consistency for trapezoidal FMPCMs is confronted with Definitions 46 and 48. In particular, it is demonstrated how the drawbacks regarding the dependence of Definition 46 on permutation of objects and violation of multiplicative-reciprocity property in Definition 48 are removed by Definition 51. Further, multiplicative weak consistency according to Definition 50 is examined.

Example 37 Let us examine the interval FMPCM $\overline{M}$ of objects o_1, o_2, o_3 given as

$$
\overline{M} = \begin{pmatrix}
1 & \left[\dfrac{2}{5}, \dfrac{2}{3}\right] & \left[\dfrac{1}{5}, \dfrac{2}{3}\right] \\[2ex]
\left[\dfrac{3}{2}, \dfrac{5}{2}\right] & 1 & \left[\dfrac{1}{2}, 1\right] \\[2ex]
\left[\dfrac{3}{2}, 5\right] & [1, 2] & 1
\end{pmatrix}
\tag{4.29}
$$

and its permutation $\overline{M}^{\pi} = P\overline{M}P^T$ given as

$$
\overline{M}^{\pi} = \begin{pmatrix}
1 & \left[\dfrac{1}{2}, 1\right] & \left[\dfrac{3}{2}, \dfrac{5}{2}\right] \\[2ex]
[1, 2] & 1 & \left[\dfrac{3}{2}, 5\right] \\[2ex]
\left[\dfrac{2}{5}, \dfrac{2}{3}\right] & \left[\dfrac{1}{5}, \dfrac{2}{3}\right] & 1
\end{pmatrix},
\tag{4.30}
$$

which is obtained from $\overline{M}$ by applying the permutation matrix

$$
P = \begin{pmatrix}
0 & 1 & 0 \\
0 & 0 & 1 \\
1 & 0 & 0
\end{pmatrix}.
\tag{4.31}
$$

As pointed out by Wang (2015a) and Wang (2015b) and mentioned in Sect. 4.2.2.1, Definitions 46 and 47 of multiplicative consistency for interval and trapezoidal FMPCMs, respectively, are not invariant under permutation of objects. In fact, the interval FMPCM (4.29) results to be multiplicatively consistent, and its permutation (4.30) results to be inconsistent according to Definition 46.

Now, let us apply Definition 51 to the interval FMPCM (4.29). By using Theorem 28, the interval FMPCM (4.29) is judged as multiplicatively consistent since it satisfies the inequalities (4.25); see Table 4.4. Also the permuted interval FMPCM (4.30) satisfies the inequalities (4.25); see Table 4.5. Therefore, it is again judged as multiplicatively consistent. Moreover, from Theorem 25 it follows that any permutation of the interval FMPCM (4.29) is multiplicatively consistent according to Definition 51.

Table 4.4 Inequality conditions (4.25) for the interval FMPCM (4.29)

$i < j$	$m_{ij}^L \geq m_{ik}^L m_{kj}^L$	$m_{ij}^U \leq m_{ik}^U m_{kj}^U$
1, 2:	$\frac{2}{5} \geq \frac{2}{5} \cdot 1$	$\frac{2}{3} \leq \frac{2}{3} \cdot 2$
1, 3:	$\frac{1}{5} \geq \frac{2}{5} \cdot \frac{1}{2}$	$\frac{2}{3} \leq \frac{2}{3} \cdot 1$
2, 3:	$\frac{1}{2} \geq \frac{3}{2} \cdot \frac{1}{3}$	$1 \leq \frac{5}{2} \cdot \frac{2}{3}$

Table 4.5 Inequality conditions (4.25) for the permuted interval FMPCM (4.30)

$i < j$:	$m_{ij}^L \geq m_{ik}^L m_{kj}^L$	$m_{ij}^U \leq m_{ik}^U m_{kj}^U$
1, 2:	$\frac{1}{2} \geq \frac{3}{2} \cdot \frac{1}{5}$	$1 \leq \frac{5}{2} \cdot \frac{2}{3}$
1, 3:	$\frac{3}{2} \geq \frac{1}{2} \cdot \frac{3}{2}$	$\frac{5}{2} \leq 1 \cdot 5$
2, 3:	$\frac{3}{2} \geq 1 \cdot \frac{3}{2}$	$5 \leq 2 \cdot \frac{5}{2}$

In Example 36, it was demonstrated that Definition 48 violates the multiplicative reciprocity of the related PCs. In fact, by using the property (iii) of Theorem 20 (more precisely a version of the theorem adapted for interval FMPCMs), the multiplicative consistency of the matrix

$$
M^L = \begin{pmatrix} 1 & \frac{2}{5} & \frac{1}{5} \\ \frac{3}{2} & 1 & \frac{1}{2} \\ \frac{3}{2} & 1 & 1 \end{pmatrix} \tag{4.32}
$$

is checked in order to verify multiplicative consistency of the interval FMPCM (4.29) according to Definition 48. The matrix (4.32) is not multiplicatively reciprocal, and thus it is not even a MPCM. Therefore, verifying its consistency is nonsensical.

According to Theorem 26, the multiplicative-reciprocity property is preserved by new Definition 51. This basically means that by taking any value from any interval PC in the interval FMPCM (4.29), there exist values in the remaining interval PCs such that they form a multiplicatively consistent MPCM. Let us examine the triplet $i = 1, j = 2, k = 3$ of indices and let us consider the value $m_{12} = \frac{1}{2} \in \left[\frac{2}{5}, \frac{2}{3}\right]$. Then, according to (4.18), there exist values $m_{13} \in \left[\frac{1}{5}, \frac{2}{3}\right]$ and $m_{32} \in [1, 2]$ such that $\frac{1}{2} = m_{13} m_{32}$. It is, for example, $m_{13} = \frac{1}{4}$, $m_{32} = 2$. The multiplicative reciprocity is clearly not violated. More interestingly, let us consider the triplet $i = 1, j = 1, k = 2$. Then, according to (4.18), there exist values $m_{12} \in \left[\frac{2}{5}, \frac{2}{3}\right]$ and $m_{21} \in \left[\frac{3}{2}, \frac{5}{2}\right]$ such that $1 = m_{12} m_{21}$. This equality is satisfied by any value $m_{12} \in \left[\frac{2}{5}, \frac{2}{3}\right]$ and the corresponding value $m_{21} \in \left[\frac{3}{2}, \frac{5}{2}\right]$ such that $m_{21} = \frac{1}{m_{12}}$, which again preserves the multiplicative reciprocity.

By verifying the inequalities (4.16) we also find out that the interval FMPCM (4.29) is multiplicatively consistent according to Definition 50; see Table 4.6. Analogously it could be shown that also the permuted interval FMPCM (4.30) is multiplicatively weakly consistent according to Definition 50. $\triangle$

Table 4.6 Inequality conditions (4.16) for the interval FMPCM (4.29)

$i < j$:	$\max_{k=1,\dots,n} \left\{ m_{ik}^L m_{kj}^L \right\} \leq \min_{k=1,\dots,n} \left\{ m_{ik}^U m_{kj}^U \right\}$
1, 2:	$\max \left\{ \frac{2}{5}, \frac{2}{5}, \frac{1}{5} \right\} \leq \min \left\{ \frac{2}{3}, \frac{2}{3}, \frac{4}{3} \right\}$
1, 3:	$\max \left\{ \frac{1}{5}, \frac{1}{5}, \frac{1}{5} \right\} \leq \min \left\{ \frac{2}{3}, \frac{2}{3}, \frac{2}{3} \right\}$
2, 3:	$\max \left\{ \frac{3}{10}, \frac{1}{2}, \frac{1}{2} \right\} \leq \min \left\{ \frac{5}{3}, 1, 1 \right\}$

In the following example, Definition 51 of multiplicative consistency is applied to an incomplete FMPCM in order to identify a missing PC.

Example 38 Let us consider the interval FMPCM in the form (4.11) with unknown interval PCs $\overline{m}_{13}$ and $\overline{m}_{31} = \frac{1}{\overline{m}_{13}}$ examined in Example 36. It was shown in Example 36 that applying Definition 48 to identify the missing PCs leads to unreasonable results.

Now let us apply Definition 51 to the interval FMPCM (4.11). By using Theorem 29, we obtain the following:

$$
\begin{array}{llll}
i = 1, j = 2: & \frac{3}{2} \geq x \cdot \frac{1}{2} = \frac{x}{2} \Rightarrow x \leq 3, & 2 \leq y \cdot \frac{2}{3} \Rightarrow y \geq 3, \\
i = 1, j = 3: & x \geq \frac{3}{2} \cdot \frac{3}{2} = \frac{9}{4} \Rightarrow x \geq \frac{9}{4}, & y \leq 2 \cdot 2 \Rightarrow y \leq 4, \\
i = 2, j = 3: & \frac{3}{2} \geq \frac{1}{2} \cdot x = \frac{x}{2} \Rightarrow x \leq 3, & 2 \leq \frac{2}{3} \cdot y \Rightarrow y \geq 3.
\end{array}
$$

Therefore, the interval FMPCM (4.11) is multiplicatively consistent according to Definition 51 if $\overline{m}_{13} = [x, y]$, $x \leq y$, is such that $x \in \left[\frac{9}{4}, 3\right]$, $y \in [3, 4]$. This means that the lowest possible intensity of preference of object o_1 over object o_3 is at least $\frac{9}{4} > 1$, i.e., object o_1 is preferred to object o_3. Moreover, the highest possible intensity of preference of object o_1 over object o_3 is 4, which is reachable under the multiplicative consistency condition for $m_{12} = 2 \in \left[\frac{3}{2}, 2\right]$, $m_{23} = 2 \in \left[\frac{3}{2}, 2\right]$.   △

In the rest of this section, some interesting properties of multiplicatively weakly consistent and multiplicatively consistent trapezoidal FMPCMs are examined. The following theorem shows the relation between Definition 51 of multiplicative consistency and Definition 50 of multiplicative weak consistency.

Theorem 30 *Let* $\widetilde{M} = \left\{\widetilde{m}_{ij}\right\}_{i,j=1}^{n}$, $\widetilde{m}_{ij} = (m_{ij}^{\alpha}, m_{ij}^{\beta}, m_{ij}^{\gamma}, m_{ij}^{\delta})$, *be a trapezoidal FMPCM. If* $\widetilde{M} = \left\{\widetilde{m}_{ij}\right\}_{i,j=1}^{n}$ *is multiplicatively consistent according to Definition 51, then it is also multiplicatively weakly consistent according to Definition 50.*

Proof The statement follows immediately from Theorem 29. In particular, the inequality (4.16) is obtained immediately from the inequalities (4.27).   □

Remark 16 According to Theorem 30, when a trapezoidal FMPCM is multiplicatively consistent according to Definition 51 then it is also automatically multiplicatively weakly consistent according to Definition 50. However, this does not hold the other way around. Clearly, the definition of multiplicative weak consistency is much weaker than the definition of multiplicative consistency; it only requires existence of one crisp multiplicatively consistent MPCM obtainable by combining particular elements from the closures of the supports of the trapezoidal fuzzy numbers in the trapezoidal FMPCM. Thus, the set of all trapezoidal FMPCMs multiplicatively consistent according to Definition 51 is a proper subset of the set of all trapezoidal FMPCMs multiplicatively weakly consistent according to Definition 50.

In the following example, the multiplicative consistency given by Definition 51 and the multiplicative weak consistency given by Definition 50 are examined.

Example 39 Let us consider the trapezoidal FMPCM

$$
\widetilde{M} = \begin{pmatrix}
1 & (2,3,4,5) & (2,3,3,4) & (1,1,2,3) \\
\left(\dfrac{1}{5},\dfrac{1}{4},\dfrac{1}{3},\dfrac{1}{2}\right) & 1 & (1,2,3,3) & \left(1,\dfrac{3}{2},2,2\right) \\
\left(\dfrac{1}{4},\dfrac{1}{3},\dfrac{1}{3},\dfrac{1}{2}\right) & \left(\dfrac{1}{3},\dfrac{1}{3},\dfrac{1}{2},1\right) & 1 & \left(\dfrac{2}{5},\dfrac{3}{5},\dfrac{4}{5},1\right) \\
\left(\dfrac{1}{3},\dfrac{1}{2},1,1\right) & \left(\dfrac{1}{2},\dfrac{1}{2},\dfrac{2}{3},1\right) & \left(1,\dfrac{5}{4},\dfrac{5}{3},\dfrac{5}{2}\right) & 1
\end{pmatrix}. \quad (4.33)
$$

$\widetilde{M}$ is not multiplicatively consistent according to new Definition 51 of multiplicative consistency since inequalities (4.27) are violated; e.g.,

$$
\max_{k=2,3}\left\{m_{1k}^{\alpha}m_{k4}^{\alpha}\right\} = \max\left\{2,\frac{4}{5}\right\} = 2 \nleq 1 = m_{14}^{\alpha}.
$$

However, $\widetilde{M}$ is multiplicatively consistent according to Definition 50 since inequalities (4.16) are satisfied; see Table 4.7. Therefore, according to Definition 50, there exists at least one multiplicatively consistent MPCM $M = \left\{\dfrac{w_i}{w_j}\right\}_{i,j=1}^{n}$, such that $m_{ij}^{\alpha} \leq \dfrac{w_i}{w_j} \leq m_{ij}^{\delta}$, $i,j = 1,\ldots,n$, $\sum_{i=1}^{n} w_i = 1$. It is, for example,

$$
M = \begin{pmatrix}
1 & 2 & 2 & 2 \\
\dfrac{1}{2} & 1 & 1 & 1 \\
\dfrac{1}{2} & 1 & 1 & 1 \\
\dfrac{1}{2} & 1 & 1 & 1
\end{pmatrix}
$$

with the priority vector $\underline{w} = (\frac{2}{5},\frac{1}{5},\frac{1}{5},\frac{1}{5})^{T}$. △

Table 4.7 Inequality conditions (4.16) for the interval FMPCM (4.33)	$i < j$:	$\max\limits_{k=1,\ldots,4}\left\{m_{ik}^{\alpha}m_{kj}^{\alpha}\right\} \leq \min\limits_{k=1,\ldots,4}\left\{m_{ik}^{\delta}m_{kj}^{\delta}\right\}$
	1, 2:	$\max\left\{2,2,\frac{2}{3},\frac{1}{2}\right\} \leq \min\{5,5,4,3\}$
	1, 3:	$\max\{2,2,2,1\} \leq \min\left\{4,15,4,\frac{15}{2}\right\}$
	1, 4:	$\max\left\{1,2,\frac{4}{5},1\right\} \leq \min\{3,10,4,3\}$
	2, 3:	$\max\left\{\frac{2}{5},1,1,1\right\} \leq \min\{2,3,3,5\}$
	2, 4:	$\max\left\{\frac{1}{5},1,\frac{2}{5},1\right\} \leq \min\left\{\frac{3}{5},2,3,2\right\}$
	3, 4:	$\max\left\{\frac{1}{4},\frac{1}{3},\frac{2}{5},\frac{2}{5}\right\} \leq \min\left\{\frac{3}{2},2,1,1\right\}$

Theorem 31 *Let $\widetilde{M}$ be a trapezoidal FMPCM multiplicatively weakly consistent according to Definition 50. A trapezoidal FMPCM $\widetilde{M}^*$ constructed by eliminating the l-th row and the l-th column, $l \in \{1, \ldots, n\}$, of $\widetilde{M}$ is again multiplicatively weakly consistent.*

Proof For $\widetilde{M}$, the inequalities (4.16) are valid for every $i, j, k = 1, \ldots, n$. After eliminating the l-th row and the l-th column of $\widetilde{M}$, the inequalities (4.16) are still valid for every remaining $i, j, k \in \{1, \ldots, n\} \setminus \{l\}$. Therefore, the new trapezoidal FMPCM $\widetilde{M}^*$ is still multiplicatively weakly consistent.   □

The same holds also for multiplicatively consistent trapezoidal FMPCMs.

Theorem 32 *Let $\widetilde{M}$ be a trapezoidal FMPCM multiplicatively consistent according to Definition 51. A trapezoidal FMPCM $\widetilde{M}^*$ constructed by eliminating the l-th row and the l-th column, $l \in \{1, \ldots, n\}$, of $\widetilde{M}$ is again multiplicatively consistent.*

Proof For $\widetilde{M}$, the inequalities (4.27) and (4.28) are valid for each $i, j, k \in \{1, \ldots, n\}$. After eliminating the l-th row and the l-th column of $\widetilde{M}$, (4.27) and (4.28) are still valid for each remaining $i, j, k \in \{1, \ldots, n\} \setminus \{l\}$. Therefore, the new trapezoidal FMPCM $\widetilde{M}^*$ is still multiplicatively consistent.   □

Remark 17 Theorems 31 and 32 are useful in situations when the set of objects compared pairwisely is being reduced. According to the theorems, elimination of one or more objects has no impact on the multiplicative or multiplicative weak consistency of fuzzy PCs of the remaining objects.

The following theorems provide results regarding aggregation of multiplicatively and multiplicatively weakly consistent trapezoidal FMPCMs, which are particularly useful in group decision making.

Theorem 33 *Let $\widetilde{M}^1 = \left\{ \widetilde{m}_{ij}^1 \right\}_{i,j=1}^n$, $\widetilde{m}_{ij}^1 = (m_{ij}^{1\alpha}, m_{ij}^{1\beta}, m_{ij}^{1\gamma}, m_{ij}^{1\delta})$, and $\widetilde{M}^2 = \left\{ \widetilde{m}_{ij}^2 \right\}_{i,j=1}^n$, $\widetilde{m}_{ij}^2 = (m_{ij}^{2\alpha}, m_{ij}^{2\beta}, m_{ij}^{2\gamma}, m_{ij}^{2\delta})$, be trapezoidal FMPCMs multiplicatively weakly consistent according to Definition 50. Then $\widetilde{M} = \left\{ \widetilde{m}_{ij} \right\}_{i,j=1}^n$, $\widetilde{m}_{ij} = (m_{ij}^{\alpha}, m_{ij}^{\beta}, m_{ij}^{\gamma}, m_{ij}^{\delta})$, such that*

$$m_{ij}^{\alpha} = (m_{ij}^{1\alpha})^{\epsilon}(m_{ij}^{2\alpha})^{1-\epsilon}, \quad m_{ij}^{\beta} = (m_{ij}^{1\beta})^{\epsilon}(m_{ij}^{2\beta})^{1-\epsilon},$$
$$m_{ij}^{\gamma} = (m_{ij}^{1\gamma})^{\epsilon}(m_{ij}^{2\gamma})^{1-\epsilon}, \quad m_{ij}^{\delta} = (m_{ij}^{1\delta})^{\epsilon}(m_{ij}^{2\delta})^{1-\epsilon},$$

is a multiplicatively weakly consistent trapezoidal FMPCM for any $\epsilon \in [0, 1]$.

Proof First, let us show that $\widetilde{M}$ is a trapezoidal FMPCM. For $i = 1, \ldots, n$, we get

$$m_{ii}^{\alpha} = (m_{ii}^{1\alpha})^{\epsilon}(m_{ii}^{2\alpha})^{1-\epsilon} = 1^{\epsilon}1^{1-\epsilon} = 1, \quad m_{ii}^{\delta} = (m_{ii}^{1\delta})^{\epsilon}(m_{ii}^{2\delta})^{1-\epsilon} = 1^{\epsilon}1^{1-\epsilon} = 1.$$

Similarly, $m_{ii}^\beta = 1$, $m_{ii}^\gamma = 1$, and thus, $\tilde{m}_{ii} = 1$, $i = 1, \ldots, n$. Further,

$$m_{ij}^\alpha = (m_{ij}^{1\alpha})^\epsilon (m_{ij}^{2\alpha})^{1-\epsilon} = \left(\frac{1}{m_{ji}^{1\delta}}\right)^\epsilon \left(\frac{1}{m_{ji}^{2\delta}}\right)^{1-\epsilon} = \frac{1}{(m_{ji}^{1\delta})^\epsilon (m_{ji}^{2\delta})^{1-\epsilon}} = \frac{1}{m_{ji}^\delta},$$

$$m_{ij}^\delta = (m_{ij}^{1\delta})^\epsilon (m_{ij}^{2\delta})^{1-\epsilon} = \left(\frac{1}{m_{ji}^{1\alpha}}\right)^\epsilon \left(\frac{1}{m_{ji}^{2\alpha}}\right)^{1-\epsilon} = \frac{1}{(m_{ji}^{1\alpha})^\epsilon (m_{ji}^{2\alpha})^{1-\epsilon}} = \frac{1}{m_{ji}^\alpha},$$

and analogously we obtain $m_{ij}^\beta = \frac{1}{m_{ji}^\gamma}$, $m_{ij}^\gamma = \frac{1}{m_{ji}^\beta}$. Therefore, $\tilde{m}_{ij} = \frac{1}{\tilde{m}_{ji}}$, $i, j = 1, \ldots, n$.

Second, let us show that $\tilde{M}$ is multiplicatively weakly consistent. It is sufficient to prove inequalities (4.16). Since (4.16) is valid for FMPCMs $\tilde{M}^1$ and $\tilde{M}^2$, we obtain

$$\max_{k=1,\ldots,n} \left\{ m_{ik}^\alpha m_{kj}^\alpha \right\} = \max_{k=1,\ldots,n} \left\{ \left(m_{ik}^{1\alpha}\right)^\epsilon \left(m_{ik}^{2\alpha}\right)^{1-\epsilon} \left(m_{kj}^{1\alpha}\right)^\epsilon \left(m_{kj}^{2\alpha}\right)^{1-\epsilon} \right\} \le$$

$$\max_{k=1,\ldots,n} \left\{ (m_{ik}^{1\alpha} m_{kj}^{1\alpha})^\epsilon \right\} \max_{k=1,\ldots,n} \left\{ (m_{ik}^{2\alpha} m_{kj}^{2\alpha})^{1-\epsilon} \right\} \le$$

$$\min_{k=1,\ldots,n} \left\{ (m_{ik}^{1\delta} m_{kj}^{1\delta})^\epsilon \right\} \min_{k=1,\ldots,n} \left\{ (m_{ik}^{2\delta} m_{kj}^{2\delta})^{1-\epsilon} \right\} \le$$

$$\min_{k=1,\ldots,n} \left\{ \left(m_{ik}^{1\delta}\right)^\epsilon \left(m_{ik}^{2\delta}\right)^{1-\epsilon} \left(m_{kj}^{1\delta}\right)^\epsilon \left(m_{kj}^{2\delta}\right)^{1-\epsilon} \right\} = \min_{k=1,\ldots,n} \left\{ m_{ik}^\delta m_{kj}^\delta \right\}$$

which proves the theorem.   □

Theorem 33 can be further extended to the aggregation of $p \ge 2$ multiplicatively weakly consistent trapezoidal FMPCMs as follows.

Theorem 34 *Let* $\tilde{M}^\tau = \left\{ \tilde{m}_{ij}^\tau \right\}_{i,j=1}^n$, $\tilde{m}_{ij}^\tau = (m_{ij}^{\tau\alpha}, m_{ij}^{\tau\beta}, m_{ij}^{\tau\gamma}, m_{ij}^{\tau\delta})$, $\tau = 1, \ldots, p$, *be trapezoidal FMPCMs multiplicatively weakly consistent according to Definition 50. Then* $\tilde{M} = \left\{ \tilde{m}_{ij} \right\}_{i,j=1}^n$, *such that*

$$\tilde{m}_{ij} = \left(m_{ij}^\alpha, m_{ij}^\beta, m_{ij}^\gamma, m_{ij}^\delta \right) = \left(\prod_{\tau=1}^p \left(m_{ij}^{\tau\alpha}\right)^{\epsilon_\tau}, \prod_{\tau=1}^p (m_{ij}^{\tau\beta})^{\epsilon_\tau}, \prod_{\tau=1}^p \left(m_{ij}^{\tau\gamma}\right)^{\epsilon_\tau}, \prod_{\tau=1}^p \left(m_{ij}^{\tau\delta}\right)^{\epsilon_\tau} \right),$$

is a multiplicatively weakly consistent trapezoidal FMPCM for any $\epsilon_\tau \in [0, 1]$, $\tau = 1, \ldots, p$, *with* $\sum_{\tau=1}^p \epsilon_\tau = 1$.

Proof The proof is analogous to the proof of Theorem 33.   □

Similar theorems are formulated also for multiplicatively consistent trapezoidal FMPCMs.

Theorem 35 *Let* $\widetilde{M}^1 = \left\{\widetilde{m}_{ij}^1\right\}_{i,j=1}^n$, $\widetilde{m}_{ij}^1 = \left(m_{ij}^{1\alpha}, m_{ij}^{1\beta}, m_{ij}^{1\gamma}, m_{ij}^{1\delta}\right)$, *and* $\widetilde{M}^2 = \left\{\widetilde{m}_{ij}^2\right\}_{i,j=1}^n$, $\widetilde{m}_{ij}^2 = \left(m_{ij}^{2\alpha}, m_{ij}^{2\beta}, m_{ij}^{2\gamma}, m_{ij}^{2\delta}\right)$, *be trapezoidal FMPCMs multiplicatively consistent according to Definition 51. Then* $\widetilde{M} = \left\{\widetilde{m}_{ij}\right\}_{i,j=1}^n$, *such that*

$$m_{ij}^\alpha = (m_{ij}^{1\alpha})^\epsilon (m_{ij}^{2\alpha})^{1-\epsilon}, \quad m_{ij}^\beta = (m_{ij}^{1\beta})^\epsilon (m_{ij}^{2\beta})^{1-\epsilon},$$
$$m_{ij}^\gamma = (m_{ij}^{1\gamma})^\epsilon (m_{ij}^{2\gamma})^{1-\epsilon}, \quad m_{ij}^\delta = (m_{ij}^{1\delta})^\epsilon (m_{ij}^{2\delta})^{1-\epsilon},$$

is a multiplicatively consistent trapezoidal FMPCM for any $\epsilon \in [0, 1]$.

Proof From Theorem 33 we already know that $\widetilde{M}$ is a FMPCM. It remains to show that $\widetilde{M}$ is multiplicatively consistent. It is sufficient to prove inequalities (4.25) and (4.26). Since (4.25) are valid for the FMPCMs $\widetilde{M}^1$ and $\widetilde{M}^2$, we obtain

$$m_{ik}^\alpha m_{kj}^\alpha = (m_{ik}^{1\alpha})^\epsilon (m_{ik}^{2\alpha})^{1-\epsilon} (m_{kj}^{1\alpha})^\epsilon (m_{kj}^{2\alpha})^{1-\epsilon} =$$
$$(m_{ik}^{1\alpha} m_{kj}^{1\alpha})^\epsilon (m_{ik}^{2\alpha} m_{kj}^{2\alpha})^{1-\epsilon} \le (m_{ij}^{1\alpha})^\epsilon (m_{ij}^{2\alpha})^{1-\epsilon} = m_{ij}^\alpha,$$
$$m_{ik}^\delta m_{kj}^\delta = (m_{ik}^{1\delta})^\epsilon (m_{ik}^{2\delta})^{1-\epsilon} (m_{kj}^{1\delta})^\epsilon (m_{kj}^{2\delta})^{1-\epsilon} =$$
$$(m_{ik}^{1\delta} m_{kj}^{1\delta})^\epsilon (m_{ik}^{2\delta} m_{kj}^{2\delta})^{1-\epsilon} \ge (m_{ij}^{1\delta})^\epsilon (m_{ij}^{2\delta})^{1-\epsilon} = m_{ij}^\delta.$$

Analogously, the validity of inequalities (4.26) is proved.   □

Theorem 35 can be further extended to the aggregation of $p \ge 2$ multiplicatively consistent trapezoidal FMPCMs as follows.

Theorem 36 *Let* $\widetilde{M}^\tau = \left\{\widetilde{m}_{ij}^\tau\right\}_{i,j=1}^n$, $\widetilde{m}_{ij}^\tau = \left(m_{ij}^{\tau\alpha}, m_{ij}^{\tau\beta}, m_{ij}^{\tau\gamma}, m_{ij}^{\tau\delta}\right)$, $\tau = 1, \ldots, p$, *be trapezoidal FMPCMs multiplicatively consistent according to Definition 51. Then* $\widetilde{M} = \left\{\widetilde{m}_{ij}\right\}_{i,j=1}^n$, *such that*

$$\widetilde{m}_{ij} = \left(m_{ij}^\alpha, m_{ij}^\beta, m_{ij}^\gamma, m_{ij}^\delta\right) = \left(\prod_{\tau=1}^p \left(m_{ij}^{\tau\alpha}\right)^{\epsilon_\tau}, \prod_{\tau=1}^p (m_{ij}^{\tau\beta})^{\epsilon_\tau}, \prod_{\tau=1}^p \left(m_{ij}^{\tau\gamma}\right)^{\epsilon_\tau}, \prod_{\tau=1}^p \left(m_{ij}^{\tau\delta}\right)^{\epsilon_\tau}\right),$$

is a multiplicatively consistent trapezoidal FMPCM for any $\epsilon_\tau \in [0, 1]$, $\tau = 1, \ldots, p$, *with* $\sum_{\tau=1}^p \epsilon_\tau = 1$.

Proof The proof is similar to the proof of Theorem 35.   □

Reaching full consistency is not always manageable in practice. Often, even when DMs are asked to reconsider their inconsistent preference information they are not able to build a consistent FMPCM. That is why the problem of measuring acceptable inconsistency of FMPCMs has been addressed in the literature.

4.2.2.3 Fuzzy Consistency Index and Fuzzy Consistency Ratio

A number of inconsistency indices for measuring an acceptable level of inconsistency of MPCMs has been proposed in the literature. One of the well-known and most often applied ones is the Consistency Index (2.9) proposed by Saaty (1980). Strangely, the problem of verifying an acceptable level of inconsistency of FMPCMs has been quite neglected in the literature. Very often, authors dealing with the fuzzy extension of PC methods do not address the issue of (in)consistency at all; see, e.g., the well-known theoretical articles by Laarhoven and Pedrycz (1983); Chang (1996); Enea and Piazza (2004). Similarly, also most real-application articles do not address this important issue.

In some real-application articles, the authors verify the acceptable inconsistency of a FMPCM by means of CR only for the crisp MPCM $M = \left\{m_{ij}^{M}\right\}_{i,j=1}^{n}$ constructed from the middle values of the triangular FMPCM $\widetilde{M} = \{\tilde{m}_{ij}\}_{i,j=1}^{n}$, $\tilde{m}_{ij} = (m_{ij}^{L}, m_{ij}^{M}, m_{ij}^{U}), i,j = 1, \ldots, n$; see, e.g., Tesfamariam and Sadiq (2006); Pan (2008); Vahidnia et al. (2009). However, by verifying acceptable inconsistency of just one particular matrix of crisp numbers obtained from the original FMPCM (in this case the middle values of the triangular fuzzy numbers) the uncertainty modeled by the fuzzy numbers in the FMPCM is neglected. This is inconsistent with the original intention to model the incompleteness of information as well as the linguistically expressed preference information by fuzzy numbers.

A similar approach was considered by Zheng et al. (2012), although they suggested to compute CR for the crisp matrix $M^{*} = \{m_{ij}^{*}\}_{i,j=1}^{n}$ whose elements are obtained from the trapezoidal FMPCM $\widetilde{M} = \left\{m_{ij}\right\}_{i,j=1}^{n}$, $\tilde{m}_{ij} = (m_{ij}^{\alpha}, m_{ij}^{\beta}, m_{ij}^{\gamma}, m_{ij}^{\delta})$, by formula

$$m_{ij}^{*} = \frac{m_{ij}^{\alpha} + 2m_{ij}^{\beta} + 2m_{ij}^{\gamma} + m_{ij}^{\delta}}{6}, \qquad i,j = 1, \ldots, n. \tag{4.34}$$

It is worth to note that the obtained matrix M^{*} is in general not multiplicatively reciprocal; for example, for $\tilde{m}_{ij} = (2, 3, 4, 6)$ and $\tilde{m}_{ji} = \left(\frac{1}{6}, \frac{1}{4}, \frac{1}{3}, \frac{1}{2}\right)$, we obtain $m_{ij}^{*} = \frac{22}{6}$ and $m_{ji}^{*} = \frac{11}{36} \neq \frac{22}{6} = \frac{1}{m_{ij}^{*}}$. Thus, verifying acceptable inconsistency of such a matrix is meaningless.

Liu (2009) proposed a method for verifying acceptable level of inconsistency (he actually calls it acceptable consistency) of interval FMPCMs. For an interval FMPCM $\overline{M} = \left\{\overline{m}_{ij}\right\}_{i,j=1}^{n}$, $\overline{m}_{ij} = [m_{ij}^{L}, m_{ij}^{U}]$, Liu (2009) constructs two MPCMs C and D by using (4.7) and verifies their acceptable inconsistency by comparing their Consistency Ratio (2.10) with the boundary value 0.1, i.e. $CR \leq 0.1$. When both MPCMs C and D are acceptably inconsistent, then also the interval FMPCM $\overline{M}$ is said to be acceptably inconsistent. When at least one of the MPCMs C and D is not acceptably inconsistent, then also the interval FMPCM $\overline{M}$ is considered as inconsistent. However, this method, similarly to Definition 46 of multiplicative consistency for interval FMPCMs proposed by Liu (2009) is not invariant under permutation of objects.

Another index of inconsistency for FMPCMs was proposed by Ramík and Korviny (2010). This index is based on the idea of measuring distance of the FMPCM from the closest fuzzy matrix of ratios of fuzzy priorities. The advantage of this approach is that, unlike the two approaches mentioned above, it takes into account the uncertainty present in the FMPCM. Since in the focus of this book is the fuzzy extension of well-known and most often applied methods based on PCMs, only CI and CR and their extension to FMPCMs are dealt with here.

As already reviewed in Sect. 2.2.2, using the maximal eigenvalue of a MPCM, Saaty defined the Consistency Index CI and the Consistency Ratio CR to verify an acceptable level of inconsistency of the matrix. In order to verify an acceptable level of inconsistency of FMPCMs, these two measures should be fuzzified properly.

In order to compute CI, it is necessary to know the maximal eigenvalue of the given MPCM. Having a FMPCM whose entries are fuzzy numbers, it is natural to compute the maximal eigenvalue of this matrix in form of a fuzzy number as well. This fuzzy maximal eigenvalue $\widetilde{\lambda}$ will then substitute the crisp maximal eigenvalue in the formula (2.9), which can be easily fuzzified in order to obtain CI in the form of a fuzzy number. Thus, for a trapezoidal FMPCM $\widetilde{M} = \{\widetilde{m}_{ij}\}_{i,j=1}^{n}$, Fuzzy Consistency Index $\widetilde{CI}$ is given as

$$\widetilde{CI} = \frac{\widetilde{\lambda} - n}{n-1} = \left(\frac{\lambda^{\alpha} - n}{n-1}, \frac{\lambda^{\beta} - n}{n-1}, \frac{\lambda^{\gamma} - n}{n-1}, \frac{\lambda^{\delta} - n}{n-1} \right), \qquad (4.35)$$

where $\widetilde{\lambda} = \left(\lambda^{\alpha}, \lambda^{\beta}, \lambda^{\gamma}, \lambda^{\delta} \right)$ is the fuzzy maximal eigenvalue of $\widetilde{M}$ (whose computation will be discussed later). Fuzzy Consistency Ratio $\widetilde{CR}$ is then given as

$$\widetilde{CR} = \frac{\widetilde{CI}}{RI} = \left(\frac{\lambda^{\alpha} - n}{RI(n-1)}, \frac{\lambda^{\beta} - n}{RI(n-1)}, \frac{\lambda^{\gamma} - n}{RI(n-1)}, \frac{\lambda^{\delta} - n}{RI(n-1)} \right). \qquad (4.36)$$

Notice that Random Index RI used in the formula (4.36) is the same as RI used for crisp MPCMs; the values of RI are given in Table 2.2. Analogously as in the case of crisp MPCMs, we need to compare $\widetilde{CR}$ with the boundary value 0.1 in order to decide whether the FMPCM is acceptably inconsistent or not. This might be done easily by defuzzifying the trapezoidal fuzzy number $\widetilde{CR}$ by the center-of-area defuzzification method using formula (3.10) first, and then comparing the center of area $COA_{\widetilde{CR}}$ of $\widetilde{CR}$ with the boundary value 0.1. Thus, the FMPCM $\widetilde{M}$ is said to be acceptably inconsistent if

$$COA_{\widetilde{CR}} \leq 0.1. \qquad (4.37)$$

The remaining question is how to obtain the fuzzy maximal eigenvalue of a FMPCM. This task is not as simple as it might seem at first sight. Several methods for deriving the fuzzy maximal eigenvalue from a FMPCM have been proposed in the literature, but, as it will be shown in the following section, these methods suffer from severe drawbacks. An appropriate method for deriving the fuzzy maximal eigenvalue is indispensable not only for verifying acceptable inconsistency of a

FMPCM by using the Fuzzy Consistency Index and the Fuzzy Consistency Ratio, but it is also necessary for the fuzzy extension of the EVM for deriving fuzzy priorities of objects from a FMPCM. Thus, particular attention will be paid to the problem of deriving properly the fuzzy maximal eigenvalue from a FMPCM in the following section.

4.2.2.4 Fuzzy Maximal Eigenvalue of a FMPCM

As reviewed in Sect. 2.2, the maximal eigenvalue of a MPCM is used in the consistency index formula (2.9) to verify acceptable multiplicative inconsistency of a MPCM and in the EVM method (2.21) to obtain normalized priorities of objects from a MPCM. In this section, extension of the maximal eigenvalue to the fuzzy maximal eigenvalue of a FMPCM is studied. The formulas for obtaining the fuzzy maximal eigenvalue of a FMPCM proposed in the literature are reviewed, deficiencies of these formulas are pointed out, and then new formulas are proposed. Subsequently, properties of the new fuzzy maximal eigenvalue are discussed. The section extends the results presented by Krejčí (2017c).

Csutora and Buckley (2001) proposed formulas for obtaining $\alpha-$cuts, $\alpha \in [0, 1]$, of the fuzzy maximal eigenvalue of a FMPCM. For the sake of simplicity, this book is limited to trapezoidal fuzzy numbers. Therefore, the formulas of Csutora and Buckley (2001) will be shown for trapezoidal fuzzy numbers here.

By applying the trapezoidal approximation, the representing values of the fuzzy maximal eigenvalue $\widetilde{\lambda}_S = (\lambda_S^\alpha, \lambda_S^\beta, \lambda_S^\gamma, \lambda_S^\delta)$ (the lower index S stands for standard fuzzy arithmetic that is applied to obtain the fuzzy maximal eigenvalue) of a trapezoidal FMPCM $\widetilde{M} = \{\widetilde{m}_{ij}\}_{i,j=1}^n$, $\widetilde{m}_{ij} = (m_{ij}^\alpha, m_{ij}^\beta, m_{ij}^\gamma, m_{ij}^\delta)$, is obtained as

$$\lambda_S^\alpha = EVM_\lambda(M^\alpha), \quad \text{where } M^\alpha = \left\{m_{ij}^\alpha\right\}_{i,j=1}^n, \tag{4.38}$$

$$\lambda_S^\beta = EVM_\lambda(M^\beta), \quad \text{where } M^\beta = \left\{m_{ij}^\beta\right\}_{i,j=1}^n, \tag{4.39}$$

$$\lambda_S^\gamma = EVM_\lambda(M^\gamma), \quad \text{where } M^\gamma = \left\{m_{ij}^\gamma\right\}_{i,j=1}^n, \tag{4.40}$$

$$\lambda_S^\delta = EVM_\lambda(M^\delta), \quad \text{where } M^\delta = \left\{m_{ij}^\delta\right\}_{i,j=1}^n. \tag{4.41}$$

This means that the lower boundary value λ_S^α of the fuzzy maximal eigenvalue $\widetilde{\lambda}_S$ of a trapezoidal FMPCM $\widetilde{M}$ is computed as the maximal eigenvalue of the matrix $M^\alpha = \left\{m_{ij}^\alpha\right\}_{i,j=1}^n$, whose elements are the lower boundary values of the trapezoidal fuzzy numbers from the trapezoidal FMPCM $\widetilde{M}$. Analogously, the upper boundary value λ_S^δ is computed as the maximal eigenvalue of the matrix of the upper boundary values of the trapezoidal fuzzy numbers from the trapezoidal FMPCM $\widetilde{M}$, similarly for the representing values λ_S^β and λ_S^γ of the resulting fuzzy maximal eigenvalue $\widetilde{\lambda}_S$.

Clearly, none of the matrices $M^\alpha, M^\beta, M^\gamma$, and M^δ is multiplicatively recip-rocal. Csutora and Buckley (2001) observed this fact, but they did not consider it to be a flaw. Also Wang and Chin (2006), who adopted formulas (4.38)–(4.41) in their method for obtaining the fuzzy priorities of objects from FMPCMs (reviewed later in Sect. 4.2.3.1), did not realize the flaw. However, as already Krejčí (2017c) emphasized, multiplicative reciprocity is a key property of MPCMs, and thus it has to be preserved also under the fuzzy extension. Unless $\widetilde{M} = \left\{ \widetilde{m}_{ij} \right\}_{i,j=1}^n$, $\widetilde{m}_{ij} = \left(m_{ij}^\alpha, m_{ij}^\beta, m_{ij}^\gamma, m_{ij}^\delta \right)$, is a crisp MPCM, it is meaningless to consider the matrix $M^\alpha = \left\{ m_{ij}^\alpha \right\}_{i,j=1}^n$ of the lower boundary values of $\widetilde{M}$ for the computation of the fuzzy maximal eigenvalue $\widetilde{\lambda}_S$ (in particular its lower boundary value λ_S^α). Since $M^\alpha = \left\{ m_{ij}^\alpha \right\}_{i,j=1}^n$ does not satisfy the multiplicative-reciprocity property (2.2), it is not even a MPCM, and thus it does not reflect the preference information provided by the DM in the original trapezoidal FMPCM $\widetilde{M}$. The same holds also for the matrices $M^\beta = \left\{ m_{ij}^\beta \right\}_{i,j=1}^n$, $M^\gamma = \left\{ m_{ij}^\gamma \right\}_{i,j=1}^n$, and $M^\delta = \left\{ m_{ij}^\delta \right\}_{i,j=1}^n$.

Despite the violation of the multiplicative reciprocity of the related PCs, we have to acknowledge that the method proposed by Csutora and Buckley (2001) is at least invariant under permutation of objects in the FMPCM $\widetilde{M}$. By permut-ing $\widetilde{M} = \left\{ \widetilde{m}_{ij} \right\}_{i,j=1}^n$ to $\widetilde{M}^\pi = P \widetilde{M} P^T$, using any permutation matrix P, the matrices $M^\alpha, M^\beta, M^\gamma$, and M^δ used in the formulas (4.38)–(4.41) are permuted in the same way, which does not have any impact on the resulting maximal eigenval-ues $\lambda_S^\alpha, \lambda_S^\beta, \lambda_S^\gamma$, and λ_S^δ. Thus, the resulting fuzzy maximal eigenvalue $\widetilde{\lambda}_S$ remains unchanged under any permutation of objects compared in the FMPCM $\widetilde{M}$.

Example 40 Let us apply the approach for obtaining the fuzzy maximal eigenvalue proposed by Csutora and Buckley (2001) to the FMPCM

$$\widetilde{M} = \begin{pmatrix} 1 & (2,2,3,4) & (2,4,5,8) \\ \left(\frac{1}{4}, \frac{1}{3}, \frac{1}{2}, \frac{1}{2} \right) & 1 & (4,5,6,7) \\ \left(\frac{1}{8}, \frac{1}{5}, \frac{1}{4}, \frac{1}{2} \right) & \left(\frac{1}{7}, \frac{1}{6}, \frac{1}{5}, \frac{1}{4} \right) & 1 \end{pmatrix}. \tag{4.42}$$

By applying the formulas (4.38)–(4.41), we actually compute the maximal eigenval-ues $\lambda^\alpha, \lambda^\beta, \lambda^\gamma$, and λ^δ of the matrices

$$M^\alpha = \begin{pmatrix} 1 & 2 & 2 \\ \frac{1}{4} & 1 & 4 \\ \frac{1}{8} & \frac{1}{7} & 1 \end{pmatrix}, \quad M^\beta = \begin{pmatrix} 1 & 2 & 4 \\ \frac{1}{3} & 1 & 5 \\ \frac{1}{5} & \frac{1}{6} & 1 \end{pmatrix}, \quad M^\gamma = \begin{pmatrix} 1 & 3 & 5 \\ \frac{1}{2} & 1 & 6 \\ \frac{1}{4} & \frac{1}{5} & 1 \end{pmatrix}, \quad M^\delta = \begin{pmatrix} 1 & 4 & 8 \\ \frac{1}{2} & 1 & 7 \\ \frac{1}{2} & \frac{1}{4} & 1 \end{pmatrix},$$
$$\tag{4.43}$$

respectively. The fuzzy maximal eigenvalue obtained by the formulas (4.38)–(4.41) is in the form $\widetilde{\lambda}_S = (2.4376, 2.8680, 3.4480, 4.4739)$.

However, as we can see from (4.43), none of the matrices $M^\alpha, M^\beta, M^\gamma$, and M^δ is multiplicatively reciprocal. This means that they are not MPCMs. Therefore, computing their maximal eigenvalues in order to verify the acceptable consistency or to derive priorities is nonsensical as these matrices do not represent the preference information contained in the FMPCM (4.42).   △

Also Ishizaka (2014) used formulas (4.38)–(4.41) to obtain the fuzzy maximal eigenvalue of a trapezoidal FMPCM. However, in his method, a particular approach for the construction of the trapezoidal FMPCM was employed. In order to distinguish the method proposed by Ishizaka (2014) from the method proposed by Csutora and Buckley (2001), the FMPCM and the fuzzy maximal eigenvalue obtained by formulas (4.38)–(4.41) in the approach of Ishizaka (2014) will be denoted $\widetilde{M}_I = \left\{\widetilde{m}_{Iij}\right\}_{i,j=1}^n$, and $\widetilde{\lambda}_I = \left(\lambda_I^\alpha, \lambda_I^\beta, \lambda_I^\gamma, \lambda_I^\delta\right)$, respectively.

Ishizaka (2014) constructed the multiplicative reciprocals of trapezoidal fuzzy numbers $\widetilde{m}_{Iij} = (m_{Iij}^\alpha, m_{Iij}^\beta, m_{Iij}^\gamma, m_{Iij}^\delta)$, $i, j = 1, \ldots, n$, $i < j$, in the trapezoidal FMPCM $\widetilde{M}_I = \left\{\widetilde{m}_{Iij}\right\}_{i,j=1}^n$ as $\widetilde{m}_{Iji} = \left(\frac{1}{m_{Iij}^\alpha}, \frac{1}{m_{Iij}^\beta}, \frac{1}{m_{Iij}^\gamma}, \frac{1}{m_{Iij}^\delta}\right)$. Thus, $\widetilde{m}_{Iji}$ does not represent a (trapezoidal) fuzzy number anymore since $\frac{1}{m_{Iij}^\alpha} \geq \frac{1}{m_{Iij}^\beta} \geq \frac{1}{m_{Iij}^\gamma} \geq \frac{1}{m_{Iij}^\delta}$; it is just a quadruple of real numbers (Krejčí 2017c). This means that Ishizaka's approach violates even the widely accepted approach to the construction of FMPCMs.

Furthermore, the resulting fuzzy maximal eigenvalue $\widetilde{\lambda}_I$ is not invariant under permutation of objects in the FMPCM $\widetilde{M}$, which will be demonstrated on an illustrative example. Moreover, because of the inappropriate construction of the reciprocals of the fuzzy numbers in the FMPCM $\widetilde{M}_I$, it is not even guaranteed that the fuzzy maximal eigenvalue $\widetilde{\lambda}_I$ obtained by formulas (4.38)–(4.41) is a fuzzy number; in general, it is just a quadruple of real numbers, similarly as for the reciprocals of $\widetilde{m}_{Iij}$, $i, j = 1, \ldots, n$, $i < j$, in $\widetilde{M}_I = \left\{\widetilde{m}_{Iij}\right\}_{i,j=1}^n$.

Example 41 Let us apply the method for obtaining the fuzzy maximal eigenvalue proposed by Ishizaka (2014) to the FMPCM $\widetilde{M}$ in the form (4.42) examined in Example 40.

The corresponding matrix $\widetilde{M}_I$ utilized in the approach of Ishizaka (2014) is given as

$$
\widetilde{M}_I = \begin{pmatrix}
1 & (2, 2, 3, 4) & (2, 4, 5, 8) \\
\left(\frac{1}{2}, \frac{1}{2}, \frac{1}{3}, \frac{1}{4}\right) & 1 & (4, 5, 6, 7) \\
\left(\frac{1}{2}, \frac{1}{4}, \frac{1}{5}, \frac{1}{8}\right) & \left(\frac{1}{4}, \frac{1}{5}, \frac{1}{6}, \frac{1}{7}\right) & 1
\end{pmatrix}.
\tag{4.44}
$$

Obviously, the elements below the main diagonal of the matrix $\widetilde{M}_I$ are not trapezoidal fuzzy numbers as they do not satisfy Definition 22. In fact, $m_{Iji}^\alpha \geq m_{Iji}^\beta \geq$

$m_{Iji}^{\gamma} \geq m_{Iji}^{\delta}$, $i, j = 1, \ldots, n$, $i < j$. The fuzzy maximal eigenvalue obtained by formulas (4.38)–(4.41) from the matrix $\widetilde{M}_I$ is in the form $\widetilde{\lambda}_I = (3.2174, 3.0940, 3.1851, 3.1769)$, which again does not satisfy Definition 22 of a trapezoidal fuzzy number.

Let us now permute the FMPCM (4.42) to $\widetilde{M}^{\pi}$ by applying the permutation matrix P in the form (4.31). The corresponding permuted matrix $\widetilde{M}_I^{\pi}$ is in the form

$$\widetilde{M}_I^{\pi} = \begin{pmatrix} 1 & (4, 5, 6, 7) & \left(\dfrac{1}{4}, \dfrac{1}{3}, \dfrac{1}{2}, \dfrac{1}{2}\right) \\ \left(\dfrac{1}{4}, \dfrac{1}{5}, \dfrac{1}{6}, \dfrac{1}{7}\right) & 1 & \left(\dfrac{1}{8}, \dfrac{1}{5}, \dfrac{1}{4}, \dfrac{1}{2}\right) \\ (4, 3, 2, 2) & (8, 5, 4, 2) & 1 \end{pmatrix}. \tag{4.45}$$

The fuzzy maximal eigenvalue $\widetilde{\lambda}_I^{\pi}$ obtained from this permuted matrix is in the form $\widetilde{\lambda}_I^{\pi} = (3.0536, 3.1356, 3.1356, 3.4357)$. We see that $\widetilde{\lambda}_I^{\pi} \neq \widetilde{\lambda}_I$, which means that the method for obtaining the fuzzy maximal eigenvalue from a FMPCM proposed by Ishizaka (2014) is not invariant under permutation of objects in FMPCMs.   △

It is obvious that the reciprocals of the fuzzy numbers in the FMPCM have to be constructed properly, as given in Definition 43 of a FMPCM. Furthermore, it is necessary to consider the multiplicative reciprocity of the related PCs in a FMPCM also in the process of deriving the fuzzy maximal eigenvalue since it is an inherent property of FMPCMs (see discussions on p. 86 and p. 98). For this it is necessary to apply constrained fuzzy arithmetic (3.45) instead of standard fuzzy arithmetic (3.36) when extending the formula (2.19) for obtaining the maximal eigenvalue of a MPCM to fuzzy numbers.

By applying simplified constrained fuzzy arithmetic (3.47), the fuzzy maximal eigenvalue $\widetilde{\lambda}_C = (\lambda_C^{\alpha}, \lambda_C^{\beta}, \lambda_C^{\gamma}, \lambda_C^{\delta})$ (the lower index C stands for applying constrained fuzzy arithmetic) of a trapezoidal FMPCM $\widetilde{M} = \{\widetilde{m}_{ij}\}_{i,j=1}^{n}$, $\widetilde{m}_{ij} = (m_{ij}^{\alpha}, m_{ij}^{\beta}, m_{ij}^{\gamma}, m_{ij}^{\delta})$, is obtained as:

$$\lambda_C^{\alpha} = \min \left\{ EVM_{\lambda}(M); \begin{array}{l} M = \{m_{rs}\}_{r,s=1}^{n}, \, m_{rs} \in \left[m_{rs}^{\alpha}, m_{rs}^{\delta}\right], \\ m_{sr} = \frac{1}{m_{rs}}, \, r, s = 1, \ldots, n \end{array} \right\}, \tag{4.46}$$

$$\lambda_C^{\beta} = \min \left\{ EVM_{\lambda}(M); \begin{array}{l} M = \{m_{rs}\}_{r,s=1}^{n}, \, m_{rs} \in \left[m_{rs}^{\beta}, m_{rs}^{\gamma}\right], \\ m_{sr} = \frac{1}{m_{rs}}, \, r, s = 1, \ldots, n \end{array} \right\}, \tag{4.47}$$

$$\lambda_C^{\gamma} = \max \left\{ EVM_{\lambda}(M); \begin{array}{l} M = \{m_{rs}\}_{r,s=1}^{n}, \, m_{rs} \in \left[m_{rs}^{\beta}, m_{rs}^{\gamma}\right], \\ m_{sr} = \frac{1}{m_{rs}}, \, r, s = 1, \ldots, n \end{array} \right\}, \tag{4.48}$$

$$\lambda_C^{\delta} = \max \left\{ EVM_{\lambda}(M); \begin{array}{l} M = \{m_{rs}\}_{r,s=1}^{n}, \, m_{rs} \in \left[m_{rs}^{\alpha}, m_{rs}^{\delta}\right], \\ m_{sr} = \frac{1}{m_{rs}}, \, r, s = 1, \ldots, n \end{array} \right\}. \tag{4.49}$$

The formulas (4.46)–(4.49) are obtained by generalizing the corresponding formulas for triangular FMPCMs introduced by Krejčí (2017c).

By using constrained fuzzy arithmetic (3.47) in the formulas (4.46)–(4.49) in order to reflect multiplicative reciprocity of the related PCs, all redundant vagueness is eliminated from the fuzzy maximal eigenvalue $\tilde{\lambda}_C$, which is the advantage over the method proposed by Csutora and Buckley (2001). Thus, $\tilde{\lambda}_C$ represents the actual fuzzy maximal eigenvalue (more precisely its best trapezoidal approximation) of a trapezoidal FMPCM. Further, unlike the method proposed by Ishizaka (2014), the new method is invariant under permutation of objects.

Theorem 37 Let $\tilde{M} = \{\tilde{m}_{ij}\}_{i,j=1}^{n}$ be a trapezoidal FMPCM. The fuzzy maximal eigenvalue $\tilde{\lambda}_C$ of $\tilde{M}$ obtained by the formulas (4.46)–(4.49) is invariant under permutation of objects in FMPCMs.

Proof We already know from Sect. 2.2.3.1 that the maximal eigenvalue $\lambda = EVM_\lambda(M)$ is invariant under permutation of objects in a MPCM M, i.e., $EVM_\lambda(M) = EVM_\lambda(PMP^T)$ for every permutation matrix P. Thus also the maximal eigenvalue of any MPCM M constructed from the elements from the closures of the supports of the trapezoidal fuzzy numbers $\tilde{m}_{ij} = (m_{ij}^\alpha, m_{ij}^\beta, m_{ij}^\gamma, m_{ij}^\delta)$ in the FMPCM $\tilde{M} = \{\tilde{m}_{ij}\}_{i,j=1}^{n}$ is invariant under permutation of objects in M. Therefore, also the minimum λ_C^α and the maximum λ_C^δ of these maximal eigenvalues obtained by the formulas (4.46) and (4.49), respectively, are invariant under permutation of objects. Similarly, also λ_C^β and λ_C^γ obtained by the formulas (4.47) and (4.48), respectively, are invariant under permutation of objects. Thus, it results that the fuzzy maximal eigenvalue $\tilde{\lambda}_C = (\lambda_C^\alpha, \lambda_C^\beta, \lambda_C^\gamma, \lambda_C^\delta)$ is invariant under permutation of objects in FMPCMs. $\square$

Using the properties of the maximal eigenvalues reviewed in Sect. 2.2.3.1, it is possible to derive some properties of the fuzzy maximal eigenvalue of a trapezoidal FMPCM obtained by the new formulas (4.46)–(4.49) as well as of the fuzzy maximal eigenvalue obtained by the formulas (4.38)–(4.41) in the approaches proposed by Csutora and Buckley (2001) and by Ishizaka (2014). Analogous properties were already derived by Krejčí (2017c) for the fuzzy maximal eigenvalues of triangular FMPCMs. The case when the trapezoidal FMPCM $\tilde{M} = \{\tilde{m}_{ij}\}_{i,j=1}^{n}$, $\tilde{m}_{ij} = \left(m_{ij}^\alpha, m_{ij}^\beta, m_{ij}^\gamma, m_{ij}^\delta\right)$, is a crisp MPCM, i.e. $m_{ij}^\alpha = m_{ij}^\delta$, $i, j = 1, \ldots, n$, is not interesting regarding the properties of the fuzzy maximal eigenvalue as this is simply a crisp number $\lambda \geq n$. Without loss of generality, let us consider trapezoidal FMPCMs $\tilde{M} = \{\tilde{m}_{ij}\}_{i,j=1}^{n}$, $\tilde{m}_{ij} = \left(m_{ij}^\alpha, m_{ij}^\beta, m_{ij}^\gamma, m_{ij}^\delta\right)$, for which $\exists k, l \in \{1, \ldots, n\}$: $m_{kl}^\alpha < m_{kl}^\beta < m_{kl}^\gamma < m_{kl}^\delta$.

Being $\tilde{M} = \{\tilde{m}_{ij}\}_{i,j=1}^{n}$, $\tilde{m}_{ij} = \left(m_{ij}^\alpha, m_{ij}^\beta, m_{ij}^\gamma, m_{ij}^\delta\right)$, a positive trapezoidal FMPCM, i.e. $m_{ij}^\alpha > 0$, $i, j = 1, \ldots, n$, the maximal eigenvalue of any matrix constructed from the elements from the closures of the supports of its fuzzy elements is positive too. Since the representing values of the fuzzy maximal eigenvalue $\widetilde{\lambda}_C = \left(\lambda_C^\alpha, \lambda_C^\beta, \lambda_C^\gamma, \lambda_C^\delta \right)$ are obtained from MPCMs that are multiplicatively reciprocal (see formulas (4.46)–(4.49)), the inequalities $\lambda_C^\alpha \geq n, \lambda_C^\beta \geq n, \lambda_C^\gamma \geq n, \lambda_C^\delta \geq n$ necessarily hold.

The lower boundary value λ_C^α and the upper boundary value λ_C^δ of the fuzzy maximal eigenvalue $\widetilde{\lambda}_C$ are obtained as the minimum and the maximum, respectively, of function EVM_λ defined on the closures of the supports of the trapezoidal fuzzy numbers in the trapezoidal FMPCM. Analogously, λ_C^β and λ_C^γ are obtained as the minimum and the maximum, respectively, of function EVM_λ defined on the cores of the trapezoidal fuzzy numbers in the trapezoidal FMPCM. Therefore, the inequalities $\lambda_C^\alpha \leq \lambda_C^\beta \leq \lambda_C^\gamma \leq \lambda_C^\delta$ necessarily hold. Thus, overall, for the fuzzy maximal eigenvalue obtained by formulas (4.46)–(4.49), the inequalities $n \leq \lambda_C^\alpha \leq \lambda_C^\beta \leq \lambda_C^\gamma \leq \lambda_C^\delta$ hold.

In the special case where there exists a multiplicatively consistent MPCM obtainable by combining particular elements from the cores of the trapezoidal fuzzy numbers $\widetilde{m}_{ij}, i,j = 1, \ldots, n$, in the trapezoidal FMPCM $\widetilde{M}$, the representing values of the fuzzy maximal eigenvalue $\widetilde{\lambda}_C = \left(\lambda_C^\alpha, \lambda_C^\beta, \lambda_C^\gamma, \lambda_C^\delta \right)$ are in the form $n = \lambda_C^\alpha = \lambda_C^\beta < \lambda_C^\gamma < \lambda_C^\delta$. In the case where there does not exist a multiplicatively consistent MPCM obtainable by combining elements from the cores of the trapezoidal fuzzy numbers $\widetilde{m}_{ij}, i,j = 1, \ldots, n$, but where there exist elements in the closures of the supports of the trapezoidal fuzzy numbers $\widetilde{m}_{ij}, i,j = 1, \ldots, n$, in $\widetilde{M} = \{\widetilde{m}_{ij}\}_{i,j=1}^n$ such that they form a multiplicatively consistent MPCM, the representing values of the fuzzy maximal eigenvalue of the FMPCM are in the form $n = \lambda_C^\alpha < \lambda_C^\beta < \lambda_C^\gamma < \lambda_C^\delta$.

Because the inequalities $m_{ij}^\alpha \leq m_{ij}^\beta \leq m_{ij}^\gamma \leq m_{ij}^\delta$ hold for $i,j = 1, \ldots, n$, and because $\exists k, l \in \{1, \ldots, n\} : m_{kl}^\alpha < m_{kl}^\beta < m_{kl}^\gamma < m_{kl}^\delta$, then, clearly, the inequalities $\lambda_S^\alpha < \lambda_S^\beta < \lambda_S^\gamma < \lambda_S^\delta$ hold for the fuzzy maximal eigenvalue $\widetilde{\lambda}_S = (\lambda_S^\alpha, \lambda_S^\beta, \lambda_S^\gamma, \lambda_S^\delta)$ obtained by the formulas (4.38)–(4.41) in Csutora and Buckley's method.

The properties mentioned above do not hold for the fuzzy maximal eigenvalue $\widetilde{\lambda}_I = \left(\lambda_I^\alpha, \lambda_I^\beta, \lambda_I^\gamma, \lambda_I^\delta \right)$ of the particular FMPCM $\widetilde{M}_I = \{\widetilde{m}_{Iij}\}_{i,j=1}^n$ obtained by the formulas (4.38)–(4.41) in Ishizaka's approach. Since there are $\frac{n^2-n}{2}$ elements $\widetilde{m}_{Iij}$ in the FMPCM $\widetilde{M}_I = \{\widetilde{m}_{Iij}\}_{i,j=1}^n$, $\widetilde{m}_{Iij} = \left(m_{Iij}^\alpha, m_{Iij}^\beta, m_{Iij}^\gamma, m_{Iij}^\delta \right)$ such that the inequalities $m_{Iij}^\alpha \leq m_{Iij}^\beta \leq m_{Iij}^\gamma \leq m_{Iij}^\delta$ do not hold for them, then also the inequalities $\lambda_I^\alpha < \lambda_I^\beta < \lambda_I^\gamma < \lambda_I^\delta$ cannot be guaranteed for the resulting fuzzy maximal eigenvalue $\widetilde{\lambda}_I$. Thus, in general, the resulting quadruple $\widetilde{\lambda}_I = \left(\lambda_I^\alpha, \lambda_I^\beta, \lambda_I^\gamma, \lambda_I^\delta \right)$ does not represent a fuzzy number, which is a very serious flaw of the method.

There exists a very interesting relation between the fuzzy maximal eigenvalues $\widetilde{\lambda}_C = (\lambda_C^\alpha, \lambda_C^\beta, \lambda_C^\gamma, \lambda_C^\delta)$ and $\widetilde{\lambda}_S = (\lambda_S^\alpha, \lambda_S^\beta, \lambda_S^\gamma, \lambda_S^\delta)$. Since λ_S^α is the maximal eigenvalue of $M^\alpha = \{m_{ij}^\alpha\}_{i,j=1}^n$ and λ_C^α is the maximal eigenvalue of a multiplicatively reciprocal matrix $M^* = \left\{ m_{ij}^* \right\}_{i,j=1}^n$, $m_{ij}^* \in \left[m_{ij}^\alpha, m_{ij}^\delta \right]$, $m_{ij}^\alpha \leq m_{ij}^*$, $i,j = 1,\ldots,n$, with at least one strict inequality, then the inequality $\lambda_S^\alpha < \lambda_C^\alpha$ follows from the Perron-Frobenius Theorem. Analogously, since λ_S^δ is the maximal eigenvalue of $M^\delta = \left\{ m_{ij}^\delta \right\}_{i,j=1}^n$ and λ_C^δ is the maximal eigenvalue of a multiplicatively reciprocal matrix $M^{**} = \left\{ m_{ij}^{**} \right\}_{i,j=1}^n$, $m_{ij}^{**} \in \left[m_{ij}^\alpha, m_{ij}^\delta \right]$, $m_{ij}^\delta \geq m_{ij}^{**}$, $i,j = 1,\ldots,n$, with at least one strict inequality, then also the inequality $\lambda_C^\delta < \lambda_S^\delta$ holds. Therefore, the support of the fuzzy maximal eigenvalue $\widetilde{\lambda}_C$ of a given FMPCM $\widetilde{M}$ is a proper subset of the support of the fuzzy maximal eigenvalue $\widetilde{\lambda}_S$ of $\widetilde{M}$, i.e. $]\lambda_C^\alpha, \lambda_C^\delta[\subset]\lambda_S^\alpha, \lambda_S^\delta[$. The same holds also for the cores of the fuzzy maximal eigenvalues $\widetilde{\lambda}_C$ and $\widetilde{\lambda}_S$, i.e. $[\lambda_C^\beta, \lambda_C^\gamma] \subset [\lambda_S^\beta, \lambda_S^\gamma]$. Thus, by employing the multiplicative-reciprocity condition in formulas (4.46)–(4.49), all unfeasible combinations of elements from the supports of the fuzzy numbers in the FMPCM are eliminated. As a consequence, the resulting fuzzy maximal eigenvalue $\widetilde{\lambda}_C$ is less vague than the original fuzzy maximal eigenvalue $\widetilde{\lambda}_S$ obtained by the formulas (4.38)–(4.41) proposed by Csutora and Buckley (2001) where the multiplicative reciprocity of the related PCs is violated.

The properties of the fuzzy maximal eigenvalues derived above are valid for a trapezoidal FMPCM $\widetilde{M} = \left\{ \widetilde{m}_{ij} \right\}_{i,j=1}^n$, $\widetilde{m}_{ij} = (m_{ij}^\alpha, m_{ij}^\beta, m_{ij}^\gamma, m_{ij}^\delta)$, such that $m_{ij}^\alpha \leq m_{ij}^\beta \leq m_{ij}^\gamma \leq m_{ij}^\delta$, $i,j = 1,\ldots,n$, and $\exists k,l \in \{1,\ldots,n\}: m_{kl}^\alpha < m_{kl}^\beta < m_{kl}^\gamma < m_{kl}^\delta$. This means that triangular FMPCMs were excluded from the analysis above. Therefore, some interesting properties that appear only for the fuzzy maximal eigenvalues of triangular FMPCMs will be shown here. Without loss of generality, let as assume a triangular FMPCM $\widetilde{M} = \left\{ \widetilde{m}_{ij} \right\}_{i,j=1}^n$, $\widetilde{m}_{ij} = (m_{ij}^L, m_{ij}^M, m_{ij}^U)$, such that $m_{ij}^L \leq m_{ij}^M \leq m_{ij}^U$, $i,j = 1,\ldots,n$, and $\exists k,l \in \{1,\ldots,n\}: m_{kl}^L < m_{kl}^M < m_{kl}^U$. Since $M^M = \left\{ m_{ij}^M \right\}_{i,j=1}^n$ is multiplicatively reciprocal, the inequality $\lambda_S^M \geq n$ holds for its maximal eigenvalue $\widetilde{\lambda}_S = (\lambda_S^L, \lambda_S^M, \lambda_S^U)$. In the special case where $M^M = \left\{ m_{ij}^M \right\}_{i,j=1}^n$ is multiplicatively consistent according to (2.4), the equality $\lambda_S^M = n$ occurs. Furthermore, in such case, the representing values of the fuzzy maximal eigenvalue $\widetilde{\lambda}_C = (\lambda_C^L, \lambda_C^M, \lambda_C^U)$ are in the form $n = \lambda_C^L = \lambda_C^M < \lambda_C^U$.

In the following three examples, the fuzzy maximal eigenvalues $\widetilde{\lambda}_C$ of three different FMPCMs are examined. In particular, based on the analysis above, three main types of the fuzzy maximal eigenvalues $\widetilde{\lambda}_C$ are identified and studied. For the simplicity of presentation, triangular FMPCMs are considered. In addition, in each example, the fuzzy maximal eigenvalues $\widetilde{\lambda}_S$ and $\widetilde{\lambda}_I$ obtained by formulas (4.38)–(4.41) in the approaches proposed by Csutora and Buckley (2001) and by Ishizaka (2014) are computed and confronted with the fuzzy maximal eigenvalue λ_C obtained by the formulas (4.46)–(4.49).

Example 42 Let us consider the triangular FMPCM of four objects o_1, o_2, o_3, and o_4 given as

$$\widetilde{M} = \begin{pmatrix} 1 & (2,3,4) & (4,5,6) & (8,9,9) \\ \left(\dfrac{1}{4}, \dfrac{1}{3}, \dfrac{1}{2}\right) & 1 & (2,3,4) & (6,7,8) \\ \left(\dfrac{1}{6}, \dfrac{1}{5}, \dfrac{1}{4}\right) & \left(\dfrac{1}{4}, \dfrac{1}{3}, \dfrac{1}{2}\right) & 1 & (4,5,6) \\ \left(\dfrac{1}{9}, \dfrac{1}{9}, \dfrac{1}{8}\right) & \left(\dfrac{1}{8}, \dfrac{1}{7}, \dfrac{1}{6}\right) & \left(\dfrac{1}{6}, \dfrac{1}{5}, \dfrac{1}{4}\right) & 1 \end{pmatrix}. \tag{4.50}$$

The fuzzy maximal eigenvalue obtained by the formulas (4.46)–(4.49) is $\widetilde{\lambda}_C = (4.0312, 4.1707, 4.4115)$. Since $\lambda_C^L = 4.0312 > n = 4$, it is clear that there does not exist a single MPCM $M^* = \left\{ m_{ij}^* \right\}_{i,j=1}^4$, $m_{ij}^* \in \left[m_{ij}^L, m_{ij}^U \right]$, $i, j = 1, \ldots, 4$, that would be multiplicatively consistent according to (2.4). According to Theorem 37, by permuting the FMPCM $\widetilde{M}$, the corresponding fuzzy maximal eigenvalue $\widetilde{\lambda}_C$ obtained by the formulas (4.46)–(4.49) remains unchanged.

Note that the triangular fuzzy number $\widetilde{\lambda}_C = (4.0312, 4.1707, 4.4115)$ is only a triangular approximation of the actual fuzzy maximal eigenvalue of the FMPCM $\widetilde{M}$. The actual fuzzy maximal eigenvalue $\widetilde{\lambda}$ obtained by applying properly constrained fuzzy arithmetic (3.45), i.e., $\widetilde{\lambda} = \bigcup_{\alpha=0}^1 \alpha[\lambda_{(\alpha)}^L, \lambda_{(\alpha)}^U]$ such that

$$\begin{aligned} \lambda_{(\alpha)}^L &= \min \left\{ EVM_{MAX}(M); \begin{array}{l} M = \{m_{rs}\}_{r,s=1}^n, \ m_{rs} \in [m_{rs(\alpha)}^L, m_{rs(\alpha)}^U], \\ m_{sr} = \frac{1}{m_{rs}}, \ r, s = 1, \ldots, n \end{array} \right\}, \\ \lambda_{(\alpha)}^U &= \max \left\{ EVM_{MAX}(M); \begin{array}{l} M = \{m_{rs}\}_{r,s=1}^n, \ m_{rs} \in [m_{rs(\alpha)}^L, m_{rs(\alpha)}^U], \\ m_{sr} = \frac{1}{m_{rs}}, \ r, s = 1, \ldots, n \end{array} \right\}, \end{aligned} \tag{4.51}$$

and its triangular approximation $\widetilde{\lambda}_C = (4.0312, 4.1707, 4.4115)$ are displayed in Fig. 4.1. It is clear from Fig. 4.1 that the actual fuzzy maximal eigenvalue $\widetilde{\lambda}$ is not triangular (but such result was expected since even a product of two triangular fuzzy numbers is not a triangular fuzzy number anymore). However, the triangular fuzzy maximal eigenvalue $\widetilde{\lambda}_C = (4.0312, 4.1707, 4.4115)$ is a sufficient approximation since the lower and upper boundary values and the middle value of the fuzzy maximal eigenvalue $\widetilde{\lambda}$ are computed correctly.

The fuzzy maximal eigenvalue $\widetilde{\lambda}_S$ obtained from the FMPCM (4.50) by the formulas (4.38)–(4.41) is in the form $\widetilde{\lambda}_S = (3.5653, 4.1707, 4.9446)$, and thus it is obviously much vaguer than $\widetilde{\lambda}_C = (4.0312, 4.1707, 4.4115)$. The huge difference in vagueness of both fuzzy maximal eigenvalues is even more noticeable from graphical representation, see Fig. 4.2. By applying the approach of Ishizaka (2014), the fuzzy maximal eigenvalue $\widetilde{\lambda}_I$ obtained by the formulas (4.38)–(4.41) from the matrix $\widetilde{M}_I$ corresponding to the FMPCM (4.50) is in the form $\widetilde{\lambda}_I = (4.0458, 4.1707, 4.3675)$. It is only a coincidence that $\widetilde{\lambda}_I$ is such that $\lambda_I^L \leq \lambda_I^M \leq \lambda_I^U$; in general, this property is not satisfied.   △

Fig. 4.1 Fuzzy maximal eigenvalues $\widetilde{\lambda}$ and $\widetilde{\lambda}_C$ of the FMPCM (4.50)

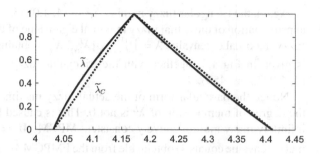

Fig. 4.2 Fuzzy maximal eigenvalues $\widetilde{\lambda}_C$ and $\widetilde{\lambda}_S$ of the FMPCM (4.50)

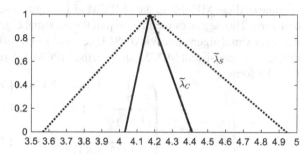

Example 43 Let us consider the FMPCM $\widetilde{M}$

$$\widetilde{M} = \begin{pmatrix} 1 & (1,2,3) & (2,3,4) & (3,4,5) \\ \left(\frac{1}{3},\frac{1}{2},1\right) & 1 & (1,2,3) & (3,4,5) \\ \left(\frac{1}{4},\frac{1}{3},\frac{1}{2}\right) & \left(\frac{1}{3},\frac{1}{2},1\right) & 1 & (2,3,4) \\ \left(\frac{1}{5},\frac{1}{4},\frac{1}{3}\right) & \left(\frac{1}{5},\frac{1}{4},\frac{1}{3}\right) & \left(\frac{1}{4},\frac{1}{3},\frac{1}{2}\right) & 1 \end{pmatrix}. \tag{4.52}$$

The fuzzy maximal eigenvalue obtained by the formulas (4.46)–(4.49) is $\widetilde{\lambda}_C = (4, 4.0875, 4.4453)$. Since $\lambda_C^L = 4 = n$, it is clear that there exists a MPCM $M^* = \left\{m_{ij}^*\right\}_{i,j=1}^4$, $m_{ij}^* \in \left[m_{ij}^L, m_{ij}^U\right]$, $i,j = 1, \ldots, n$, that is multiplicatively consistent according to (2.4). It is, for example,

$$M^* = \begin{pmatrix} 1 & 1 & 2 & 5 \\ 1 & 1 & 2 & 5 \\ \frac{1}{2} & \frac{1}{2} & 1 & 2.5 \\ \frac{1}{5} & \frac{1}{5} & \frac{1}{2.5} & 1 \end{pmatrix}, \tag{4.53}$$

but there exist infinitely many of them.

Again, the triangular fuzzy number $\widetilde{\lambda}_C = (4, 4.0875, 4.4453)$ is only a triangular approximation of the actual fuzzy maximal eigenvalue of the FMPCM $\widetilde{M}$. The actual fuzzy maximal eigenvalue $\widetilde{\lambda} = \bigcup_{\alpha=0}^{1} \alpha[\lambda_{(\alpha)}^L, \lambda_{(\alpha)}^U]$ obtainable by the formula (4.51) is given in Fig. 4.3 together with the triangular fuzzy maximal eigenvalue $\widetilde{\lambda}_C = (4, 4.0875, 4.4453)$.

Notice the particular form of the actual fuzzy maximal eigenvalue $\widetilde{\lambda}$ in Fig. 4.3; the degree of membership of λ^L is not 0. This is caused by the fact that there exist infinitely many multiplicatively consistent MPCMs $M = \{m_{ij}\}_{i,j=1}^{n}$ (i.e. their maximal eigenvalue equals 4) obtainable from the FMPCM $\widetilde{M}$. The degree of membership of some of these MPCMs to the FMPCM $\widetilde{M}$ (computed according to Definition 25) is non-zero. The degree of membership of the maximal eigenvalue $\lambda^L = 4$ to the actual fuzzy maximal eigenvalue $\widetilde{\lambda}$ is 0.1843, i.e. $\widetilde{\lambda}(\lambda^L) = 0.1843$. The corresponding multiplicatively consistent MPCM M^* (i.e. the MPCM M^* such that $\widetilde{M}(M^*) = 0.1843$) is in the form

$$M^* = \begin{pmatrix} 1 & 1.1887 & 2.1843 & 4.7681 \\ \dfrac{1}{1.1887} & 1 & 1.8372 & 4.0121 \\ \dfrac{1}{2.1843} & \dfrac{1}{1.8372} & 1 & 2.1843 \\ \dfrac{1}{4.7681} & \dfrac{1}{4.0121} & \dfrac{1}{2.1843} & 1 \end{pmatrix}. \tag{4.54}$$

The fuzzy maximal eigenvalue $\widetilde{\lambda}_S$ obtained from the FMPCM (4.52) by the formulas (4.38)–(4.41) is in the form $\widetilde{\lambda}_S = (3.1056, 4.0875, 5.5250)$, and thus it is again significantly vaguer than $\widetilde{\lambda}_C = (4, 4.0875, 4.4453)$. The huge difference in vagueness of both fuzzy maximal eigenvalues is even more noticeable from graphical representation, see Fig. 4.4. By applying the approach of Ishizaka (2014), the fuzzy maximal eigenvalue $\widetilde{\lambda}_I$ obtained by the formulas (4.38)–(4.41) from the matrix $\widetilde{M}_I$ corresponding to the FMPCM (4.52) is in the form $\widetilde{\lambda}_I = (4.1031, 4.0875, 4.1407)$. Thus, $\widetilde{\lambda}_I$ is not a fuzzy number since $\lambda_I^L > \lambda_I^M$.   △

Fig. 4.3 Fuzzy maximal eigenvalues $\widetilde{\lambda}$ and $\widetilde{\lambda}_C$ of the FMPCM (4.52)

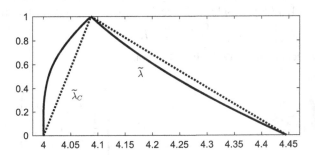

Fig. 4.4 Fuzzy maximal eigenvalues $\widetilde{\lambda}_C$ and $\widetilde{\lambda}_S$ of the FMPCM (4.52)

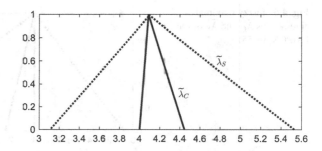

Example 44 Let us consider the FMPCM

$$
\widetilde{M} =
\begin{pmatrix}
1 & \left(\dfrac{1}{2}, 1, 2\right) & (1, 2, 3) & (5, 6, 7) \\
\left(\dfrac{1}{2}, 1, 2\right) & 1 & (1, 2, 3) & (5, 6, 6) \\
\left(\dfrac{1}{3}, \dfrac{1}{2}, 1\right) & \left(\dfrac{1}{3}, \dfrac{1}{2}, 1\right) & 1 & (2, 3, 4) \\
\left(\dfrac{1}{7}, \dfrac{1}{6}, \dfrac{1}{5}\right) & \left(\dfrac{1}{6}, \dfrac{1}{6}, \dfrac{1}{5}\right) & \left(\dfrac{1}{4}, \dfrac{1}{3}, \dfrac{1}{2}\right) & 1
\end{pmatrix}.
\tag{4.55}
$$

The fuzzy maximal eigenvalue obtained by the formulas (4.46)–(4.49) is $\widetilde{\lambda}_C = (4, 4, 4.2961)$. Since $\lambda_C^L = \lambda_C^M = 4 = n$, it is clear that MPCM $M^M = \left\{m_{ij}^M\right\}_{i,j=1}^4$ is multiplicatively consistent according to (2.4). However, there exist infinitely many multiplicatively consistent MPCMs obtainable from the FMPCM (4.55).

Again, the triangular fuzzy number $\widetilde{\lambda}_C = (4, 4, 4.2961)$ is only a sufficient triangular approximation of the actual fuzzy maximal eigenvalue $\widetilde{\lambda}$ of the FMPCM $\widetilde{M}$. The actual fuzzy maximal eigenvalue $\widetilde{\lambda}$ obtainable by the formula (4.51) is given together with its triangular approximation $\widetilde{\lambda}_C$ in Fig. 4.5.

The fuzzy maximal eigenvalue $\widetilde{\lambda}_S$ obtained from the FMPCM (4.52) by the formulas (4.38)–(4.41) is in the form $\widetilde{\lambda}_S = (3.0815, 4, 5.6300)$, and thus it is obviously much vaguer than $\widetilde{\lambda}_C = (4, 4, 4.2961)$. The huge difference in vagueness of both

Fig. 4.5 Fuzzy maximal eigenvalues $\widetilde{\lambda}$ and $\widetilde{\lambda}_C$ of the FMPCM (4.55)

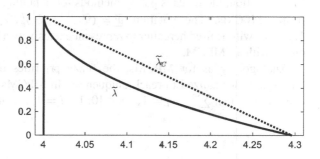

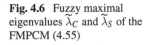

Fig. 4.6 Fuzzy maximal eigenvalues $\widetilde{\lambda}_C$ and $\widetilde{\lambda}_S$ of the FMPCM (4.55)

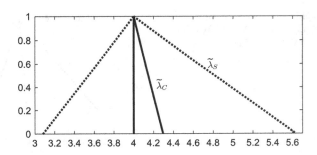

fuzzy maximal eigenvalues is even more noticeable from graphical representation, see Fig. 4.6. By applying the approach of Ishizaka (2014), the fuzzy maximal eigenvalue $\widetilde{\lambda}_I$ obtained by the formulas (4.38)–(4.41) from the matrix $\widetilde{M}_I$ corresponding to the FMPCM (4.52) is in the form $\widetilde{\lambda}_I = (4.1674, 4, 4.0972)$. Thus, as in the previous example, $\widetilde{\lambda}_I$ is again not a fuzzy number.   △

Since the fuzzy maximal eigenvalue $\widetilde{\lambda}_C$ obtained by the formulas (4.46)–(4.49) has all desired properties, it can be used in the formulas (4.35) and (4.36) for computing $\widetilde{CI}$ and $\widetilde{CR}$, respectively, in order to verify acceptable inconsistency of FMPCMs.

Theorem 38 *Fuzzy consistency index $\widetilde{CI}$ given by the formula (4.35) with the fuzzy maximal eigenvalue $\widetilde{\lambda}_C$ given by the formulas (4.46)–(4.49) is invariant under permutation of objects in FMPCMs.*

Proof According to Theorem 37, the fuzzy maximal eigenvalue $\widetilde{\lambda}_C = (\lambda_C^\alpha, \lambda_C^\beta, \lambda_C^\gamma, \lambda_C^\delta)$ obtained by the formulas (4.46)–(4.49) is invariant under permutation of objects in FMPCMs. Thus, also the expressions $\frac{\lambda_C^\alpha - n}{n-1}, \frac{\lambda_C^\beta - n}{n-1}, \frac{\lambda_C^\gamma - n}{n-1}$, and $\frac{\lambda_C^\delta - n}{n-1}$ in the formula (4.35) are invariant under permutation of objects. This means that $\widetilde{CI}$ given by the formula (4.35) is invariant under permutation of objects in FMPCMs.   □

4.2.3 Deriving Priorities from FMPCMs

In this section, the focus is put on methods for obtaining fuzzy priorities of objects from FMPCMs. The notation $\widetilde{\underline{w}} = (\widetilde{w}_1, \ldots, \widetilde{w}_n)^T, \widetilde{w}_i = (w_i^\alpha, w_i^\beta, w_i^\gamma, w_i^\delta), i = 1, \ldots, n$, will be used hereafter to represent exclusively a fuzzy priority vector associated with a FMPCM.

Analogously as for MPCMs, the fuzzy priorities obtained from a FMPCM are usually normalized to reach uniqueness. In MPCMs theory, the normalization condition (2.18), $\sum_{i=1}^{n} w_i = 1, w_i \in [0, 1], i = 1, \ldots, n$, is usually applied to the priorities. This normalization condition is usually extended to the fuzzy priorities $\tilde{w}_i = (w_i^\alpha, w_i^\beta, w_i^\gamma, w_i^\delta)$, $i = 1, \ldots, n$, as

$$
w_i^\alpha + \sum_{\substack{j=1 \\ j \neq i}}^n w_j^\delta \geq 1, \quad w_i^\delta + \sum_{\substack{j=1 \\ j \neq i}}^n w_j^\alpha \leq 1, \quad w_i^\beta + \sum_{\substack{j=1 \\ j \neq i}}^n w_j^\gamma \geq 1, \quad w_i^\gamma + \sum_{\substack{j=1 \\ j \neq i}}^n w_j^\beta \leq 1,
$$

(4.56)

which is in fact Definition 28 of the normalized fuzzy vector.

There exist several well-known methods for deriving priorities of objects from FMPCMs. Laarhoven and Pedrycz (1983) proposed the fuzzy logarithmic least squares method for obtaining fuzzy priorities of objects from FMPCMs. Later, fuzzy extension of the GMM was proposed by Buckley (1985a) in order to obtain fuzzy priorities. Chang (1996) introduced the extent analysis method to obtain crisp priorities from FMPCMs. However, this method has been severely criticized, especially because the priorities determined by this method do not represent the relative importance of objects; see, e.g., Wang et al. (2008). Despite this criticism, the extent analysis method seems to be the most popular in practice, mainly because of its computational simplicity; see Kubler et al. (2016). Csutora and Buckley (2001) proposed a fuzzy extension of the EVM in order to obtain fuzzy priorities of objects. Mikhailov (2003) proposed a fuzzy preference programming method for obtaining crisp priorities from FMPCMs.

In this book only the methods based on the fuzzy extension of well-known methods originally developed for PCMs are of interest. In Sect. 2.2.3, two famous methods for obtaining priorities from MPCMs were reviewed—the EVM and the GMM. The fuzzy extension of these two methods to FMPCMs is studied in detail in the following two subsections.

4.2.3.1 Fuzzy Extension of the Eigenvector Method

This section focuses on the methods for obtaining the fuzzy maximal eigenvector corresponding to the fuzzy maximal eigenvalue of a FMPCM. The methods proposed by Csutora and Buckley (2001), Wang and Chin (2006), and Ishizaka (2014) are reviewed, and their drawbacks regarding the violation of the multiplicative reciprocity of the related PCs and the invariance under permutation of objects are pointed out. Afterwards, a new method for deriving the fuzzy maximal eigenvector corresponding to the fuzzy maximal eigenvalue obtained by the formulas proposed in the previous section is introduced. The section mainly reviews the results of Krejčí (2018).

Csutora and Buckley (2001) proposed a procedure for obtaining the lower and upper boundary values of α-cuts of the fuzzy maximal eigenvector corresponding to the fuzzy maximal eigenvalue $\tilde{\lambda}_S$ of a given FMPCM $\tilde{M} = \{\tilde{m}_{ij}\}_{i,j=1}^n$, $\tilde{m}_{ij} = (m_{ij}^\alpha, m_{ij}^\beta, m_{ij}^\gamma, m_{ij}^\delta)$. Simplification to trapezoidal representation is used here to review their method.

Csutora and Buckley (2001) computed the normalized maximal eigenvectors $\underline{w}^{\alpha*}$, $\underline{w}^{\beta*}$, $\underline{w}^{\gamma*}$, $\underline{w}^{\delta*}$ corresponding to the representing values of the fuzzy maximal eigenvalue $\widetilde{\lambda}_S = (\lambda_S^{\alpha}, \lambda_S^{\beta}, \lambda_S^{\gamma}, \lambda_S^{\delta})$ obtained by the formulas (4.38)–(4.41) as follows:

$$\underline{w}^{\alpha*} = \left(w_1^{\alpha*}, \ldots, w_n^{\alpha*}\right)^T: \quad \underline{w}^{\alpha*} = EVM_{\underline{w}}(M^{\alpha}),\ M^{\alpha} = \left\{m_{ij}^{\alpha}\right\}_{i,j=1}^n, \tag{4.57}$$

$$\underline{w}^{\beta*} = (w_1^{\beta*}, \ldots, w_n^{\beta*})^T: \quad \underline{w}^{\beta*} = EVM_{\underline{w}}(M^{\beta}),\ M^{\beta} = \left\{m_{ij}^{\beta}\right\}_{i,j=1}^n, \tag{4.58}$$

$$\underline{w}^{\gamma*} = \left(w_1^{\gamma*}, \ldots, w_n^{\gamma*}\right)^T: \quad \underline{w}^{\gamma*} = EVM_{\underline{w}}(M^{\gamma}),\ M^{\gamma} = \left\{m_{ij}^{\gamma}\right\}_{i,j=1}^n, \tag{4.59}$$

$$\underline{w}^{\delta*} = \left(w_1^{\delta*}, \ldots, w_n^{\delta*}\right)^T: \quad \underline{w}^{\delta*} = EVM_{\underline{w}}(M^{\delta}),\ M^{\delta} = \left\{m_{ij}^{\delta}\right\}_{i,j=1}^n. \tag{4.60}$$

Since the maximal eigenvectors $\underline{w}^{\alpha*}$, $\underline{w}^{\beta*}$, $\underline{w}^{\gamma*}$, $\underline{w}^{\delta*}$ are normalized, it is clear that the resulting fuzzy vector $\widetilde{w} = (\underline{w}^{\alpha*}, \underline{w}^{\beta*}, \underline{w}^{\gamma*}, \underline{w}^{\delta*})$ is not a vector of trapezoidal fuzzy numbers; the inequalities $w_i^{\alpha*} \leq w_i^{\beta*} \leq w_i^{\gamma*} \leq w_i^{\delta*}$ are not satisfied for each $i = 1, \ldots, n$ (unless $\widetilde{M}$ is a crisp MPCM). Thus, Csutora and Buckley (2001) proposed to adjust the resulting maximal eigenvectors in the following way. First, the normalized maximal eigenvector $\underline{w}^M = (w_1^M \ldots, w_n^M)^T$, $\sum_{i=1}^n w_i^M = 1$, of the MPCM $M^M = \left\{m_{ij}^M\right\}_{i,j=1}^n$, $m_{ij}^M = \sqrt{m_{ij}^{\beta} m_{ij}^{\gamma}}$, is computed. Afterwards, the fuzzy maximal eigenvector $\widetilde{w} = (\underline{w}^{\alpha}, \underline{w}^{\beta}, \underline{w}^{\gamma}, \underline{w}^{\delta})$, $\underline{w}^{\alpha} = \left(w_1^{\alpha}, \ldots, w_n^{\alpha}\right)^T$, $\underline{w}^{\beta} = (w_1^{\beta}, \ldots, w_n^{\beta})^T$, $\underline{w}^{\gamma} = \left(w_1^{\gamma}, \ldots, w_n^{\gamma}\right)^T$, $\underline{w}^{\delta} = \left(w_1^{\delta}, \ldots, w_n^{\delta}\right)^T$ corresponding to the fuzzy maximal eigenvalue $\widetilde{\lambda}_S$ is obtained as

$$k^{\beta} = \min\left\{\frac{w_i^M}{w_i^{\beta*}}; i = 1, \ldots, n\right\} \rightarrow \underline{w}^{\beta} = k^{\beta}\underline{w}^{\beta*}, \tag{4.61}$$

$$k^{\gamma} = \max\left\{\frac{w_i^M}{w_i^{\gamma*}}; i = 1, \ldots, n\right\} \rightarrow \underline{w}^{\gamma} = k^{\gamma}\underline{w}^{\gamma*}, \tag{4.62}$$

$$k^{\alpha} = \min\left\{\frac{w_i^{\beta}}{w_i^{\alpha*}}; i = 1, \ldots, n\right\} \rightarrow \underline{w}^{\alpha} = k^{\alpha}\underline{w}^{\alpha*}, \tag{4.63}$$

$$k^{\delta} = \max\left\{\frac{w_i^{\gamma}}{w_i^{\delta*}}; i = 1, \ldots, n\right\} \rightarrow \underline{w}^{\delta} = k^{\delta}\underline{w}^{\delta*}. \tag{4.64}$$

Note that the fuzzy maximal eigenvector $\widetilde{w} = \left(\underline{w}^{\alpha}, \underline{w}^{\beta}, \underline{w}^{\gamma}, \underline{w}^{\delta}\right)$ is a vector of trapezoidal fuzzy numbers $\widetilde{w}_i = \left(w_i^{\alpha}, w_i^{\beta}, w_i^{\gamma}, w_i^{\delta}\right)$, $i = 1, \ldots, n$, i.e., it can be written as $\widetilde{w} = (\widetilde{w}_1, \ldots, \widetilde{w}_n)^T$. However, for the convenience of representation, the different notation is used in the method given by the formulas (4.57)–(4.64).

Example 45 Let us examine the method for obtaining the fuzzy maximal eigenvector proposed by Csutora and Buckley (2001) on the trapezoidal FMPCM

$$\tilde{M} = \begin{pmatrix} 1 & (1,1,2,3) & (2,2.5,3,4) & (4,6,7,8) \\ \left(\frac{1}{3},\frac{1}{2},1,1\right) & 1 & (3,4,4,5) & (4,5,6,6) \\ \left(\frac{1}{4},\frac{1}{3},\frac{1}{2.5},\frac{1}{2}\right) & \left(\frac{1}{5},\frac{1}{4},\frac{1}{4},\frac{1}{3}\right) & 1 & (1,2,2,3) \\ \left(\frac{1}{8},\frac{1}{7},\frac{1}{6},\frac{1}{4}\right) & \left(\frac{1}{6},\frac{1}{6},\frac{1}{5},\frac{1}{4}\right) & \left(\frac{1}{3},\frac{1}{2},\frac{1}{2},1\right) & 1 \end{pmatrix} \tag{4.65}$$

of four objects o_1, o_2, o_3, and o_4.

The fuzzy maximal eigenvalue obtained from the FMPCM (4.65) by the formulas (4.38)–(4.41) is $\tilde{\lambda}_S = (3.1250, 3.7752, 4.4053, 5.5024)$. The fuzzy priorities of objects obtained by the formulas (4.57)–(4.64) are given as

$$\begin{aligned} \tilde{w}_1 &= (0.3747, 0.3838, 0.4843, 0.5574), \\ \tilde{w}_2 &= (0.3284, 0.3657, 0.4057, 0.4057), \\ \tilde{w}_3 &= (0.1053, 0.1254, 0.1254, 0.1497), \\ \tilde{w}_4 &= (0.0643, 0.0643, 0.0659, 0.0867). \end{aligned} \tag{4.66}$$

$\triangle$

Ishizaka (2014) proposed another approach for obtaining the fuzzy maximal eigenvector of a trapezoidal FMPCM. In this approach, the fuzzy maximal eigenvector $\underline{\tilde{w}}_I = \left(\underline{w}_I^\alpha, \underline{w}_I^\beta, \underline{w}_I^\gamma, \underline{w}_I^\delta\right)$, similarly to the fuzzy maximal eigenvalue $\tilde{\lambda}_I$ in Sect. 4.2.2.4, is obtained from the fuzzy matrix $\tilde{M}_I = \{\tilde{m}_{Iij}\}_{i,j=1}^n$, $\tilde{m}_{Iij} = \left(m_{Iij}^\alpha, m_{Iij}^\beta, m_{Iij}^\gamma, m_{Iij}^\delta\right)$, such that $\tilde{m}_{Iji} = \left(\frac{1}{m_{Iij}^\alpha}, \frac{1}{m_{Iij}^\beta}, \frac{1}{m_{Iij}^\gamma}, \frac{1}{m_{Iij}^\delta}\right)$, $i < j$. In particular, the fuzzy maximal eigenvector $\underline{\tilde{w}}_I = \left(\underline{w}_I^\alpha, \underline{w}_I^\beta, \underline{w}_I^\gamma, \underline{w}_I^\delta\right)$ is obtained as

$$\underline{w}_I^\alpha = \left(w_{I1}^\alpha, \ldots, w_{In}^\alpha\right)^T: \quad \underline{w}_I^\alpha = EVM_{\underline{w}}(M_I^\alpha), \ M_I^\alpha = \left\{m_{Iij}^\alpha\right\}_{i,j=1}^n, \tag{4.67}$$

$$\underline{w}_I^\beta = \left(w_{I1}^\beta, \ldots, w_{In}^\beta\right)^T: \quad \underline{w}_I^\beta = EVM_{\underline{w}}(M_I^\beta), \ M_I^\beta = \left\{m_{Iij}^\beta\right\}_{i,j=1}^n, \tag{4.68}$$

$$\underline{w}_I^\gamma = \left(w_{I1}^\gamma, \ldots, w_{In}^\gamma\right)^T: \quad \underline{w}_I^\gamma = EVM_{\underline{w}}(M_I^\gamma), \ M_I^\gamma = \left\{m_{Iij}^\gamma\right\}_{i,j=1}^n, \tag{4.69}$$

$$\underline{w}_I^\delta = \left(w_{I1}^\delta, \ldots, w_{In}^\delta\right)^T: \quad \underline{w}_I^\delta = EVM_{\underline{w}}(M_I^\delta), \ M_I^\delta = \left\{m_{Iij}^\delta\right\}_{i,j=1}^n. \tag{4.70}$$

However, similarly as for $\tilde{\lambda}_I$ obtained from the FMPCM $\tilde{M}_I$, also the elements $\tilde{w}_{Ii} = \left(w_{Ii}^\alpha, w_{Ii}^\beta, w_{Ii}^\gamma, w_{Ii}^\delta\right)$, $i = 1, \ldots, n$, of $\underline{\tilde{w}}_I$ are not even fuzzy numbers in general, just quadruples of real numbers. This is caused not only by the inappropriate form of the fuzzy maximal eigenvalue $\tilde{\lambda}_I$ used in the formulas (4.67)–(4.70) but also by the inappropriate normalization of the maximal eigenvectors $\underline{w}_I^\alpha$, $\underline{w}_I^\beta$, $\underline{w}_I^\gamma$, and $\underline{w}_I^\delta$. Since $\sum_{i=1}^n w_{Ii}^\alpha = 1$, $\sum_{i=1}^n w_{Ii}^\beta = 1$, $\sum_{i=1}^n w_{Ii}^\gamma = 1$, $\sum_{i=1}^n w_{Ii}^\delta = 1$, then clearly the inequalities $w_{Ii}^\alpha \leq w_{Ii}^\beta \leq w_{Ii}^\gamma \leq w_{Ii}^\delta$, $i = 1, \ldots, n$, cannot be guaranteed. Moreover, since Ishizaka's method for obtaining the fuzzy maximal eigenvalue $\tilde{\lambda}_I$ reviewed in Sect. 4.2.2.4 is not invariant under permutation of objects in the FMPCM, it is obvious that also the method for obtaining the fuzzy maximal eigenvector $\tilde{w}_I$ is not invariant under permutation of objects.

Example 46 Let us examine the method for obtaining the fuzzy maximal eigenvector proposed by Ishizaka (2014) on the trapezoidal FMPCM (4.65) from Example 45. The corresponding matrix $\tilde{M}_I$ used in Ishizaka's approach is

$$\tilde{M}_I = \begin{pmatrix} 1 & (1,1,2,3) & (2,2.5,3,4) & (4,6,7,8) \\ \left(1,1,\dfrac{1}{2},\dfrac{1}{3}\right) & 1 & (3,4,4,5) & (4,5,6,6) \\ \left(\dfrac{1}{2},\dfrac{1}{2.5},\dfrac{1}{3},\dfrac{1}{4}\right) & \left(\dfrac{1}{3},\dfrac{1}{4},\dfrac{1}{4},\dfrac{1}{5}\right) & 1 & (1,2,2,3) \\ \left(\dfrac{1}{4},\dfrac{1}{6},\dfrac{1}{7},\dfrac{1}{8}\right) & \left(\dfrac{1}{4},\dfrac{1}{5},\dfrac{1}{6},\dfrac{1}{6}\right) & \left(1,\dfrac{1}{2},\dfrac{1}{2},\dfrac{1}{3}\right) & 1 \end{pmatrix}. \quad (4.71)$$

The fuzzy maximal eigenvalue obtained from (4.71) by the formulas (4.38)–(4.41) is $\tilde{\lambda}_I = (4.0458, 4.0319, 4.0863, 4.2184)$. Obviously, $\tilde{\lambda}_I$ is not a fuzzy number since $\lambda_I^\alpha = 4.0458 \not\leq 4.0319 = \lambda_I^\beta$. The corresponding fuzzy priorities of objects obtained by the formulas (4.67)–(4.70) are given as

$$\begin{aligned} \tilde{w}_{I1} &= (0.3606, 0.3837, 0.4753, 0.5379), \\ \tilde{w}_{I2} &= (0.3946, 0.4151, 0.3466, 0.3088), \\ \tilde{w}_{I3} &= (0.1376, 0.1311, 0.1182, 0.1055), \\ \tilde{w}_{I4} &= (0.1072, 0.0701, 0.0599, 0.0478). \end{aligned} \quad (4.72)$$

Notice that $\tilde{w}_{I2}$, $\tilde{w}_{I3}$, and $\tilde{w}_{I4}$ are not trapezoidal fuzzy numbers since the inequalities $w_{Ii}^\alpha \leq w_{Ii}^\beta \leq w_{Ii}^\gamma \leq w_{Ii}^\delta$, $i = 2, 3, 4$, are not satisfied.

Let us now permute the trapezoidal FMPCM (4.65) to $\tilde{M}^\pi = P\tilde{M}P^T$ by using the permutation matrix

$$P = \begin{pmatrix} 0 & 0 & 0 & 1 \\ 0 & 0 & 1 & 0 \\ 1 & 0 & 0 & 0 \\ 0 & 1 & 0 & 0 \end{pmatrix}. \quad (4.73)$$

The corresponding matrix in Ishizaka's approach is

$$
\tilde{M}_I^\pi =
\begin{pmatrix}
1 & \left(\dfrac{1}{3},\dfrac{1}{2},\dfrac{1}{2},1\right) & \left(\dfrac{1}{8},\dfrac{1}{7},\dfrac{1}{6},\dfrac{1}{4}\right) & \left(\dfrac{1}{6},\dfrac{1}{6},\dfrac{1}{5},\dfrac{1}{4}\right) \\[2mm]
(3,2,2,1) & 1 & \left(\dfrac{1}{4},\dfrac{1}{3},\dfrac{1}{2.5},\dfrac{1}{2}\right) & \left(\dfrac{1}{5},\dfrac{1}{4},\dfrac{1}{4},\dfrac{1}{3}\right) \\[2mm]
(8,7,6,4) & (4,3,2.5,2) & 1 & (1,1,2,3) \\[2mm]
(6,6,5,4) & (5,4,4,3) & \left(1,1,\dfrac{1}{2},\dfrac{1}{3}\right) & 1
\end{pmatrix}.
\tag{4.74}
$$

The fuzzy maximal eigenvalue obtained from (4.74) by the formulas (4.38)–(4.41) is $\tilde{\lambda}_I^\pi = (4.0720, 4.0133, 4.1193, 4.2606)$. The corresponding fuzzy priorities of objects obtained by the formulas (4.67)–(4.70) are given as

$$
\begin{aligned}
\tilde{w}_{I\pi(1)}^\pi &= (0.4171, 0.4017, 0.4560, 0.4738), \\
\tilde{w}_{I\pi(2)}^\pi &= (0.4191, 0.4162, 0.3472, 0.4738), \\
\tilde{w}_{I\pi(3)}^\pi &= (0.1119, 0.1201, 0.1295, 0.1332), \\
\tilde{w}_{I\pi(4)}^\pi &= (0.0519, 0.0620, 0.0674, 0.0997).
\end{aligned}
\tag{4.75}
$$

Again, $\tilde{\lambda}_I^\pi$, $\tilde{w}_{I1}^\pi$, $\tilde{w}_{I2}^\pi$ are not fuzzy numbers. Further, we see that $\tilde{\lambda}_I^\pi \neq \tilde{\lambda}_I$ and $\tilde{w}_{I\pi(i)}^\pi \neq \tilde{w}_{Ii}$, $i = 1\ldots, 4$, which means that the method is not invariant under permutation of objects. $\triangle$

Wang and Chin (2006) revised the method for obtaining the fuzzy maximal eigenvector proposed by Csutora and Buckley (2001). They argued that the fuzzy maximal eigenvector obtained by formulas (4.57)–(4.64) is not normalized according to Definition 29, and they proposed a new procedure for obtaining the normalized fuzzy maximal eigenvector corresponding to the fuzzy maximal eigenvalue $\tilde{\lambda}_S$ obtained by formulas (4.38)–(4.41). First, formulas (4.57)–(4.60) for obtaining the normalized maximal eigenvectors $\underline{w}^{\alpha*}$, $\underline{w}^{\beta*}$, $\underline{w}^{\gamma*}$, $\underline{w}^{\delta*}$ corresponding to the maximal eigenvalues $\lambda_S^\alpha, \lambda_S^\beta, \lambda_S^\gamma, \lambda_S^\delta$ obtained by formulas (4.38)–(4.41), respectively, are applied. Afterwards, the normalization constants $k^\alpha, k^\beta, k^\gamma$, and k^δ are searched for to obtain $\underline{w}^\alpha = k^\alpha \underline{w}^{\alpha*}$, $\underline{w}^\beta = k^\beta \underline{w}^{\beta*}$, $\underline{w}^\gamma = k^\gamma \underline{w}^{\gamma*}$, and $\underline{w}^\delta = k^\delta \underline{w}^{\delta*}$ so that $\tilde{\underline{w}} = \left(\underline{w}^\alpha, \underline{w}^\beta, \underline{w}^\gamma, \underline{w}^\delta\right)$ is a normalized fuzzy vector according to (3.15), i.e., the inequalities

$$
k^\alpha w_i^{\alpha*} + \sum_{j=1, j\neq i}^n k^\delta w_j^{\delta*} \geq 1, \quad k^\delta w_i^{\delta*} + \sum_{j=1, j\neq i}^n k^\alpha w_j^{\alpha*} \leq 1,
$$

$$
k^\beta w_i^{\beta*} + \sum_{j=1, j\neq i}^n k^\gamma w_j^{\gamma*} \geq 1, \quad k^\gamma w_i^{\gamma*} + \sum_{j=1, j\neq i}^n k^\beta w_j^{\beta*} \leq 1,
$$

hold for every $i = 1, \ldots, n$. The inequalities can be further written as

$$k^\alpha w_i^{\alpha*} + k^\delta \left(1 - w_i^{\delta*}\right) \geq 1, \qquad k^\delta w_i^{\delta*} + k^\alpha \left(1 - w_i^{\alpha*}\right) \leq 1,$$

$$k^\beta w_i^{\beta*} + k^\gamma \left(1 - w_i^{\gamma*}\right) \geq 1, \qquad k^\gamma w_i^{\gamma*} + k^\beta (1 - w_i^{\beta*}) \leq 1.$$

Obviously, $0 \leq k^\alpha \leq k^\beta \leq 1 \leq k^\gamma \leq k^\delta$.

In order not to lose any information obtained in the fuzzy maximal eigenvector, parameters $k^\alpha, k^\beta, k^\gamma$, and k^δ are chosen in such a form that the supports of the trapezoidal fuzzy numbers in the fuzzy maximal eigenvector are as wide as possible. It means that the parameters k^α and k^β are minimized and the parameters k^γ and k^δ are maximized. The parameters $k^\alpha, k^\beta, k^\gamma$, and k^δ can be obtained as the solution of the following linear programming problem (Wang and Chin 2006):

$$J^* = \max k^\alpha + \delta_1 + \delta_2 + \delta_3$$

$$s.t. \left. \begin{array}{l} k^\delta \left(1 - w_i^{\delta*}\right) + k^\alpha w_i^{\alpha*} \geq 1, \\ k^\delta w_i^{\delta*} + k^\alpha \left(1 - w_i^{\alpha*}\right) \leq 1, \\ k^\gamma \left(1 - w_i^{\gamma*}\right) + k^\beta w_i^{\beta*} \geq 1, \\ k^\gamma w_i^{\gamma*} + k^\beta (1 - w_i^{\beta*}) \leq 1, \\ k^\delta w_i^{\delta*} - k^\gamma w_i^{\gamma*} - \delta_1 \geq 0, \\ k^\gamma w_i^{\gamma*} - k^\beta w_i^{\beta*} - \delta_2 \geq 0, \\ k^\beta w_i^{\beta*} - k^\alpha w_i^{\alpha*} - \delta_3 \geq 0, \end{array} \right\} \quad i = 1, \ldots, n. \qquad (4.76)$$

$$k^\alpha, \delta_1, \delta_2, \delta_3 \geq 0,$$

Readers can refer to Wang and Chin (2006) and Krejčí (2017c) for more details. The notation $\underline{\widetilde{w}}_S = (\widetilde{w}_{S1}, \ldots, \widetilde{w}_{Sn})^T$, $\widetilde{w}_{Si} = (w_{Si}^\alpha, w_{Si}^\beta, w_{Si}^\gamma, w_{Si}^\delta)$, will be used hereafter to refer to the normalized fuzzy maximal eigenvector obtained by the formulas (4.57)–(4.60) and (4.76) corresponding to the fuzzy maximal eigenvalue $\widetilde{\lambda}_S$ (the lower index S stands for standard fuzzy arithmetic that is used in the formulas (4.38)–(4.41)).

It is true that the supports of the trapezoidal fuzzy numbers $\widetilde{w}_{Si} = (w_{Si}^\alpha, w_{Si}^\beta, w_{Si}^\gamma, w_{Si}^\delta)$, $i = 1, \ldots, n$, in the fuzzy vector $\underline{\widetilde{w}}_S$ obtained by the method proposed by Wang and Chin (2006) are as wide as possible, still meeting the condition (3.15) of normalized fuzzy numbers. Moreover, the method for deriving the normalized fuzzy maximal eigenvector proposed by Wang and Chin (2006) is invariant under permutation of objects. However, the fuzzy vector $\underline{\widetilde{w}}_S$ does not represent the actual normalized fuzzy maximal eigenvector of the FMPCM $\widetilde{M}$. First of all, as already shown in Sect. 4.2.2.4, the fuzzy maximal eigenvalue $\widetilde{\lambda}_S$ obtained by formulas (4.38)–(4.41) does not represent the actual fuzzy maximal eigenvalue of a FMPCM $\widetilde{M}$ since it violates the multiplicative reciprocity of the related PCs. Thus, it should not be used in the process of deriving the fuzzy maximal eigenvectors of FMPCMs. However, just simply replacing the fuzzy maximal eigenvalue $\widetilde{\lambda}_S$ by the fuzzy maximal eigenvalue $\widetilde{\lambda}_C$ in the method for deriving the normalized fuzzy maximal eigenvector is not sufficient.

A severe drawback independent of the formulas for obtaining the fuzzy maximal eigenvalue is that the normalized fuzzy vector $\widetilde{w}_S$ obtained by the formulas (4.57)–(4.60) and (4.76) does not "consist" of normalized eigenvectors. For example, the eigenvector $\underline{w}^\alpha = k^\alpha \underline{w}^{\alpha*} = (k^\alpha w_1^{\alpha*}, \ldots, k^\alpha w_n^{\alpha*})^T$ obtained as the maximal eigenvector of the matrix $M^\alpha = \left\{ m_{ij}^\alpha \right\}_{i,j=1}^n$ corresponding to the maximal eigenvalue λ_S^α is not normalized, $\sum_{i=1}^n k^\alpha w_i^{\alpha*} < 1$. This drawback would occur even if the eigenvector $\underline{w}^\alpha$ was obtained as the maximal eigenvector of the MPCM $M^* = \left\{ m_{ij} \right\}_{i,j=1}^n$ corresponding to the maximal eigenvalue λ_C^α obtainable by the formula (4.46). The same drawback occurs also for the eigenvector $\underline{w}^\beta = k^\beta \underline{w}^{\beta*} = (k^\beta w_1^{\beta*}, \ldots, k^\beta w_n^{\beta*})^T$, i.e. $\sum_{i=1}^n k^\beta w_i^{\beta*} < 1$, unless $M^\beta = M^\gamma$. Analogously, also the eigenvectors $\underline{w}^\delta = k^\delta \underline{w}^{\delta*}$ and $\underline{w}^\gamma = k^\gamma \underline{w}^{\gamma*}$ are such that $\sum_{i=1}^n k^\delta w_i^{\delta*} > 1$ and $\sum_{i=1}^n k^\gamma w_i^{\gamma*} > 1$, unless $M^\beta = M^\gamma$.

Example 47 Let us examine the method for obtaining the fuzzy maximal eigenvector proposed by Wang and Chin (2006) on the trapezoidal FMPCM (4.65) from Example 45. The fuzzy priorities obtained by the formulas (4.57)–(4.60) and (4.76) are

$$
\begin{aligned}
\widetilde{w}_{S1} &= (0.2584, 0.2969, 0.5704, 0.6565), \\
\widetilde{w}_{S2} &= (0.2265, 0.2829, 0.4778, 0.4778), \\
\widetilde{w}_{S3} &= (0.0726, 0.0970, 0.1477, 0.1763), \\
\widetilde{w}_{S4} &= (0.0444, 0.0497, 0.0777, 0.1022).
\end{aligned}
\tag{4.77}
$$

It can be easily verified by using (3.15) that the fuzzy priorities are normalized.

In the approach of Wang and Chin (2006), as well as in the original approach of Csutora and Buckley (2001), the representing values of the fuzzy priorities $\widetilde{w}_{Si}$ are obtained from matrices that are not multiplicatively reciprocal. For example, the lower boundary values w_{Si}^α, $i = 1, \ldots, 4$, are obtained as the components of the maximal eigenvector of matrix

$$
M^\alpha = \begin{pmatrix}
1 & 1 & 2 & 4 \\
\dfrac{1}{3} & 1 & 3 & 4 \\
\dfrac{1}{4} & \dfrac{1}{5} & 1 & 1 \\
\dfrac{1}{8} & \dfrac{1}{6} & \dfrac{1}{3} & 1
\end{pmatrix}
\tag{4.78}
$$

that does not keep multiplicative reciprocity of the related PCs. The meaning of the maximal eigenvectors obtained from such matrices is questionable. This violation of multiplicative reciprocity of the related PCs occurs in the approach of Wang and Chin (2006) as well as in the approach of Csutora and Buckley (2001).

Further, the fuzzy vector $\widetilde{\underline{w}}_S = (\widetilde{w}_{S1}, \widetilde{w}_{S2}, \widetilde{w}_{S3}, \widetilde{w}_{S4})^T$ with the components given as (4.77) does not really represent a normalized fuzzy maximal eigenvector of the trapezoidal FMPCM (4.65) since it does not consist of normalized eigenvectors. For example, the vector $\underline{w}_S^\alpha = (0.2584, 0.2265, 0.0726, 0.0444)^T$ obtained as the maximal eigenvector of the matrix (4.78) is not normalized; $\sum_{i=1}^4 w_{Si}^\alpha = 0.6019 < 1$. $\triangle$

As emphasized repeatedly, it is necessary to consider the multiplicative reciprocity of the related PCs when performing operations on the elements of a FMPCM in order to reflect properly the preference information contained in the FMPCM. Thus, similarly to the formulas for obtaining the fuzzy maximal eigenvalue $\tilde{\lambda}_C$ of a FMPCM, it is necessary to apply constrained fuzzy arithmetic to the fuzzy extension of the formula (2.21) in order to obtain the normalized fuzzy maximal eigenvector of a FMPCM. For a trapezoidal FMPCM $\tilde{M} = \{\tilde{m}_{ij}\}_{i,j=1}^n$, $\tilde{m}_{ij} = (m_{ij}^\alpha, m_{ij}^\beta, m_{ij}^\gamma, m_{ij}^\delta)$, the components $\tilde{w}_{Ci} = (w_{Ci}^\alpha, w_{Ci}^\beta, w_{Ci}^\gamma, w_{Ci}^\delta)$, $i = 1, \ldots, n$, of the normalized fuzzy maximal eigenvector $\tilde{\underline{w}}_C$ (the lower index C stands for the applied concept of constrained fuzzy arithmetic 3.47) should be obtained as Krejčí (2018):

$$w_{Ci}^\alpha = \min\left\{w_i \; ; \; \begin{array}{l} \underline{w} = (w_1, \ldots, w_i, \ldots, w_n)^T, \;\; \underline{w} = EVM_{\underline{w}}(M), \;\; M = \{m_{rs}\}_{r,s=1}^n, \\ m_{rs} \in [m_{rs}^\alpha, m_{rs}^\delta], \;\; m_{sr} = \frac{1}{m_{rs}}, \;\; r, s = 1, \ldots, n \end{array}\right\},$$
$$(4.79)$$

$$w_{Ci}^\beta = \min\left\{w_i \; ; \; \begin{array}{l} \underline{w} = (w_1, \ldots, w_i, \ldots, w_n)^T, \;\; \underline{w} = EVM_{\underline{w}}(M), \;\; M = \{m_{rs}\}_{r,s=1}^n, \\ m_{rs} \in [m_{rs}^\beta, m_{rs}^\gamma], \;\; m_{sr} = \frac{1}{m_{rs}}, \;\; r, s = 1, \ldots, n \end{array}\right\},$$
$$(4.80)$$

$$w_{Ci}^\gamma = \max\left\{w_i \; ; \; \begin{array}{l} \underline{w} = (w_1, \ldots, w_i, \ldots, w_n)^T, \;\; \underline{w} = EVM_{\underline{w}}(M), \;\; M = \{m_{rs}\}_{r,s=1}^n, \\ m_{rs} \in [m_{rs}^\beta, m_{rs}^\gamma], \;\; m_{sr} = \frac{1}{m_{rs}}, \;\; r, s = 1, \ldots, n \end{array}\right\},$$
$$(4.81)$$

$$w_{Ci}^\delta = \max\left\{w_i \; ; \; \begin{array}{l} \underline{w} = (w_1, \ldots, w_i, \ldots, w_n)^T, \;\; \underline{w} = EVM_{\underline{w}}(M), \;\; M = \{m_{rs}\}_{r,s=1}^n, \\ m_{rs} \in [m_{rs}^\alpha, m_{rs}^\delta], \;\; m_{sr} = \frac{1}{m_{rs}}, \;\; r, s = 1, \ldots, n \end{array}\right\}.$$
$$(4.82)$$

Theorem 39 (Krejčí 2018) *The fuzzy priorities* $\tilde{w}_{Ci} = (w_{Ci}^\alpha, w_{Ci}^\beta, w_{Ci}^\gamma, w_{Ci}^\delta)$, $i = 1, \ldots, n$, *obtained from a FMPCM* $\tilde{M}$ *by the formulas* (4.79)–(4.82) *are normalized.*

Proof It is sufficient to prove that the fuzzy priorities $\tilde{w}_{Ci}, i = 1, \ldots, n$, satisfy the inequalities (3.15). From the formula (4.79) it follows that w_{Ci}^α was obtained as the i-th component of the normalized maximal eigenvector of one particular MPCM $M^{\alpha i} = \{m_{pq}\}_{p,q=1}^n$, $m_{pq} \in [m_{pq}^\alpha, m_{pq}^\delta]$, $p, q = 1, \ldots, n$. Let us denote by $w_k^{\alpha i}$ the priorities of objects $o_k, k \neq i$, obtainable from the same MPCM $M^{\alpha i}$, i.e., $(w_1^{\alpha i}, \ldots, w_{Ci}^\alpha, \ldots, w_n^{\alpha i})^T$ is the normalized maximal eigenvector of $M^{\alpha i}$.

Obviously, $w_{Ci}^{\alpha} + \sum_{\substack{k=1 \\ k \neq i}}^{n} w_k^{\alpha i} = 1$ and $w_k^{\alpha i} \in [w_{Ck}^{\alpha}, w_{Ck}^{\delta}], k \neq i$. From this it follows that $w_{Ci}^{\alpha} + \sum_{\substack{k=1 \\ k \neq i}}^{n} w_{Ck}^{\delta} \geq 1$. The remaining inequalities in (3.15) are proved analogously. □

Remark 18 According to Theorem 39, the fuzzy maximal eigenvector $\widetilde{w}_C$ obtained from a FMPCM by the formulas (4.79)–(4.82) is normalized in the sense of Definition 29, i.e., we can really call $\widetilde{w}_C$ the normalized fuzzy maximal eigenvector. Notice that the normality of the fuzzy maximal eigenvector was reached naturally by just properly applying constrained fuzzy arithmetic to the fuzzy extension of the formula (2.21) for obtaining the normalized maximal eigenvector of a MPCM; no forced normalization was done as in the case of normalizing the fuzzy maximal eigenvector in the method proposed by Wang and Chin (2006). Therefore, the normalized fuzzy vector given by Definition 29 is a natural counterpart of the normalized crisp vector given by (2.18).

Theorem 40 (Krejčí 2018) *The fuzzy extension of the EVM based on the formulas (4.79)–(4.82) is invariant under permutation of objects in FMPCMs.*

Proof It is sufficient to show that for a given object $o_i, i \in \{1, \ldots, n\}$, its priority $\widetilde{w}_{Ci}$ corresponding to the i–th component of the normalized maximal eigenvector $\widetilde{w}$ does not change under permutation of objects in the FMPCM $\widetilde{M}$.

From the invariance of the EVM reviewed in Sect. 2.2.3.1 it follows that the normalized maximal eigenvector w of the given MPCM M does not change under any permutation $M^{\pi} = PMP^T$ of M, but it is just permuted accordingly to the normalized maximal eigenvector w^{π}. This means that each component $w_i, i \in \{1, \ldots, n\}$, of the normalized maximal eigenvector w is equal to the corresponding component $w_{\pi(i)}^{\pi}$ of the permuted normalized maximal eigenvector w^{π}. Therefore, neither the minimum w_{Ci}^{α} nor the maximum w_{Ci}^{δ} of the component w_i of the normalized maximal eigenvector w over all MPCMs obtainable from the closures of the supports of the trapezoidal fuzzy numbers in the trapezoidal FMPCM $\widetilde{M} = \{\widetilde{m}_{ij}\}_{i,j=1}^n$, $\widetilde{m}_{ij} = (m_{ij}^{\alpha}, m_{ij}^{\beta}, m_{ij}^{\gamma}, m_{ij}^{\delta})$, change under permutation of objects in the FMPCM $\widetilde{M}$. Analogously, also the minimum w_{Ci}^{β} and the maximum w_{Ci}^{γ} of the component w_i of the normalized maximal eigenvector w over all MPCMs obtainable from the cores of the trapezoidal fuzzy numbers in the trapezoidal FMPCM $\widetilde{M} = \{\widetilde{m}_{ij}\}_{i,j=1}^n$ do not change under permutation. Therefore, the fuzzy priority $\widetilde{w}_{Ci} = (w_{Ci}^{\alpha}, w_{Ci}^{\beta}, w_{Ci}^{\gamma}, w_{Ci}^{\delta})$ obtained by the formulas (4.79)–(4.82) does not change under permutation of objects in FMPCMs (it is only permuted accordingly), which concludes the proof. □

Example 48 Let us apply the new method given by formulas (4.79)–(4.82) to the FMPCM (4.65). The fuzzy maximal eigenvalue obtained by the formulas (4.46)–(4.49) is $\widetilde{\lambda}_C = (4, 4.0104, 4.1245, 4.4747)$ and the normalized fuzzy maximal eigenvector $\widetilde{w}_C$ obtained by the formulas (4.79)–(4.82) is given as

$$\widetilde{w}_{C1} = (0.3256, 0.3786, 0.4795, 0.5711),$$
$$\widetilde{w}_{C2} = (0.2551, 0.3367, 0.4267, 0.4639),$$
$$\widetilde{w}_{C3} = (0.0821, 0.1182, 0.1313, 0.1781),$$
$$\widetilde{w}_{C4} = (0.0478, 0.0599, 0.0701, 0.1072).$$
(4.83)

The fuzzy priorities are normalized, i.e., they satisfy the inequalities (3.15).

The normalized fuzzy maximal eigenvector $\widetilde{w}_C$ given by (4.83) differs significantly from the fuzzy maximal eigenvector $\widetilde{w}_S$ given by (4.77) that was obtained by the formulas (4.57)–(4.60) and (4.76). For easier comparison, both fuzzy maximal eigenvectors are displayed in Fig. 4.7.

Let us now examine the lower boundary value $w_{C1}^{\alpha} = 0.3256$ of $\widetilde{w}_{C1}$. It was obtained as the solution of the optimization problem (4.79), in particular as the first component of the normalized maximal eigenvector $\underline{w} = (0.3256, 0.4631, 0.1451, 0.0662)^T$ of the MPCM

$$M^{\alpha} = \begin{pmatrix} 1 & 1 & 2 & 4 \\ 1 & 1 & 5 & 6 \\ \dfrac{1}{2} & \dfrac{1}{5} & 1 & 3 \\ \dfrac{1}{4} & \dfrac{1}{6} & \dfrac{1}{3} & 1 \end{pmatrix}.$$
(4.84)

Similarly, any other element from the closure of the support of any fuzzy priority $\widetilde{w}_{Ci}, i = 1, \dots, 4$, is an element of a normalized maximal eigenvector corresponding to a MPCM obtainable from the closures of the supports of the trapezoidal fuzzy numbers in the FMPCM (4.65). $\triangle$

Krejčí (2018) provided an interesting comparative study, where the approaches to the computation of the fuzzy maximal eigenvector reviewed in this section are compared in terms of distance of the fuzzy priorities from the fuzzy priorities obtained by the fuzzy extension of the geometric mean method based on constrained fuzzy

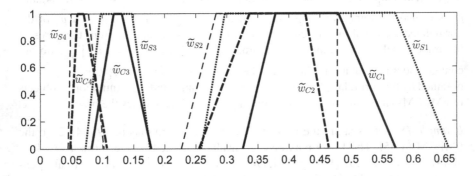

Fig. 4.7 Fuzzy maximal eigenvectors $\widetilde{w}_C$ and $\widetilde{w}_S$ of the FMPCM (4.65)

arithmetic that is going to be introduced in the following section. The fuzzy priorities obtained by the fuzzy maximal eigenevector method resulted to be closest.

4.2.3.2 Fuzzy Extension of the Geometric-Mean Method

In this section, the fuzzy extension of the GMM to FMPCMs is dealt with. The methods proposed by Buckley (1985a) and by Liu (2009) are reviewed and their drawbacks regarding the violation of multiplicative reciprocity and of invariance under permutation are pointed out. Afterwards, the formulas proposed by Enea and Piazza (2004) based on constrained fuzzy arithmetic are analyzed and some interesting properties are derived. The section extends the results presented by Krejčí et al. (2017).

Buckley (1985a) proposed a fuzzy extension of the GMM to compute $\alpha-$cuts of the fuzzy priorities of objects from FMPCMs. Trapezoidal representation is again used here to review the method.

Buckley (1985a) first computed the geometric mean $\widetilde{g}_i = (g_i^\alpha, g_i^\beta, g_i^\gamma, g_i^\delta)$, $i = 1, \ldots, n$, of the elements in each row of the FMPCM $\widetilde{M} = \{\widetilde{m}_{ij}\}_{i,j=1}^n$, $\widetilde{m}_{ij} = (m_{ij}^\alpha, m_{ij}^\beta, m_{ij}^\gamma, m_{ij}^\delta)$, by applying standard fuzzy arithmetic to the fuzzy extension of the formula (2.23). Thus, the representing values of the geometric means $\widetilde{g}_i$, $i = 1, \ldots, n$, of the elements in the rows of the FMPCM $\widetilde{M}$ are computed as

$$g_i^\alpha = \sqrt[n]{\prod_{j=1}^n m_{ij}^\alpha}, \quad g_i^\beta = \sqrt[n]{\prod_{j=1}^n m_{ij}^\beta}, \quad g_i^\gamma = \sqrt[n]{\prod_{j=1}^n m_{ij}^\gamma}, \quad g_i^\delta = \sqrt[n]{\prod_{j=1}^n m_{ij}^\delta}. \qquad (4.85)$$

The geometric means $\widetilde{g}_i = (g_i^\alpha, g_i^\beta, g_i^\gamma, g_i^\delta)$, $i = 1, \ldots, n$, represent the non-normalized fuzzy priorities of objects compared in the FMPCM $\widetilde{M}$. Notice that simplified standard fuzzy arithmetic (3.38) is used in (4.85).

Afterwards, Buckley (1985a) divided each geometric mean by their sum, analogously to the formula (2.24), in order to normalize the fuzzy priorities of objects standard fuzzy arithmetic was again used for this purpose. Thus, according to Buckley (1985a), the trapezoidal fuzzy priorities $\widetilde{w}_{Si} = (w_{Si}^\alpha, w_{Si}^\beta, w_{Si}^\gamma, w_{Si}^\delta)$, $i = 1, \ldots, n$, (the lower index S stands for the standard fuzzy arithmetic that is applied to the formulas) are obtained from a trapezoidal FMPCM $\widetilde{M} = \{\widetilde{m}_{ij}\}_{i,j=1}^n$, $\widetilde{m}_{ij} = (m_{ij}^\alpha, m_{ij}^\beta, m_{ij}^\gamma, m_{ij}^\delta)$, as

$$w_{Si}^\alpha = \frac{\sqrt[n]{\prod_{j=1}^n m_{ij}^\alpha}}{\sum_{k=1}^n \sqrt[n]{\prod_{j=1}^n m_{kj}^\delta}}, \qquad (4.86)$$

$$w_{Si}^{\beta} = \frac{\sqrt[n]{\prod_{j=1}^{n} m_{ij}^{\beta}}}{\sum_{k=1}^{n} \sqrt[n]{\prod_{j=1}^{n} m_{kj}^{\gamma}}}, \tag{4.87}$$

$$w_{Si}^{\gamma} = \frac{\sqrt[n]{\prod_{j=1}^{n} m_{ij}^{\gamma}}}{\sum_{k=1}^{n} \sqrt[n]{\prod_{j=1}^{n} m_{kj}^{\beta}}}, \tag{4.88}$$

$$w_{Si}^{\delta} = \frac{\sqrt[n]{\prod_{j=1}^{n} m_{ij}^{\delta}}}{\sum_{k=1}^{n} \sqrt[n]{\prod_{j=1}^{n} m_{kj}^{\alpha}}}. \tag{4.89}$$

It is a well-known fact that the fuzzy extension of the GMM given by the formulas (4.86)–(4.89) is invariant under permutation of objects.

The fuzzy priorities $\widetilde{w}_{Si}, i = 1, \ldots, n$, obtained by the formulas (4.86)–(4.89) are not normalized according to Definition 28 as they do not satisfy the inequalities (3.15). In fact, they are not even constrained to the interval $[0, 1]$. Thus, Buckley (1985a) suggested to multiply all fuzzy priorities $\widetilde{w}_{Si}, i = 1, \ldots, n$, by a suitable normalization constant $c < 1$ in order to limit them to the interval $[0, 1]$. However, such fuzzy priorities still do not satisfy the inequalities (3.15), i.e., they are not normalized.

Furthermore, the representing values $w_{Si}^{\alpha}, w_{Si}^{\beta}, w_{Si}^{\gamma}, w_{Si}^{\delta}$ given by (4.86)–(4.89), respectively, are not obtained from MPCMs. In particular, in the formula (4.86), the upper boundary values m_{kj}^{δ} of all PCs $\widetilde{m}_{kj}, k, j = 1, \ldots, n$, are used. This violates the multiplicative reciprocity of the related PCs since $m_{kj}^{\delta} \neq \frac{1}{m_{jk}^{\delta}}$ (unless $\widetilde{m}_{kj}$ is a crisp number). In addition, also the lower boundary values m_{ij}^{α} of the PCs $\widetilde{m}_{ij}, j = 1, \ldots, n$, in the i−th row of the FMPCM are present in the formula. This even violates the extension principle (3.3) since two different values, in particular m_{ij}^{α} and $m_{ij}^{\delta}, j = 1, \ldots, n$, of one variable are used in the formula at the same time. The formulas (4.87), (4.88), and (4.89) suffer from the same drawbacks.

Liu (2009) proposed the following extension of the GMM to interval FMPCMs. For an interval FMPCM $\overline{M} = \{\overline{m}_{ij}\}_{i,j=1}^{n}, \overline{m}_{ij} = [m_{ij}^{L}, m_{ij}^{U}]$, he constructed two MPCMs $C = \{c_{ij}\}_{i,j=1}^{n}$ and $D = \{d_{ij}\}_{i,j=1}^{n}$ by applying (4.7). Afterwards, he derived non-normalized priorities $w_i(C)$ and $w_i(D), i = 1, \ldots, n$, of objects from these MPCMs C and D, respectively, by using the formula (2.23). The interval priorities $\overline{w}_i = [w_i^{L}, w_i^{U}], i = 1, \ldots, n$, were then determined as

$$w_i^{L} = \min\{w_i(C), w_i(D)\}, \qquad w_i^{U} = \max\{w_i(C), w_i(D)\}. \tag{4.90}$$

This method, similarly to Definitions 46 and 47 of multiplicative consistency for interval and triangular FMPCMs proposed by Liu (2009) and Liu et al. (2014), respectively, reviewed already in Sect. 4.2.2.1, is not invariant under permutation of objects. This drawback is illustrated on the following example.

Example 49 Let us apply the method for obtaining interval priorities proposed by Liu (2009) to the interval FMPCM (4.29). The interval priorities of objects obtained by the formula (4.90) are

$$\overline{w}_1 = [0.4309, 0.7631], \quad \overline{w}_2 = [1.0772, 1.1447], \quad \overline{w}_3 = [1.1447, 2.1544].$$

By applying the formula (4.90) to the permuted interval FMPCM (4.30), the interval priorities of objects are obtained as

$$\overline{w}_{\pi(1)}^{\pi} = [0.4309, 0.7631], \quad \overline{w}_{\pi(2)}^{\pi} = [0.9086, 1.3572], \quad \overline{w}_{\pi(3)}^{\pi} = [1.4422, 1.7100].$$

We see that $\overline{w}_{\pi(2)}^{\pi} \neq \overline{w}_2$ and $\overline{w}_{\pi(3)}^{\pi} \neq \overline{w}_3$, which demonstrates that the method for obtaining interval priorities from interval FMPCMs proposed by Liu (2009) is not invariant under permutation of objects.

The non-normalized interval priorities obtained from the interval FMPCM (4.29) by the formulas (4.85) are

$$\overline{w}_1 = [0.4309, 0.7631], \quad \overline{w}_2 = [0.9086, 1.3572], \quad \overline{w}_3 = [1.1447, 2.1544].$$

The same interval priorities are obtained by the formulas (4.85) also from the permuted interval FMPCM (4.30), i.e. $\overline{w}_{\pi(i)}^{\pi} = \overline{w}_i$, $i = 1, 2, 3$.   △

Note that the formulas (4.85) for obtaining non-normalized fuzzy priorities do not violate the multiplicative-reciprocity property as well as the invariance under permutation of objects. That follows from the absence of mutually reciprocal PCs in the formulas. Therefore, use of standard fuzzy arithmetic instead of constrained fuzzy arithmetic is sufficient here. Contrarily, the formulas (4.86)–(4.89) violate the multiplicative reciprocity and the formulas (4.90) violate the invariance under permutation of objects. The reason is that the formulas (4.86)–(4.89) are based on standard fuzzy arithmetic instead of constrained fuzzy arithmetic, which would be in this case indispensable, whereas the formulas (4.90) are not even based on standard fuzzy arithmetic.

In order to handle properly the multiplicative reciprocity of PCs, and thus automatically ensuring also the invariance under permutation, it is necessary to follow the same approach as in the previous section where the fuzzy extension of the EVM was dealt with. This means that constrained fuzzy arithmetic has to be applied to the fuzzy extension of the formula (2.24) instead of standard fuzzy arithmetic in order to respect the multiplicative reciprocity of the related PCs in the FMPCM. Enea and Piazza (2004) realized this necessity and proposed appropriate fuzzy extension of the formula (2.24) to triangular FMPCM.

Extending the method proposed by Enea and Piazza (2004) to trapezoidal FMPCMs $\widetilde{M} = \{\widetilde{m}_{ij}\}_{i,j=1}^{n}$, $\widetilde{m}_{ij} = (m_{ij}^{\alpha}, m_{ij}^{\beta}, m_{ij}^{\gamma}, m_{ij}^{\delta})$, fuzzy priorities $\widetilde{w}_{Ci} = \left(w_{Ci}^{\alpha}, w_{Ci}^{\beta}, w_{Ci}^{\gamma}, w_{Ci}^{\delta}\right)$, $i = 1, \ldots, n$, (the lower index C stands for the applied concept of constrained fuzzy arithmetic (3.47)) are obtained in the following form:

$$w_{Ci}^{\alpha} = \min \left\{ \frac{\sqrt[n]{\prod_{j=1}^{n} m_{ij}}}{\sum_{k=1}^{n} \sqrt[n]{\prod_{j=1}^{n} m_{kj}}} \; ; \; \begin{array}{l} m_{rs} \in \left[m_{rs}^{\alpha}, m_{rs}^{\delta} \right], \\ m_{sr} = \frac{1}{m_{rs}}, \\ r, s = 1, \ldots, n \end{array} \right\}, \tag{4.91}$$

$$w_{Ci}^{\beta} = \min \left\{ \frac{\sqrt[n]{\prod_{j=1}^{n} m_{ij}}}{\sum_{k=1}^{n} \sqrt[n]{\prod_{j=1}^{n} m_{kj}}} \; ; \; \begin{array}{l} m_{rs} \in \left[m_{rs}^{\beta}, m_{rs}^{\gamma} \right], \\ m_{sr} = \frac{1}{m_{rs}}, \\ r, s = 1, \ldots, n \end{array} \right\}, \tag{4.92}$$

$$w_{Ci}^{\gamma} = \max \left\{ \frac{\sqrt[n]{\prod_{j=1}^{n} m_{ij}}}{\sum_{k=1}^{n} \sqrt[n]{\prod_{j=1}^{n} m_{kj}}} \; ; \; \begin{array}{l} m_{rs} \in \left[m_{rs}^{\beta}, m_{rs}^{\gamma} \right], \\ m_{sr} = \frac{1}{m_{rs}}, \\ r, s = 1, \ldots, n \end{array} \right\}, \tag{4.93}$$

$$w_{Ci}^{\delta} = \max \left\{ \frac{\sqrt[n]{\prod_{j=1}^{n} m_{ij}}}{\sum_{k=1}^{n} \sqrt[n]{\prod_{j=1}^{n} m_{kj}}} \; ; \; \begin{array}{l} m_{rs} \in \left[m_{rs}^{\alpha}, m_{rs}^{\delta} \right], \\ m_{sr} = \frac{1}{m_{rs}}, \\ r, s = 1, \ldots, n \end{array} \right\}. \tag{4.94}$$

Theorem 41 *The fuzzy priorities $\widetilde{w}_{Ci} = (w_{Ci}^{\alpha}, w_{Ci}^{\beta}, w_{Ci}^{\gamma}, w_{Ci}^{\delta}), i = 1, \ldots, n,$ obtained from a FMPCM $\widetilde{M}$ by the formulas (4.91)–(4.94) are normalized.*

Proof It is sufficient to prove that the fuzzy priorities $\widetilde{w}_{Ci}, i = 1, \ldots, n,$ satisfy the inequalities (3.15). From the formula (4.91) it follows that w_{Ci}^{α} was obtained by applying the formula (2.24) to one particular MPCM $M^{\alpha i} = \left\{ m_{pq} \right\}_{p,q=1}^{n}, m_{pq} \in [m_{pq}^{\alpha}, m_{pq}^{\delta}], p, q = 1, \ldots, n.$ Let $w_k^{\alpha i}$ denote the priorities of objects $o_k, k \neq i$, obtainable by the formula (2.24) from the same MPCM $M^{\alpha i}$. Obviously, $w_{Ci}^{\alpha} + \sum_{\substack{k=1 \\ k \neq i}}^{n} w_k^{\alpha i} = 1$ and $w_k^{\alpha i} \in [w_{Ck}^{\alpha}, w_{Ck}^{\delta}], k \neq i.$ From this it follows that $w_{Ci}^{\alpha} + \sum_{\substack{k=1 \\ k \neq i}}^{n} w_{Ck}^{\delta} \geq 1.$ The remaining inequalities in (3.15) are proved analogously. $\square$

Remark 19 According to Theorem 41, the fuzzy priorities $\widetilde{w}_{Ci}, i = 1, \ldots, n,$ obtained from a FMPCM by the formulas (4.91)–(4.94) are normalized in the sense of Definition 29. Notice that the normality of the fuzzy priorities was reached naturally by just properly applying constrained fuzzy arithmetic to the fuzzy extension of the formula (2.24) for obtaining normalized priorities from a MPCM; no forced normalization was needed, unlike in the case of the "normalization" of the fuzzy priorities (4.86)–(4.89) as suggested by Buckley (1985a).

Theorem 42 *The fuzzy extension of the GMM based on the formulas (4.91)–(4.94) is invariant under permutation of objects in FMPCMs.*

Proof It is sufficient to show that for a given object $o_i, i \in \{1, \ldots, n\}$, its priority $\tilde{w}_{Ci}$ obtained by the formulas (4.91)–(4.94) does not change under permutation of objects in a FMPCM $\tilde{M}$.

From the invariance of the GMM reviewed in Sect. 2.2.3.2 it follows that the priority w_i of object o_i determined by the formula (2.24) from the given MPCM M does not change under any permutation $M^\pi = PMP^T$ of M, it is just permuted accordingly. This means that the priority w_i obtained from M is equal to the corresponding priority $w_{\pi(i)}^\pi$ obtained from M^π. Therefore, also the minimum w_{Ci}^α and the maximum w_{Ci}^δ of the priority w_i of object o_i obtained by (2.24) over all MPCMs obtainable from the closures of the supports of the trapezoidal fuzzy numbers in the trapezoidal FMPCM $\tilde{M} = \{\tilde{m}_{ij}\}_{i,j=1}^n$, $\tilde{m}_{ij} = (m_{ij}^\alpha, m_{ij}^\beta, m_{ij}^\gamma, m_{ij}^\delta)$, do not change. Analogously also the minimum w_{Ci}^β and the maximum w_{Ci}^γ of the priority w_i of object o_i obtained by (2.24) over all MPCMs obtainable from the cores of the trapezoidal fuzzy numbers in the trapezoidal FMPCM $\tilde{M} = \{\tilde{m}_{ij}\}_{i,j=1}^n$ do not change. Therefore, the fuzzy priority $\tilde{w}_{Ci} = (w_{Ci}^\alpha, w_{Ci}^\beta, w_{Ci}^\gamma, w_{Ci}^\delta)$ obtained by the formulas (4.91)–(4.94) does not change under permutation of objects in FMPCMs (it is only permuted accordingly), which concludes the proof. □

In the following example, the difference between the fuzzy priorities obtained by formulas (4.86)–(4.89) and by formulas (4.91)–(4.94) is illustrated, and the drawbacks in the formulas (4.86)–(4.89) are demonstrated.

Example 50 Let us consider the FMPCM $\tilde{M}$ given by (4.65). The fuzzy priorities of objects obtained from this FMPCM by the fuzzy extension of the GMM proposed both by Buckley (1985a) and by Enea and Piazza (2004) are given in Table 4.8 and graphically presented in Fig. 4.8. The fuzzy priorities obtained by formulas (4.91)–(4.94) are significantly less uncertain than the fuzzy priorities obtained by formulas (4.86)–(4.89). Moreover, the closures of the supports of the fuzzy priorities obtained by formulas (4.91)–(4.94) are the proper subsets of the closures of the supports of the fuzzy priorities obtained by formulas (4.86)–(4.89). Analogously, the cores of the fuzzy priorities obtained by formulas (4.91)–(4.94) are proper subsets of the cores of the fuzzy priorities obtained by formulas (4.86)–(4.89). This is caused by using constrained fuzzy arithmetic (that preserves the multiplicative reciprocity of the related PCs) in the formulas (4.91)–(4.94).

Further, let us illustrate inappropriateness of the formulas (4.86)–(4.89) for computing the fuzzy priorities of objects from a FMPCM. For this purpose, let us see how the lower boundary value w_{S1}^α of the fuzzy priority $\tilde{w}_{S1}$ was obtained. The intensities of preference figuring in the formula (4.86) for obtaining the lower boundary value $w_{S1}^\alpha = 0.2469$ are highlighted in bold in the FMPCM.

Table 4.8 Fuzzy priorities of objects obtained from the FMPCM (4.65)

Fuzzy priorities obtained by formulas (4.86)–(4.89)	Fuzzy priorities obtained by formulas (4.91)–(4.94)
$\widetilde{w}_{S1} = (0.2469, 0.3401, 0.5399, 0.8114)$	$\widetilde{w}_{C1} = (0.3294, 0.3789, 0.4795, 0.5689)$
$\widetilde{w}_{S2} = (0.2076, 0.3073, 0.4694, 0.6067)$	$\widetilde{w}_{C2} = (0.2539, 0.3350, 0.4262, 0.4647)$
$\widetilde{w}_{S3} = (0.0694, 0.1104, 0.1418, 0.2180)$	$\widetilde{w}_{C3} = (0.0821, 0.1188, 0.1309, 0.1767)$
$\widetilde{w}_{S4} = (0.0424, 0.0571, 0.0762, 0.1296)$	$\widetilde{w}_{C4} = (0.0496, 0.0614, 0.0703, 0.1069)$

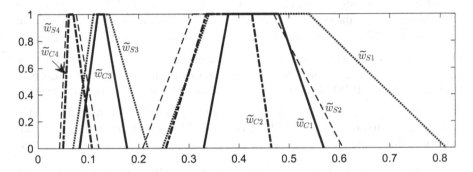

Fig. 4.8 Fuzzy priority vectors $\widetilde{\underline{w}}_C$ and $\widetilde{\underline{w}}_S$ of the FMPCM (4.65)

$$
\widetilde{M} = \begin{pmatrix}
1 & (1,1,2,3) & (2,2.5,3,4) & (4,6,7,8) \\
\left(\dfrac{1}{3},\dfrac{1}{2},1,1\right) & 1 & (3,4,4,5) & (4,5,6,6) \\
\left(\dfrac{1}{4},\dfrac{1}{3},\dfrac{1}{2.5},\dfrac{1}{2}\right) & \left(\dfrac{1}{5},\dfrac{1}{4},\dfrac{1}{4},\dfrac{1}{3}\right) & 1 & (1,2,2,3) \\
\left(\dfrac{1}{8},\dfrac{1}{7},\dfrac{1}{6},\dfrac{1}{4}\right) & \left(\dfrac{1}{6},\dfrac{1}{6},\dfrac{1}{5},\dfrac{1}{4}\right) & \left(\dfrac{1}{3},\dfrac{1}{2},\dfrac{1}{2},1\right) & 1
\end{pmatrix}. \quad (4.95)
$$

From (4.95) we see that two different intensities of preference of object o_1 over the other objects are used in the formula (4.86) at the same time. Moreover, the multiplicative reciprocity of the related PCs in the matrix is violated. This adds redundant vagueness to the fuzzy priorities, which leads to the distortion of the information contained in the FMPCM (4.65).

Contrarily, the lower boundary value $w_{C1}^{\alpha} = 0.3294$ of the fuzzy priority $\widetilde{w}_{C1}$ was obtained from a multiplicatively reciprocal matrix, in particular from the MPCM

$$
M = \begin{pmatrix}
1 & 1 & 2 & 4 \\
1 & 1 & 5 & 6 \\
\dfrac{1}{2} & \dfrac{1}{5} & 1 & 3 \\
\dfrac{1}{4} & \dfrac{1}{6} & \dfrac{1}{3} & 1
\end{pmatrix}. \quad (4.96)
$$

In the same way, it could be shown that all representing values of all four fuzzy priorities were obtained from MPCMs by the formulas (4.91)–(4.94). △

In general, the optimization problems solved in (4.91)–(4.94) have $n^2 - n$ variables and $\frac{n^2-n}{2}$ multiplicative-reciprocity constraints (the number of variables and multiplicative-reciprocity constraints gets reduced when crisp numbers are present above and below the main diagonal of the FMPCM). Thus, the computational complexity of the optimization problems increases rapidly with an increasing dimension n. However, the following theorem (that is based on the theorem introduced by Krejčí et al. (2017) for triangular FMPCMs) shows that the optimization problems (4.91)–(4.94) can be simplified significantly. First, the multiplicative-reciprocity constraints can be incorporated into the objective functions. Second, when w_{Ci}^{α} is computed, the variables $m_{ij}, j = 1, \ldots, n$, can be fixed as the lower boundary values of the trapezoidal fuzzy numbers in the i-th row of the FMPCM, i.e. as $m_{ij} := m_{ij}^{\alpha}$. Analogously also for the representing values w_{Ci}^{β}, w_{Ci}^{γ}, and w_{Ci}^{δ}. In this way, the number of variables is reduced from $n^2 - n$ to $\frac{n^2-n}{2} - (n-1)$.

Theorem 43 Let $\widetilde{M} = \{\widetilde{m}_{ij}\}_{i,j=1}^{n}$, $\widetilde{m}_{ij} = \left(m_{ij}^{\alpha}, m_{ij}^{\beta}, m_{ij}^{\gamma}, m_{ij}^{\delta}\right)$, be a trapezoidal FMPCM. The optimization problems (4.91)–(4.94) can be simplified for $i = 1, \ldots, n$ in the following way:

$$w_{Ci}^{\alpha} = \frac{\sqrt[n]{\prod_{j=1}^{n} m_{ij}^{\alpha}}}{\sqrt[n]{\prod_{j=1}^{n} m_{ij}^{\alpha}} + \max \left\{ \sum_{\substack{k=1 \\ k \neq i}}^{n} \sqrt[n]{\frac{1}{m_{ik}^{\alpha}} \prod_{\substack{l=1 \\ l \neq i}}^{k-1} \frac{1}{m_{lk}} \prod_{\substack{l=k+1 \\ l \neq i}}^{n} m_{kl}} \; ; \; \begin{array}{l} m_{rs} \in \left[m_{rs}^{\alpha}, m_{rs}^{\delta}\right], \\ r = 1, \ldots, n-1, \\ s = r+1, \ldots, n, \\ r, s \neq i \end{array} \right\}}, \quad (4.97)$$

$$w_{Ci}^{\beta} = \frac{\sqrt[n]{\prod_{j=1}^{n} m_{ij}^{\beta}}}{\sqrt[n]{\prod_{j=1}^{n} m_{ij}^{\beta}} + \max \left\{ \sum_{\substack{k=1 \\ k \neq i}}^{n} \sqrt[n]{\frac{1}{m_{ik}^{\beta}} \prod_{\substack{l=1 \\ l \neq i}}^{k-1} \frac{1}{m_{lk}} \prod_{\substack{l=k+1 \\ l \neq i}}^{n} m_{kl}} \; ; \; \begin{array}{l} m_{rs} \in \left[m_{rs}^{\beta}, m_{rs}^{\gamma}\right], \\ r = 1, \ldots, n-1, \\ s = r+1, \ldots, n, \\ r, s \neq i \end{array} \right\}}, \quad (4.98)$$

$$w_{Ci}^{\gamma} = \frac{\sqrt[n]{\prod_{j=1}^{n} m_{ij}^{\gamma}}}{\sqrt[n]{\prod_{j=1}^{n} m_{ij}^{\gamma}} + \min \left\{ \sum_{\substack{k=1 \\ k \neq i}}^{n} \sqrt[n]{\frac{1}{m_{ik}^{\gamma}} \prod_{\substack{l=1 \\ l \neq i}}^{k-1} \frac{1}{m_{lk}} \prod_{\substack{l=k+1 \\ l \neq i}}^{n} m_{kl}} \; ; \; \begin{array}{l} m_{rs} \in \left[m_{rs}^{\beta}, m_{rs}^{\gamma}\right], \\ r = 1, \ldots, n-1, \\ s = r+1, \ldots, n, \\ r, s \neq i \end{array} \right\}}, \quad (4.99)$$

$$w_{Ci}^{\delta} = \frac{\sqrt[n]{\prod\limits_{j=1}^{n} m_{ij}^{\delta}}}{\sqrt[n]{\prod\limits_{j=1}^{n} m_{ij}^{\delta}} + \min\left\{\sum\limits_{\substack{k=1\\k\neq i}}^{n} \sqrt[n]{\frac{1}{m_{ik}^{\delta}} \prod\limits_{\substack{l=1\\l\neq i}}^{k-1} \frac{1}{m_{lk}} \prod\limits_{\substack{l=k+1\\l\neq i}}^{n} m_{kl}} \; ; \; \begin{array}{l} m_{rs} \in \left[m_{rs}^{\alpha}, m_{rs}^{\delta}\right], \\ r = 1, \ldots, n-1, \\ s = r+1, \ldots, n, \\ r, s \neq i \end{array}\right\}}. \quad (4.100)$$

Proof First, let us show that the formulas (4.91) and (4.97) are identical. For any $i \in \{1, \ldots, n\}$, the formula (4.91) can be written in the following way:

$$w_{Ci}^{\alpha} = \min\left\{ \frac{\sqrt[n]{\prod\limits_{j=1}^{n} m_{ij}}}{\sqrt[n]{\prod\limits_{j=1}^{n} m_{ij}} + \sum\limits_{\substack{k=1\\k\neq i}}^{n} \sqrt[n]{\frac{1}{m_{ik}} \prod\limits_{\substack{j=1\\j\neq i}}^{n} m_{kj}}} \; ; \; \begin{array}{l} m_{rs} \in \left[m_{rs}^{\alpha}, m_{rs}^{\delta}\right], \\ m_{sr} = \frac{1}{m_{rs}}, \\ r, s = 1, \ldots, n, \end{array} \right\}.$$

Let us denote $\quad x_i := \sqrt[n]{\prod\limits_{j=1}^{n} m_{ij}} \quad$ and $\quad y_i := \sum\limits_{\substack{k=1\\k\neq i}}^{n} \sqrt[n]{\frac{1}{m_{ik}} \prod\limits_{\substack{j=1\\j\neq i}}^{n} m_{kj}}.$

Obviously, $x_i > 0$ for $m_{is} \in \left[m_{is}^{\alpha}, m_{is}^{\delta}\right]$, $s = 1, \ldots, n$, and $y_i > 0$ for $m_{rs} \in \left[m_{rs}^{\alpha}, m_{rs}^{\delta}\right]$, $m_{sr} = 1/m_{rs}, r, s = 1, \ldots, n$. Further, let us denote $f_i := \frac{x_i}{x_i+y_i}$. Then $\frac{\partial f_i}{\partial x_i} = \frac{y_i}{(x_i+y_i)^2} > 0$ and $\frac{\partial f_i}{\partial y_i} = \frac{-x_i}{(x_i+y_i)^2} < 0$. Hence, f_i is an increasing function of x_i and a decreasing function of y_i. It means that for minimizing the function f_i, we have to minimize x_i and maximize y_i. The function x_i is increasing in all the variables. Therefore,

$$x_i' := \min\left\{ x_i \; ; \; m_{ij} \in \left[m_{ij}^{\alpha}, m_{ij}^{\delta}\right], j = 1, \ldots, n \right\} = \sqrt[n]{\prod\limits_{j=1}^{n} m_{ij}^{\alpha}}.$$

The function y_i is decreasing in the variables $m_{i1}, \ldots, m_{in}$. Therefore,

$$y_i' := \max\left\{ y_i \; ; \; \begin{array}{l} m_{rs} \in \left[m_{rs}^{\alpha}, m_{rs}^{\delta}\right], m_{sr} = \frac{1}{m_{rs}}, \\ r, s = 1, \ldots, n \end{array} \right\}$$

$$= \left\{ \sum\limits_{\substack{k=1\\k\neq i}}^{n} \sqrt[n]{\frac{1}{m_{ik}^{\alpha}} \prod\limits_{\substack{j=1\\j\neq i}}^{n} m_{kj}} \; ; \; \begin{array}{l} m_{rs} \in \left[m_{rs}^{\alpha}, m_{rs}^{\delta}\right], m_{sr} = \frac{1}{m_{rs}}, \\ r, s = 1, \ldots, n \end{array} \right\}.$$

Finally, thanks to the reciprocity of $\widetilde{M}$, we can also replace all the elements m_{sr}, $r, s = 1, \ldots, n$, $r < s$, i.e., the elements below the main diagonal, by the reciprocals $1/m_{rs}$ of the corresponding elements m_{rs} above the main diagonal. By that we obtain formula (4.97).

Analogously, it can be demonstrated that (4.92) is equivalent to (4.98), (4.93) is equivalent to (4.99), and (4.94) is equivalent (4.100). □

In Example 50, it was pointed out that the cores and the closures of the supports of the fuzzy priorities in Table 4.8 obtained by formulas (4.91)–(4.94) are proper subsets of the cores and closures of the supports of the fuzzy priorities obtained by formulas (4.86)–(4.89), respectively. The following theorem (based on the theorem introduced by Krejčí et al. (2017) for triangular FMPCMs) shows that this property is valid in general for trapezoidal FMPCMs with at least one entry that is not a crisp number.

Theorem 44 *Let* $\widetilde{M} = \{\widetilde{m}_{ij}\}_{i,j=1}^{n}$, $\widetilde{m}_{ij} = \left(m_{ij}^{\alpha}, m_{ij}^{\beta}, m_{ij}^{\gamma}, m_{ij}^{\delta}\right)$, *be a trapezoidal FMPCM. Further, let* $\widetilde{w}_{S1}, \ldots, \widetilde{w}_{Sn}$ *be trapezoidal fuzzy priorities obtained by formulas (4.86)–(4.89), and* $\widetilde{w}_{C1}, \ldots, \widetilde{w}_{Cn}$ *be trapezoidal fuzzy priorities obtained by formulas (4.91)–(4.94). If there exists at least one* $\widetilde{m}_{kl}$, $k, l \in \{1, \ldots, n\}$, *such that* $m_{kl}^{\alpha} < m_{kl}^{\delta}$, *then*

$$[w_{Ci}^{\alpha}, w_{Ci}^{\delta}] \subset [w_{Si}^{\alpha}, w_{Si}^{\delta}] \quad and \quad [w_{Ci}^{\beta}, w_{Ci}^{\gamma}] \subseteq [w_{Si}^{\beta}, w_{Si}^{\gamma}], \quad i = 1, \ldots, n. \quad (4.101)$$

The equality

$$[w_{Ci}^{\beta}, w_{Ci}^{\gamma}] = [w_{Si}^{\beta}, w_{Si}^{\gamma}], \quad i = 1, \ldots, n, \quad (4.102)$$

occurs only when $m_{ij}^{\beta} = m_{ij}^{\gamma}$ *for all* $i, j = 1, \ldots, n$.

Proof First, let us demonstrate the validity of $[w_{Ci}^{\alpha}, w_{Ci}^{\delta}] \subset [w_{Si}^{\alpha}, w_{Si}^{\delta}]$, $i = 1 \ldots, n$. Presence of at least one fuzzy number $\widetilde{m}_{kl} = (m_{kl}^{\alpha}, m_{kl}^{\beta}, m_{kl}^{\gamma}, m_{kl}^{\delta})$ that is not a crisp number, i.e. $m_{kl}^{\alpha} < m_{kl}^{\delta}$, implies the strict inequalities

$$\sum_{\substack{k=1 \\ k \neq i}}^{n} \sqrt[n]{\prod_{j=1}^{n} m_{kj}^{\alpha}} < \sum_{\substack{k=1 \\ k \neq i}}^{n} \sqrt[n]{\frac{1}{m_{ik}^{\alpha}} \prod_{\substack{l=1 \\ l \neq i}}^{k-1} \frac{1}{m_{lk}} \prod_{\substack{l=k+1 \\ l \neq i}}^{n} m_{kl}} < \sum_{\substack{k=1 \\ k \neq i}}^{n} \sqrt[n]{\prod_{j=1}^{n} m_{kj}^{\delta}}, \quad \begin{array}{l} m_{kl} \in [m_{kl}^{\alpha}, m_{kl}^{\delta}], \\ m_{lk} = \frac{1}{m_{kl}}. \end{array}$$

$$(4.103)$$

Thus, for any $i \in \{1, \ldots, n\}$:

$$w_{Si}^{\alpha} \overset{(4.86)}{=} \frac{\sqrt[n]{\prod_{j=1}^{n} m_{ij}^{\alpha}}}{\sqrt[n]{\prod_{j=1}^{n} m_{ij}^{\delta}} + \sum_{k=1,\,k\neq i}^{n} \sqrt[n]{\prod_{j=1}^{n} m_{kj}^{\delta}}} \overset{(4.103)}{<}$$

$$\frac{\sqrt[n]{\prod_{j=1}^{n} m_{ij}^{\alpha}}}{\sqrt[n]{\prod_{j=1}^{n} m_{ij}^{\alpha}} + \max \left\{ \sum_{\substack{k=1 \\ k\neq i}}^{n} \sqrt[n]{\frac{1}{m_{ik}^{\alpha}} \prod_{\substack{l=1 \\ l\neq i}}^{k-1} \frac{1}{m_{lk}} \prod_{\substack{l=k+1 \\ l\neq i}}^{n} m_{kl}} \; ; \; \begin{array}{l} m_{rs} \in \left[m_{rs}^{\alpha}, m_{rs}^{\beta} \right], \\ r = 1, \ldots, n-1, \\ s = r+1, \ldots, n, \\ r, s \neq i \end{array} \right\}} \overset{(4.97)}{=} w_{Ci}^{\alpha}$$

and

$$w_{Si}^{\delta} \overset{(4.88)}{=} \frac{\sqrt[n]{\prod_{j=1}^{n} m_{ij}^{\delta}}}{\sqrt[n]{\prod_{j=1}^{n} m_{ij}^{\alpha}} + \sum_{k=1,\,k\neq i}^{n} \sqrt[n]{\prod_{j=1}^{n} m_{kj}^{\alpha}}} \overset{(4.103)}{>}$$

$$\frac{\sqrt[n]{\prod_{j=1}^{n} m_{ij}^{\delta}}}{\sqrt[n]{\prod_{j=1}^{n} m_{ij}^{\delta}} + \min \left\{ \sum_{\substack{k=1 \\ k\neq i}}^{n} \sqrt[n]{\frac{1}{m_{ik}^{\delta}} \prod_{\substack{l=1 \\ l\neq i}}^{k-1} \frac{1}{m_{lk}} \prod_{\substack{l=k+1 \\ l\neq i}}^{n} m_{kl}} \; ; \; \begin{array}{l} m_{rs} \in \left[m_{rs}^{\alpha}, m_{rs}^{\beta} \right], \\ r = 1, \ldots, n-1, \\ s = r+1, \ldots, n, \\ r, s \neq i \end{array} \right\}} \overset{(4.99)}{=} w_{Ci}^{\delta}.$$

Analogously, validity of $[w_{Ci}^{\beta}, w_{Ci}^{\gamma}] \subseteq [w_{Si}^{\beta}, w_{Si}^{\gamma}]$, $i = 1, \ldots, n$, is demonstrated.

Further, let us prove the validity of (4.102). The equality $m_{ij}^{\beta} = m_{ij}^{\gamma}$, $i, j = 1, \ldots, n$, means that the trapezoidal FMPCM $\tilde{M}$ is reduced to a triangular FMPCM. The domains of all variables in the formulas (4.92) and (4.93) are thus reduced to singletons. This basically means that the formula (4.92) is reduced to the formula (4.87) and the formula (4.93) is reduced to the formula (4.88). Thus, $w_{Ci}^{\beta} = w_{Si}^{\beta}$ and $w_{Ci}^{\gamma} = w_{Si}^{\gamma}$, $i = 1, \ldots, n$. Furthermore, the formulas (4.87) and (4.88) give the same results since they have the same entries $m_{ij}^{\beta} = m_{ij}^{\gamma}$. Thus, $w_{Ci}^{\beta} = w_{Si}^{\beta} = w_{Ci}^{\gamma} = w_{Si}^{\gamma}$, $i = 1, \ldots, n$. $\qquad\qquad\square$

4.3   Fuzzy Additive Pairwise Comparison Matrices

Similarly as in the case of MPCMs, also the PCs in APCMs are done either expertly or by using linguistic terms from a predefined scale. Especially when information about the decision-making problem is incomplete or imprecise it is more appropriate to provide the expert numerical judgments in form of fuzzy numbers rather than crisp numbers. The use of fuzzy numbers is more natural also when modeling the meaning of linguistic terms from a predefined scale (see the discussion in Sect. 4.2.1). Thus, in this section, the fuzzy extension of APCMs is dealt with. In particular, in Sect. 4.3.1, the construction of fuzzy APCMs is studied and the fuzzy extension of APCMs-A and APCMs-M is defined. Fuzzy APCMs-A are then studied in detail in Sect. 4.3.2, and Sect. 4.3.3 is focused on fuzzy APCMs-M.

4.3.1 Construction of FAPCMs

In Sect. 4.2, the fuzzy extension of MPCMs was dealt with. As it is obvious from the literature review provided in Sect. 4.2, MPCMs are most often extended to fuzzy numbers (usually triangular or trapezoidal), less often to intervals. That is probably because fuzzy numbers allow for more subtle modeling of preference intensities by using different degrees of membership. Intervals do not provide this option; all elements in the given interval have the same degree of membership.

The situation is different with APCMs. In the literature, the trend is to extend APCMs to intervals rather than to fuzzy numbers in general. The reason behind this might be the fact that APCMs are already perceived as fuzzy having the elements defined on interval [0, 1]. Thus, the intensity of preference of each PC in the APCM is looked at as a degree of membership to a fuzzy set, and modeling a degree of membership by a fuzzy number is not a common practice.

As already mentioned in Sect. 2.3.1, APCMs are often called "*fuzzy* preference relations" in the literature. Therefore, another reason why APCMs are usually extended to intervals rather than to fuzzy numbers in general might be the issue of terminology. Interval extension of a fuzzy preference relation is simply called "interval fuzzy preference relation" which probably seems to be acceptable (something "fuzzy" is extended to intervals). Extending fuzzy preference relations to fuzzy numbers sounds somehow wrong; "*fuzzy* preference relations" are already *fuzzy* and by extending them to fuzzy numbers we would obtain "*fuzzy fuzzy* preference relations"?

As shown in Sect. 2.4, MPCMs, APCMs-A, and APCMs-M are equivalent; each representation can be transformed into the others. This means that an APCM defined on interval [0, 1] can be transformed into a MPCM defined on interval $[\frac{1}{9}, 9]$ (or $[\frac{1}{S}, S]$, $S > 1$, in general) and vice versa. It will be shown later that there exist transformations also between the fuzzy extensions of MPCMs, APCMs-A, and APCMs-M. Therefore, using fuzzy numbers for one representation of preference information (in this case MPCMs), it does not seem reasonable to avoid using fuzzy numbers for other representations of preference information (APCMs-A and APCMs-M). If there is a need for more subtle modeling of preference intensities in MPCMs by using fuzzy numbers, why should these not be used for modeling preference intensities in APCMs as well? Why should the imprecision of information and vagueness of human judgment be modeled solely by intervals rather than by fuzzy numbers in general in the case of APCMs?

In this section, the methods related to APCMs will be extended not only to interval FAPCMs but, more in general, to fuzzy APCMs.

Definition 52 A fuzzy additive pairwise comparison matrix (FAPCM) of n objects $o_1, \ldots, o_n$ is a square matrix $\widetilde{A} = \{\widetilde{a}_{ij}\}_{i,j=1}^{n}$, whose elements $\widetilde{a}_{ij}, i, j = 1, \ldots, n$, are fuzzy numbers defined on interval $[0, 1]$. Further, the matrix is additively reciprocal, i.e.,

$$\widetilde{a}_{ij} = 1 - \widetilde{a}_{ji}, \qquad i, j = 1, \ldots, n, \tag{4.104}$$

and

$$\widetilde{a}_{ii} = 0.5, \qquad i, j = 1, \ldots, n. \tag{4.105}$$

Definition 52 of a FAPCM is very general; elements $\widetilde{a}_{ij}$ of $\widetilde{A}$ are arbitrary fuzzy numbers satisfying the additive-reciprocity condition (4.104). In practice, these fuzzy numbers are usually intervals, less often triangular or trapezoidal fuzzy numbers. In general, more types of fuzzy numbers and intervals can be present in a FAPCM at the same time. Even an APCM $A = \{a_{ij}\}_{i,j=1}^{n}$ given by Definition 7 is a FAPCM since crisp numbers are a special case of fuzzy numbers and since an APCM A satisfies (4.105).

As already mentioned above, intervals are most often used in the literature to model intensities of preference in FAPCMs. Interval representation will be preserved when reviewing the existing approaches to interval extension of APCMs. However, in order to be coherent with Sect. 4.2 on the fuzzy extension of MPCMs, the new approaches and formulas related to the fuzzy extension of APCMs proposed in this book will be presented for trapezoidal fuzzy numbers. Since intervals are a special case of trapezoidal fuzzy numbers, there will be no difficulties in confronting the new approaches and formulas presented for trapezoidal fuzzy numbers with the original approaches and formulas presented for intervals.

Being $\widetilde{a}_{ij} = (a_{ij}^{\alpha}, a_{ij}^{\beta}, a_{ij}^{\gamma}, a_{ij}^{\delta})$ a trapezoidal fuzzy number, then also $\widetilde{a}_{ji} = 1 - \widetilde{a}_{ij}$ is a trapezoidal fuzzy number. Thus, unlike in the case of FMPCMs, there is no need for simplified fuzzy arithmetic when constructing a FAPCM $\widetilde{A}$.

When constructing a FAPCM by using a scale of predefined linguistic terms it is necessary to model appropriately the meaning of the linguistic term "equal preference". Similarly as in the case of FMPCMs we have to distinguish whether o_i and o_j are the same objects or not. For $i = j$ it is necessary to set $\widetilde{a}_{ii} = 0.5, i = 1, \ldots, n$, because there is no vagueness in the PC; we compare one object with itself. On the other hand, when two different objects are assessed as "equally preferred", there is very likely some vagueness contained in such a PC. Therefore, in this case, "equal preference" should be modeled by a fuzzy number "about 0.5", i.e. $\widetilde{0.5}$, not necessarily by crisp number 0.5. Furthermore, similarly as in the case of FMPCMs, it is again necessary to preserve $\widetilde{0.5} = 1 - \widetilde{0.5}$ (a detailed reasoning behind this requirement is given in Sect. 4.2.1). Therefore, the fuzzy number $\widetilde{0.5} := (c^\alpha, c^\beta, c^\gamma, c^\delta)$ has to satisfy

$$(c^\alpha, c^\beta, c^\gamma, c^\delta) = 1 - (c^\alpha, c^\beta, c^\gamma, c^\delta). \qquad (4.106)$$

By solving (4.106), we obtain $\widetilde{0.5}$ defined as

$$\widetilde{0.5} = (1 - c, 1 - b, b, c), \qquad 0.5 \leq b \leq c \leq 1. \qquad (4.107)$$

Note 8 *From now on, by a FAPCM will be meant a FAPCM given by Definition 52 satisfying the indispensable condition $\widetilde{0.5} = 1 - \widetilde{0.5}$ for the fuzzy number $\widetilde{0.5}$ modeling the meaning of the linguistic term "equal preference", i.e., (4.107) in the case of trapezoidal fuzzy numbers.*

Similarly as in the case of FMPCMs, also for FAPCMs it seems to be reasonable to define the fuzzy numbers modeling the meanings of linguistic terms in a predefined scale in such a way that they form Ruspini's fuzzy partition of interval [0, 1]. Defining a particular fuzzy scale is, however, out of the scope of the book as this should be done in cooperation with the particular DM. In the illustrative examples in this book only expertly defined fuzzy PCs without assigned linguistic terms will be used.

Analogously to APCMs, also in the case of FAPCMs it is necessary to distinguish between FAPCMs with additive representation and FAPCMs with multiplicative representation, depending on the representation used when constructing a FAPCM.

Definition 53 A FAPCM with additive representation (FAPCM-A) is a FAPCM $\widetilde{R} = \{\widetilde{r}_{ij}\}_{i,j=1}^{n}$, where $\widetilde{r}_{ij} - \widetilde{r}_{ji}$ indicates the difference of preference intensity of object o_i and of object o_j.

Definition 54 A FAPCM with multiplicative representation (FAPCM-M) is a FAPCM $\widetilde{Q} = \{\widetilde{q}_{ij}\}_{i,j=1}^{n}$, $\widetilde{q}_{ij} \in]0, 1[$, where $\frac{\widetilde{q}_{ij}}{\widetilde{q}_{ji}}$ indicates the ratio of preference intensity of object o_i to that of object o_j, i.e., o_i is $\frac{\widetilde{q}_{ij}}{\widetilde{q}_{ji}}$ —times as good as o_j.

4.3.2 Fuzzy Additive Pairwise Comparison Matrices with Additive Representation

In this Section, the fuzzy extension of the methods related to APCMs-A reviewed in Sect. 2.3.2 is dealt with. In particular, Sect. 4.3.2.1 is dedicated to the extension of additive-consistency condition (2.28) to FAPCMs-A, and Sect. 4.3.2.2 is focused on methods for obtaining fuzzy priorities from FAPCMs-A.

4.3.2.1 Additive Consistency of FAPCMs-A

In this section, additive consistency of FAPCMs-A is studied. First, in Sect. "Review of Fuzzy Extensions of Additive Consistency", definitions of additive consistency for interval FAPCMs-A based on Tanino's characterization (2.32) proposed in the literature are reviewed and some drawbacks of the definitions are pointed out. Afterwards, in Sect."New Fuzzy Extension of Additive Consistency", two new definitions of additive consistency for FAPCMs-A are proposed.

Review of Fuzzy Extensions of Additive Consistency

Many definitions of consistency for interval FAPCMs-A have been proposed in the literature. These definitions are most often based on interval extension of additive-transitivity property (2.28) and related Tanino's characterization (2.32), or alternatively, on interval extension of the more general characterization (2.46) reviewed in Remark 2 of Sect. 2.3.2.2.

Xu (2007b) introduced a weak version of additive consistency for interval FAPCMs-A based on Tanino's characterization (2.32). Wang and Li (2012) proposed another definition of additive consistency based on Tanino's additive-transitivity property (2.28). Qian et al. (2014) proposed a method for constructing additively consistent interval FAPCMs-A satisfying definition proposed by Wang and Li (2012) from inconsistent interval FAPCMs-A. Another definition of additive consistency for interval FAPCMs-A based on Tanino's additive-transitivity property (2.28) was proposed by Liu et al. (2012a). Xu et al. (2014a) proposed another definition of additive consistency for interval FAPCMs-A and introduced a method for completing incomplete interval FAPCMs-A based on this additive-consistency condition. Wang (2014) pointed out that the additive consistency defined by Xu et al. (2014a) is not invariant under permutation of objects. Further, they approved the definition of additive consistency proposed by Wang and Li (2012) and derived some further properties. Wang et al. (2012) introduced a definition of consistency using a particular characterization based on logarithms. However, the interpretation of such characterization based on logarithms was not clarified.

As discussed in Remark 2, priorities (2.47) corresponding to more general characterization (2.46) lack intuitive representation. Therefore, in this section, the focus is put only on traditional Tanino's characterization (2.32). Definitions of additively consistent interval FAPCMs-A based on interval extension of Tanino's characterization (2.32) are reviewed in detail and some drawbacks are identified, drawing from the critical review done by Krejčí (2017a). In particular, it will be shown that some definitions are not invariant under permutation of objects in interval FAPCMs-A and some violate additive reciprocity of the related PCs of objects. Afterwards, it will be shown that the drawbacks can be eliminated by employing the constrained fuzzy arithmetic.

Xu (2007b) defined the additive consistency of interval FAPCMs-A as follows:

Definition 55 (Xu 2007b) Let $\bar{R} = \{\bar{r}_{ij}\}_{i,j=1}^n$, $\bar{r}_{ij} = \left[r_{ij}^L, r_{ij}^U\right]$, be an interval FAPCM-A. If there exists a vector $\underline{v} = (v_1, \ldots, v_n)^T$, $\sum_{i=1}^n v_i = 1, v_i \geq 0, i = 1, \ldots, n$, such that

$$r_{ij}^L \leq 0.5(v_i - v_j + 1) \leq r_{ij}^U, \qquad i,j = 1, \ldots, n, \tag{4.108}$$

then $\bar{R}$ is called an additively consistent interval FAPCM-A.

Xu and Chen (2008a) formulated the following theorem in order to verify additive consistency of interval FAPCMs-A according to Definition 55.

Theorem 45 (Xu and Chen 2008a) *An interval FAPCM-A* $\bar{R} = \{\bar{r}_{ij}\}_{i,j=1}^n$, $\bar{r}_{ij} = \left[r_{ij}^L, r_{ij}^U\right]$, *is additively consistent according to Definition 55 if and only if the solution of the optimization model*

$$
\begin{aligned}
J^* = \min &\sum_{i=1}^{n-1} \sum_{j=i+1}^n (d_{ij}^- + d_{ij}^+) \\
s.t. \; &0.5(v_i - v_j + 1) + d_{ij}^L \geq r_{ij}^L, \; i,j = 1, \ldots, n, \; i < j, \\
&0.5(v_i - v_j + 1) - d_{ij}^U \leq r_{ij}^U, \; i,j = 1, \ldots, n, \; i < j, \\
&\sum_{i=1}^n v_i = 1, \; v_i \geq 0, \qquad\qquad\quad i = 1, \ldots, n, \\
&d_{ij}^L, d_{ij}^U \geq 0, \qquad\qquad\qquad\quad i,j = 1, \ldots, n, \; i < j,
\end{aligned}
\tag{4.109}
$$

is $J^* = 0$.

Definition 55 of additive consistency for interval FAPCMs-A is clearly based on Proposition 2 for APCMs-A. According to the definition, an interval FAPCM-A $\bar{R} = \{\bar{r}_{ij}\}_{i,j=1}^n$, $\bar{r}_{ij} = \left[r_{ij}^L, r_{ij}^U\right]$, is additively consistent if there exists a vector $\underline{v} = (v_1, \ldots, v_n)^T$ using which we can construct an additively consistent APCM-A $R^* = \{r_{ij}^*\}_{i,j=1}^n$, $r_{ij}^* = 0.5(v_i - v_j + 1)$, such that $r_{ij}^* \in [r_{ij}^L, r_{ij}^U], i,j = 1, \ldots, n$. Thus Definition 55 does not violate the additive reciprocity of the related PCs and is also invariant under permutation of objects. Nevertheless, extension of Proposition 2 to Definition 55 is not done appropriately. The vector $\underline{v} = (v_1, \ldots, v_n)^T$ in Proposition 2 is such that $|v_i - v_j| \leq 1, i,j = 1, \ldots, n$, while in Definition 55, the vector should satisfy the normalization condition

$$\sum_{i=1}^n v_i = 1, \; v_i \geq 0, \; i = 1, \ldots, n. \tag{4.110}$$

The same normalization condition is employed also in the optimization model (4.109).

The incompatibility of the normalization condition (4.110) and Tanino's characterization (2.32) for APCMs-A was demonstrated by Fedrizzi and Brunelli (2009). They showed that for some additively consistent APCMs-A, there exists no vector satisfying the normalization condition (4.110). The inappropriateness of Definition 55 of additive consistency is demonstrated on the following illustrative example.

Example 51 (Krejčí 2017a) Let us examine the interval FAPCM-A

$$\bar{R} = \begin{pmatrix} 0.5 & [0.5, 0.6] & [0.8, 0.9] & [0.9, 1] \\ [0.4, 0.5] & 0.5 & [0.6, 0.7] & [0.6, 0.9] \\ [0.1, 0.2] & [0.3, 0.4] & 0.5 & [0.5, 0.7] \\ [0, 0.1] & [0.1, 0.4] & [0.3, 0.5] & 0.5 \end{pmatrix}. \tag{4.111}$$

By solving the optimization model (4.109), we obtain $J^* = 0.1$. Thus, based on Theorem 45, the interval FAPCM-A $\bar{R}$ is not additively consistent according to Definition 55. This means that there does not exist an APCM-A $R^* = \left\{ r_{ij}^* \right\}_{i,j=1}^n$ obtainable from the interval FAPCM-A (4.111) that would be additively consistent according to Definition 9. However, this conclusion is wrong.

There do exist additively consistent APCMs-A obtainable from (4.111); one of them is, for example,

$$R^* = \begin{pmatrix} 0.5 & 0.6 & 0.8 & 1 \\ 0.4 & 0.5 & 0.7 & 0.9 \\ 0.2 & 0.3 & 0.5 & 0.7 \\ 0 & 0.1 & 0.3 & 0.5 \end{pmatrix} \tag{4.112}$$

with the corresponding priority vector $\underline{v} = (1.45, 1.25, 0.85, 0.45)^T$ satisfying $|v_i - v_j| \leq 1, i, j = 1, \ldots, n$. Because there does not exist a priority vector corresponding to the APCM-A R^* that would satisfy the inappropriate normalization condition (4.110), it could not have been revealed by solving the optimization model (4.109). Thus, by utilizing Definition 55 in real-life applications, a DM who provided the interval FAPCM-A (4.111) would be judged as inconsistent in his or her preferences even though the opposite is true.   △

Because the normalization condition (4.110) is not compatible with Tanino's characterization for APCMs-A, it is clearly not reasonable to employ it in the definition of additively consistent interval FAPCMs-A (APCMs-A being a special case of interval FAPCMs-A where $r_{ij}^L = r_{ij}^U$, $i, j = 1, \ldots, n$).

Liu et al. (2012a) defined additive consistency in the following way.

Definition 56 (Liu et al. 2012a) Let $\bar{R} = \left\{ \bar{r}_{ij} \right\}_{i,j=1}^n, \bar{r}_{ij} = \left[r_{ij}^L, r_{ij}^U \right]$, be an interval FAPCM-A. $\bar{R}$ is called additively consistent if the APCMs-A $C = \left\{ c_{ij} \right\}_{i,j=1}^n$ and $D = \left\{ d_{ij} \right\}_{i,j=1}^n$ such that

$$c_{ij} = \begin{cases} r_{ij}^L, & i < j \\ 0.5, & i = j \\ r_{ij}^U, & i > j \end{cases}, \quad d_{ij} = \begin{cases} r_{ij}^U, & i < j \\ 0.5, & i = j \\ r_{ij}^L, & i > j \end{cases}, \quad i, j = 1, \ldots, n, \quad (4.113)$$

are additively consistent according to Definition 9.

However, Definition 56 is not invariant under permutation of objects (Krejčí 2017a). This serious drawback is demonstrated on the following example.

Example 52 (Krejčí 2017a) Let us consider the interval FAPCM-A $\overline{R}$ in the form

$$\overline{R} = \begin{pmatrix} 0.5 & [0.6, 0.7] & [0.8, 1] \\ [0.3, 0.4] & 0.5 & [0.7, 0.8] \\ [0, 0.2] & [0.2, 0.3] & 0.5 \end{pmatrix}. \quad (4.114)$$

The corresponding APCMs-A C and D given by (4.113) are in the form

$$C = \begin{pmatrix} 0.5 & 0.6 & 0.8 \\ 0.4 & 0.5 & 0.7 \\ 0.2 & 0.3 & 0.5 \end{pmatrix}, \quad D = \begin{pmatrix} 0.5 & 0.7 & 1 \\ 0.3 & 0.5 & 0.8 \\ 0 & 0.2 & 0.5 \end{pmatrix}.$$

Both C and D satisfy the additive-transitivity condition (2.28), which means that they are additively consistent according to Definition 9. Therefore, according to Definition 56, the interval FAPCM-A $\overline{R}$ is additively consistent.

Now, let us permute the interval FAPCM-A (4.114) by using the permutation matrix

$$P = \begin{pmatrix} 0 & 0 & 1 \\ 1 & 0 & 0 \\ 0 & 1 & 0 \end{pmatrix} \quad (4.115)$$

to the interval FAPCM-A $\overline{R}^\pi = P\overline{R}P^T$:

$$\overline{R}^\pi = \begin{pmatrix} 0.5 & [0, 0.2] & [0.2, 0.3] \\ [0.8, 1] & 0.5 & [0.6, 0.7] \\ [0.7, 0.8] & [0.3, 0.4] & 0.5 \end{pmatrix}. \quad (4.116)$$

The corresponding APCMs-A C^π and D^π given by (4.113) are in the form

$$C^\pi = \begin{pmatrix} 0.5 & 0 & 0.2 \\ 1 & 0.5 & 0.6 \\ 0.8 & 0.4 & 0.5 \end{pmatrix}, \quad D^\pi = \begin{pmatrix} 0.5 & 0.2 & 0.3 \\ 0.8 & 0.5 & 0.7 \\ 0.7 & 0.3 & 0.5 \end{pmatrix}.$$

We can see that neither C^π nor D^π satisfies the additive-transitivity condition (2.28), and thus, according to Definition 56, the interval FAPCM-A $\overline{R}^\pi$ is not additively consistent.

Obviously, the PCs of objects in the interval FAPCMs-A $\overline{R}$ and $\overline{R}^{\pi}$ are the same; they are just provided in different orders. Therefore, also the conclusion about the additive consistency should be the same for both interval FAPCMs-A. However, as demonstrated, according to Definition 56, the interval FAPCM-A $\overline{R}$ results to be additively consistent while the interval FAPCM-A $\overline{R}^{\pi}$ results to be additively inconsistent.   △

Another definition of additive consistency for interval FAPCMs-A was proposed by Wang and Li (2012).

Definition 57 (Wang and Li 2012) Let $\overline{R} = \{\overline{r}_{ij}\}_{i,j=1}^{n}$, $\overline{r}_{ij} = \left[r_{ij}^{L}, r_{ij}^{U} \right]$, be an interval FAPCM-A. $\overline{R}$ is called additively consistent if it satisfies the additive-transitivity condition

$$\overline{r}_{ij} + \overline{r}_{jk} + \overline{r}_{ki} = \overline{r}_{kj} + \overline{r}_{ji} + \overline{r}_{ik}, \qquad i, j, k = 1, \ldots, n. \tag{4.117}$$

Wang and Li (2012) defined the additive-transitivity condition (4.117) by using the standard interval arithmetic, i.e., addition is done according to formula (3.22). Therefore, the Eq. (4.117) is nothing else but the equations

$$\begin{aligned} r_{ij}^{L} + r_{jk}^{L} + r_{ki}^{L} &= r_{kj}^{L} + r_{ji}^{L} + r_{ik}^{L}, \\ r_{ij}^{U} + r_{jk}^{U} + r_{ki}^{U} &= r_{kj}^{U} + r_{ji}^{U} + r_{ik}^{U}, \end{aligned} \qquad i, j, k = 1, \ldots, n. \tag{4.118}$$

Further, Wang and Li (2012) formulated the following proposition.

Proposition 11 (Wang and Li 2012) *Let $\overline{R} = \{\overline{r}_{ij}\}_{i,j=1}^{n}$ be an interval FAPCM-A. If there exists a normalized interval vector $\overline{v} = (\overline{v}_1, \ldots, \overline{v}_n)^T$, $\overline{v}_i = [v_i^L, v_i^U]$, $i = 1, \ldots, n$, such that*

$$\overline{r}_{ij} = [0.5(v_i^L - v_j^U + 1), 0.5(v_i^U - v_j^L + 1)],$$

then $\overline{R}$ is additively consistent according to Definition 57.

According to Proposition 11, a crisp APCM-A $R = \{r_{ij}\}_{i,j=1}^{n}$, which is a particular case of interval FAPCMs-A, is additively consistent according to Definition 2 if there exists a priority vector $\underline{v} = (v_1, \ldots, v_n)^T$ satisfying Tanino's characterization (2.32) and the normalization condition (4.110). However, the normalization condition (4.110) is incompatible with Tanino's Proposition 2 (see Fedrizzi and Brunelli 2009).

Furthermore, Wang and Li (2012) pointed out that "due to the fact that $\overline{a} - \overline{a}$ does not always yield 0, we cannot derive $\overline{r}_{ij} + \overline{r}_{ji} = 1$ any more" (Wang and Li 2012, p.183). For example, for $\overline{r}_{ij} = [0.7, 0.8]$, we obtain $\overline{r}_{ij} + \overline{r}_{ji} = [0.7, 0.8] + [0.2, 0.3] = [0.9, 1.1] \neq 1$. Wang and Li (2012) conclude that "due to the possibility of $\overline{a} - \overline{a} \neq 0$, which makes it impossible to manipulate an interval-valued equation by moving terms from one side to the other, (4.117) may not necessarily be able to produce equation

$$\overline{r}_{ij} = \overline{r}_{ik} - \overline{r}_{jk} + 0.5 \tag{4.119}$$

in contrast to the case of regular APCMs-A where these two expressions are equivalent" (Wang and Li 2012, p.183). This is why Wang and Li (2012) defined the additive consistency of interval FAPCMs-A (Definition 57) by extending the property (2.30) of APCMs-A.

However, it is necessary to point out here that Wang and Li's assertion that "we cannot derive $\bar{r}_{ij} + \bar{r}_{ji} = 1$" is not true. At the end of this section, it will be demonstrated that the validity of equation $\bar{r}_{ij} + \bar{r}_{ji} = 1$ can be easily achieved by applying appropriately the constrained fuzzy arithmetic instead of the standard fuzzy arithmetic (or equivalently by applying constrained interval arithmetic (Lodwick and Jenkins 2013) instead of the standard interval arithmetic in this case). Furthermore, it will be shown that Definition 57 of additive consistency is inappropriate since it violates the additive reciprocity of the related PCs. Before doing that, let us finalize the literature review.

Definition 58 (Xu et al. 2014a) Let $\bar{R} = \{\bar{r}_{ij}\}_{i,j=1}^{n}$, $\bar{r}_{ij} = \left[r_{ij}^{L}, r_{ij}^{U}\right]$, be an interval FAPCM-A. $\bar{R}$ is called an additively consistent interval FAPCM-A if

$$\bar{r}_{ij} + \bar{r}_{jk} = \bar{r}_{ik} + [0.5, 0.5], \quad i < j < k, \ i,j,k = 1, \ldots, n. \quad (4.120)$$

Wang (2014) demonstrated that Definition 58 of additive consistency is dependent on objects labeling, i.e., it is not invariant under permutation of objects. Further, Wang (2014) adopted Definition 57 of additive consistency since it is invariant under permutation of objects and derived the following theorem.

Theorem 46 (Wang 2014) *Let* $\bar{R} = \{\bar{r}_{ij}\}_{i,j=1}^{n}$, $\bar{r}_{ij} = \left[r_{ij}^{L}, r_{ij}^{U}\right]$, *be an interval FAPCM-A. $\bar{R}$ is additively consistent (according to Definition 57) if and only if*

$$r_{ij}^{L} + r_{ij}^{U} - (r_{ik}^{L} + r_{ik}^{U}) = r_{lj}^{L} + r_{lj}^{U} - (r_{lk}^{L} + r_{lk}^{U}), \quad i,j,k,l = 1, \ldots, n. \quad (4.121)$$

Based on Theorem 46, Wang (2014) formulated also the following proposition.

Proposition 12 (Wang 2014) *Let* $\bar{R} = \{\bar{r}_{ij}\}_{i,j=1}^{n}$, $\bar{r}_{ij} = \left[r_{ij}^{L}, r_{ij}^{U}\right]$, *be an interval FAPCM-A. $\bar{R}$ is additively consistent (according to Definition 57) if and only if*

$$r_{ij}^{L} + r_{ij}^{U} + r_{jk}^{L} + r_{jk}^{U} + r_{ki}^{L} + r_{ki}^{U} = 3, \quad i,j,k = 1, \ldots, n. \quad (4.122)$$

In order to demonstrate the inappropriateness of Definition 57 and thus also of Theorem 46 and Proposition 12, let us analyze in more detail the requirement of additive reciprocity of the related PCs in APCMs and interval FAPCMs. The following analysis is based on the original analysis of Krejčí (2017a). As stated in Sect. 2.3.1, the additive reciprocity of the related PCs is an inherent property of APCMs. Because of the additive-reciprocity property, the PCs r_{ii}, $i = 1, \ldots, n$, in an APCM are always equal to 0.5 standing for equal preference. This result is very natural since the PC r_{ii} expresses the intensity of preference of object o_i over itself (clearly, any object has to be equally preferred to itself). Further, because of the additive reciprocity of the related PCs, the additive-transitivity property (2.28) for APCMs-A is equivalent to statements (ii) and (iii) in Theorem 3.

Conception of additive reciprocity becomes more complicated when extended to intervals. For an interval FAPCM $\bar{R} = \{\bar{r}_{ij}\}_{i,j=1}^{n}$, $\bar{r}_{ij} = \left[r_{ij}^{L}, r_{ij}^{U}\right]$, the additive reciprocity is defined as $\bar{r}_{ji} = 1 - \bar{r}_{ij} = [1 - r_{ij}^{U}, 1 - r_{ij}^{L}]$. According to this property, when, e.g., the highest possible intensity of preference r_{ij}^{U} of object o_i over object o_j is $r_{ij}^{U} = 0.9$, this means that the lowest possible intensity of preference $r_{ji}^{L} = 1 - r_{ij}^{U}$ of object o_j over object o_i is automatically $r_{ji}^{L} = 0.1$. However, this is not all.

FAPCMs carry more information about the preference intensities. In particular, any value $r_{ij}^{*} \in \bar{r}_{ij} = \left[r_{ij}^{L}, r_{ij}^{U}\right]$ expressing a possible intensity of preference of object o_i over object o_j is associated with a corresponding intensity of preference $r_{ji}^{*} \in \bar{r}_{ji} = \left[r_{ji}^{L}, r_{ji}^{U}\right]$ such that $r_{ji}^{*} = 1 - r_{ij}^{*}$; r_{ij}^{*} and r_{ji}^{*} express the same preference information about o_i and o_j. This property results naturally from the meaning of PCs in an interval FAPCM. The same holds for trapezoidal FAPCMs in general.

Wang and Li (2012) and Wang (2014), similarly to other researchers whose work was reviewed in this section, applied the standard interval arithmetic to the computations with intervals. This means that the additive-consistency condition (4.117) is equivalent to Eqs. (4.118) which are equivalent to (4.121) and (4.122). However, the Eqs. (4.118), (4.121), and (4.122) do not preserve the additive reciprocity of the related PCs. In Eqs. (4.118), each intensity of preference r_{pq}^{L} or r_{pq}^{U}, $p, q = 1, \ldots, n$, always appears in a pair with the intensity of preference r_{qp}^{L} or r_{qp}^{U}, respectively. And since $r_{pq}^{L} = 1 - r_{qp}^{U} \neq 1 - r_{qp}^{L}$ (unless $r_{pq}^{L} = r_{pq}^{U}$), the additive reciprocity of the related PCs is violated. Similarly, in Eqs. (4.121) and (4.122), each intensity of preference r_{pq}^{L}, $p, q = 1, \ldots, n$, appears in a pair with the intensity of preference r_{pq}^{U}. This again violates the additive reciprocity since $r_{pq}^{U} = 1 - r_{qp}^{L}$ and $r_{pq}^{L} \neq 1 - r_{qp}^{L}$. The problem is for better understanding illustrated on the following example.

Example 53 (Krejčí 2017a) Let us examine the additive consistency of the interval FAPCM-A $\bar{R}$ given by (4.114) by applying Definition 57. The expressions (4.118) mean that we construct matrices $R^{L} = \left\{r_{ij}^{L}\right\}_{i=1}^{n}$ and $R^{U} = \left\{r_{ij}^{U}\right\}_{i=1}^{n}$ from the interval FAPCM-A (4.114) as

$$R^{L} = \begin{pmatrix} 0.5 & 0.6 & 0.8 \\ 0.3 & 0.5 & 0.7 \\ 0 & 0.2 & 0.5 \end{pmatrix}, \qquad R^{U} = \begin{pmatrix} 0.5 & 0.7 & 1 \\ 0.4 & 0.5 & 0.8 \\ 0.2 & 0.3 & 0.5 \end{pmatrix},$$

and we verify their additive consistency by utilizing the property (2.30). However, we can easily see that neither R^{L} nor R^{U} is additively reciprocal, which means that both R^{L} and R^{U} are not even APCMs-A according to Definition 8. Therefore, it is nonsensical to verify their "additive consistency".   △

For reader's convenience the properties regarding the reciprocity of the related PCs and invariance under permutations of the definitions of additive consistency reviewed in this section are listed in Table 4.9. Notice that Definition 55 of additive consistency proposed by Xu (2007b) for interval FAPCMs-A satisfies both the invariance under permutation and the additive reciprocity of the related PCs. Nevertheless, as demonstrated on p. 156, Definition 55 is inappropriate as it is not compatible with Tanino's characterization (2.32). All other reviewed definitions of additive consistency violate either the invariance under permutation or the additive reciprocity of the related PCs.

As already mentioned in the discussion following Definition 57, Wang and Li (2012) claim that it is not possible to derive the equality $\overline{r}_{ij} + \overline{r}_{ji} = 1$ due to the fact that $\overline{a} - \overline{a}$ does not always yield 0. Obviously, for an interval FAPCM-A $\overline{R} = \{\overline{r}_{ij}\}_{i,j=1}^{n}$, $\overline{r}_{ij} = [r_{ij}^{L}, r_{ij}^{U}]$, by applying the standard fuzzy arithmetic, we obtain

$$\overline{r}_{ij} + \overline{r}_{ji} = [r_{ij}^{L}, r_{ij}^{U}] + [r_{ji}^{L}, r_{ji}^{U}] = [r_{ij}^{L}, r_{ij}^{U}] + [1 - r_{ij}^{U}, 1 - r_{ij}^{L}] =$$
$$= [1 + r_{ij}^{L} - r_{ij}^{U}, 1 + r_{ij}^{U} - r_{ij}^{L}] \neq 1,$$

unless $r_{ij}^{L} = r_{ij}^{U}$. Similarly for $\overline{a} = [a^{L}, a^{U}]$, we obtain

$$\overline{a} - \overline{a} = [a^{L}, a^{U}] - [a^{L}, a^{U}] = [a^{L} - a^{U}, a^{U} - a^{L}] \neq 0, \qquad \text{unless } a^{L} = a^{U}.$$

However, applying the standard fuzzy arithmetic in this case is not appropriate; the constrained fuzzy arithmetic needs to be applied whenever there are any interactions among operands. Obviously, there is an interaction between interval PCs $\overline{r}_{ij}$ and $\overline{r}_{ji}$; from the additive reciprocity of the related PCs it follows that any intensity of preference $r_{ij}^{*} \in \overline{r}_{ij}$ of object o_i over object o_j is associated with the corresponding intensity of preference $r_{ji}^{*} = 1 - r_{ij}^{*}$ of object o_j over object o_i. Thus, $\overline{b} = \overline{r}_{ij} + \overline{r}_{ji}$ should be correctly computed according to the constrained fuzzy arithmetic (3.42) as $\overline{b} = [b^{L}, b^{U}]$:

$$b^{L} = \min\left\{r_{ij} + r_{ji}; r_{ij} \in [r_{ij}^{L}, r_{ij}^{U}], r_{ji} \in [r_{ji}^{L}, r_{ji}^{U}], r_{ji} = 1 - r_{ij}\right\} =$$
$$= \min\left\{r_{ij} + 1 - r_{ij}; r_{ij} \in [r_{ij}^{L}, r_{ij}^{U}]\right\} = 1,$$
$$b^{U} = \max\left\{r_{ij} + r_{ji}; r_{ij} \in [r_{ij}^{L}, r_{ij}^{U}], r_{ji} \in [r_{ji}^{L}, r_{ji}^{U}], r_{ji} = 1 - r_{ij}\right\} =$$
$$= \max\left\{r_{ij} + 1 - r_{ij}; r_{ij} \in [r_{ij}^{L}, r_{ij}^{U}]\right\} = 1,$$

Table 4.9 Properties of the definitions of additive consistency for FAPCMs-A

Definition	Authors	Invariance under permutation	Additive reciprocity
Definition 55	Xu (2007b)	✓	✓
Definition 56	Liu et al. (2012a)	✗	✓
Definition 57	Wang and Li (2012)	✓	✗
Definition 58	Xu et al. (2014a)	✗	✓

where the constraint $g(r_{ij}, r_{ji}) = 0$ is in the form $r_{ji} = 1 - r_{ij}$ (equivalently written as $r_{ij} + r_{ji} - 1 = 0$).

Analogously, $\overline{c} = \overline{a} - \overline{a}$ should be computed as $\overline{c} = [c^L, c^U]$:

$$
\begin{aligned}
c^L &= \min\left\{a_1 - a_2; a_1 \in [a^L, a^U], a_2 \in [a^L, a^U], a_1 = a_2\right\} = \\
&= \min\left\{a_1 - a_1; a_1 \in [a^L, a^U]\right\} = 0, \\
c^U &= \max\left\{a_1 - a_2; a_1 \in [a^L, a^U], a_2 \in [a^L, a^U], a_1 = a_2\right\} = \\
&= \max\left\{a_1 - a_1; a_1 \in [a^L, a^U]\right\} = 0.
\end{aligned}
$$

Keeping in mind the importance of the additive-reciprocity property of PCs in FAPCMs-A, additive consistency needs to be defined accordingly so that it does not violate the additive reciprocity. In the following section, two definitions of additively consistent trapezoidal FAPCMs-A respecting the additive reciprocity of the related PCs and invariant under permutation of objects are provided.

New Fuzzy Extension of Additive Consistency

In this section, additive weak consistency for trapezoidal FAPCMs-A is defined by removing the mistake in Definition 55 of additive consistency for interval FAPCMs-A. Further, another definition of additive consistency much stronger than the definition of additive weak consistency is proposed. Tools for verifying both the additive weak consistency and additive consistency are provided and some properties of both additively weakly consistent and additively consistent FAPCMs-A are derived. Both new definitions are invariant under permutation and preserve the additive reciprocity of the related PCs. The findings summarized in this section are mostly based on the original paper of Krejčí (2017a) on the additive consistency for interval FAPCMs-A.

An additively weakly consistent trapezoidal FAPCM-A is defined as follows.

Definition 59 Let $\widetilde{R} = \{\widetilde{r}_{ij}\}_{i,j=1}^n$, $\widetilde{r}_{ij} = \left(r_{ij}^\alpha, r_{ij}^\beta, r_{ij}^\gamma, r_{ij}^\delta\right)$, be a trapezoidal FAPCM-A. $\widetilde{R}$ is said to be additively weakly consistent if there exists a vector $\underline{v} = (v_1, \ldots, v_n)^T$, $|v_i - v_j| \le 1$, $i, j = 1, \ldots, n$, such that

$$
r_{ij}^\alpha \le 0.5(v_i - v_j + 1) \le r_{ij}^\delta, \qquad i, j = 1, \ldots, n. \tag{4.123}
$$

Definition 59 is obtained by correcting the error concerning the normalization of vector $\underline{v} = (v_1, \ldots, v_n)^T$ in Definition 55. As already mentioned in the previous section, the normalization condition (4.110) used in Definition 55 is not compatible with Tanino's characterization (2.32). Therefore, the original normalization condition $\underline{v} = (v_1, \ldots, v_n)^T$, $|v_i - v_j| \le 1$, $i, j = 1, \ldots, n$, used in Tanino's characterization was employed in Definition 59 instead. Definition 59 is confronted with Definition 55 in the following example.

Example 54 Let us examine the additive weak consistency of the interval FAPCM-A $\widetilde{R}$ given by (4.111). It was shown in Example 51 that $\widetilde{R}$ is wrongly judged as additively inconsistent according to Definition 55. By applying Definition 59, we can find out that there exists a vector $\underline{v} = (v_1, v_2, v_3, v_4)^T$ satisfying the appropriate normalization condition $|v_i - v_j| \leq 1, i, j = 1, \ldots, 4$. It is, for example, the vector $\underline{v} = (1.45, 1.25, 0.85, 0.45)^T$ with the corresponding additively consistent APCM-A given as (4.112). The reader can easily verify that this vector satisfies the inequalities (4.123). Thus, the interval FAPCM-A $\widetilde{R}$ given by (4.111) is correctly judged as additively weakly consistent. $\triangle$

Based on Proposition 2 for FAPCMs-A, the following proposition can be formulated for interval FAPCMs-A.

Proposition 13 *Let* $\widetilde{R} = \{\widetilde{r}_{ij}\}_{i,j=1}^n$, $\widetilde{r}_{ij} = \left(r_{ij}^\alpha, r_{ij}^\beta, r_{ij}^\gamma, r_{ij}^\delta\right)$, *be a trapezoidal FAPCM-A.* $\widetilde{R} = \{\widetilde{r}_{ij}\}_{i,j=1}^n$ *is additively weakly consistent according to Definition 59 if and only if there exist elements* $r_{ij}^* \in \left[r_{ij}^\alpha, r_{ij}^\delta\right]$, $i, j = 1, \ldots, n$, *such that* $R^* = \left\{r_{ij}^*\right\}_{i,j=1}^n$ *is an APCM-A additively consistent according to Definition 9.*

Proof First, let $\widetilde{R} = \{\widetilde{r}_{ij}\}_{i,j=1}^n$, $\widetilde{r}_{ij} = \left(r_{ij}^\alpha, r_{ij}^\beta, r_{ij}^\gamma, r_{ij}^\delta\right)$, be a trapezoidal FAPCM-A additively weakly consistent according to Definition 59. Let us denote $r_{ij}^* := 0.5(v_i - v_j + 1)$. From (4.123) it follows that $r_{ij}^* \in \left[r_{ij}^\alpha, r_{ij}^\delta\right]$, $i, j = 1, \ldots, n$. Further, we have $r_{ii}^* = 0.5(v_i - v_i + 1) = 0.5$, $i = 1, \ldots, n$, and $r_{ji}^* = 0.5(v_j - v_i + 1) = 1 - 0.5(v_i - v_j + 1) = 1 - r_{ij}^*$, $i, j = 1, \ldots, n$. From $[r_{ij}^\alpha, r_{ij}^\delta] \subseteq [0, 1], i, j = 1, \ldots, n$, it follows that $r_{ij}^* \in [0, 1]$, $i, j = 1, \ldots, n$. Therefore, $R^* = \left\{r_{ij}^*\right\}_{i,j=1}^n$ is an APCM-A. Finally, $r_{ik}^* + r_{kj}^* - 0.5 = 0.5(v_i - v_k + 1) + 0.5(v_k - v_j + 1) - 0.5 = 0.5(v_i - v_j + 1) = r_{ij}^*$, $i, j, k = 1, \ldots, n$, which means that $R^* = \left\{r_{ij}^*\right\}_{i,j=1}^n$ is additively consistent according to (2.28).

In the opposite direction, let $\widetilde{R} = \{\widetilde{r}_{ij}\}_{i,j=1}^n$, $\widetilde{r}_{ij} = \left(r_{ij}^\alpha, r_{ij}^\beta, r_{ij}^\gamma, r_{ij}^\delta\right)$, be a trapezoidal FAPCM-A, and let $R^* = \left\{r_{ij}^*\right\}_{i,j=1}^n$, $r_{ij}^* \in \left[r_{ij}^\alpha, r_{ij}^\delta\right]$, $i, j = 1, \ldots, n$, be an APCM-A additively consistent according to Definition 9. Then, from Proposition 2 it follows that there exists a vector $\underline{v} = (v_1, \ldots, v_n)^T$, $|v_i - v_j| \leq 1, i, j = 1, \ldots, n$, such that $r_{ij}^* = 0.5(v_i - v_j + 1), i, j = 1, \ldots, n$. Because, $r_{ij}^* \in \left[r_{ij}^\alpha, r_{ij}^\delta\right], i, j = 1, \ldots, n$, then clearly (4.123) holds. $\square$

Remark 20 According to Proposition 13 and its proof, a trapezoidal FAPCM-A $\widetilde{R} = \{\widetilde{r}_{ij}\}_{i,j=1}^n$, $\widetilde{r}_{ij} = \left(r_{ij}^\alpha, r_{ij}^\beta, r_{ij}^\gamma, r_{ij}^\delta\right)$, is additively weakly consistent if and only if there exists an additively consistent APCM-A $R^* = \left\{r_{ij}^*\right\}_{i,j=1}^n$ such that $r_{ij}^* \in \left[r_{ij}^\alpha, r_{ij}^\delta\right]$. This consistency condition is quite easy to reach. That is why the consistency according to Definition 59 is called weak. Later in this section also a much stronger definition of additive consistency for trapezoidal FAPCMs-A will be given.

Definition 59 of additive weak consistency satisfies two desirable properties—invariance under permutation of objects and additive reciprocity of the related PCs in interval FAPCMs-A.

Theorem 47 *Definition 59 of additive weak consistency is invariant under permutation of objects in trapezoidal FAPCMs-A.*

Proof There exists a priority vector $\underline{v} = (v_1, \ldots, v_n)^T$, $|v_i - v_j| \leq 1$, $i, j = 1, \ldots, n$, such that the inequality $r_{ij}^\alpha \leq 0.5(v_i - v_j + 1) \leq r_{ij}^\delta$ is required to hold for every single PC $\widetilde{r}_{ij} = \left(r_{ij}^\alpha, r_{ij}^\beta, r_{ij}^\gamma, r_{ij}^\delta \right)$ in an additively consistent trapezoidal FAPCM-A $\widetilde{R} = \left\{ \widetilde{r}_{ij} \right\}_{i,j=1}^n$. By permuting the FAPCM-A $\widetilde{R}$ to $\widetilde{R}^\pi = P\widetilde{R}P^T$, the original PC $\widetilde{r}_{ij}$ in the i–th row and in the j–th column of $\widetilde{R}$ is moved to the $\pi(i)$–th row and the $\pi(j)$–th column of the permuted trapezoidal FAPCM-A $\widetilde{R}^\pi$ as $\widetilde{r}_{\pi(i)\pi(j)}^\pi$, but still keeping $\widetilde{r}_{ij} = \widetilde{r}_{\pi(i)\pi(j)}^\pi$, $i, j = 1, \ldots, n$. Thus, there exists a vector $\underline{v}^\pi = (v_1^\pi, \ldots, v_n^\pi)^T$ obtained by permuting the vector $\underline{v}$, i.e. $\underline{v}^\pi = P\underline{v}$, with the components satisfying the inequalities $r_{ij}^{\pi\alpha} \leq 0.5(v_i^\pi - v_j^\pi + 1) \leq r_{ij}^{\pi\delta}$ as well as the normalization condition $|v_i^\pi - v_j^\pi| \leq 1$ for every $i, j = 1, \ldots, n$. $\square$

Theorem 48 *Definition 59 of additive weak consistency does not violate the additive reciprocity of the related PCs in a trapezoidal FAPCM-A $\widetilde{R} = \left\{ \widetilde{r}_{ij} \right\}_{i,j=1}^n$, $\widetilde{r}_{ij} = (r_{ij}^\alpha, r_{ij}^\beta, r_{ij}^\gamma, r_{ij}^\delta)$, in the sense that any fixed value $r_{ij} \in [r_{ij}^\alpha, r_{ij}^\delta]$, $i, j \in \{1, \ldots, n\}$, representing the intensity of preference of object o_i over object o_j is associated with the corresponding value $r_{ji} \in [r_{ji}^\alpha, r_{ji}^\delta]$ representing the intensity of preference of object o_j over object o_i such that $r_{ji} = 1 - r_{ij}$.*

Proof The existence of the priority vector $\underline{v} = (v_1, \ldots, v_n)^T$ satisfying the inequalities (4.123) means that there exists an APCM-A $R = \left\{ r_{ij} \right\}_{i,j=1}^n$, $r_{ij} \in [r_{ij}^\alpha, r_{ij}^\delta]$, such that $r_{ij} = 0.5(v_i - v_j + 1)$, $i, j = 1, \ldots, n$. R is additively reciprocal from the definition, i.e., every PC r_{ij} is associated with the PC r_{ji} such that $r_{ji} = 1 - r_{ij}$. $\square$

Remark 21 Note that Theorem 48 does not simply state that a FAPCM-A $\widetilde{R} = \left\{ \widetilde{r}_{ij} \right\}_{i,j=1}^n$ additively weakly consistent according to Definition 59 is additively reciprocal, i.e. $\widetilde{r}_{ji} = 1 - \widetilde{r}_{ij}$, $i, j = 1, \ldots, n$. The validity of this property automatically follows from Definition 52 of a FAPCM; every FAPCM is additively reciprocal, and thus, also a FAPCM-A that is additively weakly consistent according to Definition 59 is additively reciprocal.

As explained on p. 160, the extension of the additive-reciprocity property from APCMs to FAPCMs does not concern only the "simple" additive reciprocity of the related fuzzy PCs $\widetilde{r}_{ij}$ and $\widetilde{r}_{ji}$ in the sense that $\widetilde{r}_{ji} = 1 - \widetilde{r}_{ij}$, $i, j = 1, \ldots, n$. The conception of the additive reciprocity becomes more complex for FAPCMs. In particular, every possible intensity of preference $r_{ij}^* \in \widetilde{r}_{ij}$ of object o_i over object o_j is associated inseparably with the corresponding possible intensity of preference $r_{ji}^* \in \widetilde{r}_{ji}$ such that $r_{ji}^* = 1 - r_{ij}^*$ since both r_{ij}^* and r_{ji}^* have to express the same preference information about the objects o_i and o_j. Theorem 48 states that Definition 59 is in accordance with this conception of additive reciprocity, i.e., that only additively reciprocal PCs are involved in Definition 59 of additive weak consistency.

The following theorems provide useful tools for verifying additive weak consistency of trapezoidal FAPCMs-A.

Theorem 49 *A trapezoidal FAPCM-A* $\tilde{R} = \{\tilde{r}_{ij}\}_{i,j=1}^{n}$, $\tilde{r}_{ij} = \left(r_{ij}^{\alpha}, r_{ij}^{\beta}, r_{ij}^{\gamma}, r_{ij}^{\delta}\right)$, *is additively weakly consistent according to Definition 59 if and only if*

$$\max_{k=1,\ldots,n} \left\{ r_{ik}^{\alpha} + r_{kj}^{\alpha} - 0.5 \right\} \leq \min_{k=1,\ldots,n} \left\{ r_{ik}^{\delta} + r_{kj}^{\delta} - 0.5 \right\}, \qquad i,j = 1,\ldots,n. \quad (4.124)$$

Proof From the inequalities (4.123) it follows that $r_{ik}^{\alpha} \leq 0.5(v_i - v_k + 1) \leq r_{ik}^{\delta}$ and $r_{kj}^{\alpha} \leq 0.5(v_k - v_j + 1) \leq r_{kj}^{\delta}$. Thus, for every $k \in \{1, \ldots, n\}$:

$$r_{ik}^{\alpha} + r_{kj}^{\alpha} \leq 0.5(v_i - v_k + 1) + 0.5(v_k - v_j + 1) \leq r_{ik}^{\delta} + r_{kj}^{\delta}$$

$$r_{ik}^{\alpha} + r_{kj}^{\alpha} - 0.5 \leq 0.5(v_i - v_j + 1) \leq r_{ik}^{\delta} + r_{kj}^{\delta} - 0.5.$$

From this we obtain

$$\max_{k=1,\ldots,n} \left\{ r_{ik}^{\alpha} + r_{kj}^{\alpha} - 0.5 \right\} \leq 0.5(v_i - v_j + 1) \leq \min_{k=1,\ldots,n} \left\{ r_{ik}^{\delta} + r_{kj}^{\delta} - 0.5 \right\},$$

and thus (4.124) holds for every $i, j = 1, \ldots, n$.

In the opposite direction, let (4.124) hold. Then, for every $i, j, k \in \{1, \ldots, n\}$:

$$r_{ij}^{\alpha} \leq \max_{k=1,\ldots,n} \left\{ r_{ik}^{\alpha} + r_{kj}^{\alpha} - 0.5 \right\} \leq \min_{k=1,\ldots,n} \left\{ r_{ik}^{\delta} + r_{kj}^{\delta} - 0.5 \right\} \leq r_{ij}^{\delta}.$$

Thus, for every $i, j, k \in \{1, \ldots, n\}$:

$$\exists r_{ij}^{*} \in \left[\max_{k=1,\ldots,n} \left\{ r_{ik}^{\alpha} + r_{kj}^{\alpha} - 0.5 \right\}, \min_{k=1,\ldots,n} \left\{ r_{ik}^{\delta} + r_{kj}^{\delta} - 0.5 \right\} \right] \subseteq [r_{ij}^{\alpha}, r_{ij}^{\delta}]$$

$$\wedge \exists r_{ik}^{*} \in [r_{ik}^{\alpha}, r_{ik}^{\delta}] \wedge \exists r_{kj}^{*} \in [r_{kj}^{\alpha}, r_{kj}^{\delta}] : r_{ij}^{*} = r_{ik}^{*} + r_{kj}^{*} - 0.5.$$

This means that $R^{*} = \{r_{ij}^{*}\}_{i,j=1}^{n}$ is an APCM-A. Thus, according to Proposition 2, there exists a vector $\underline{v} = (v_1, \ldots, v_n)^T$, $|v_i - v_j| \leq 1$, $i, j = 1, \ldots, n$, such that $r_{ij}^{*} = 0.5(v_i - v_j + 1)$. Since $r_{ij}^{*} \in [r_{ij}^{\alpha}, r_{ij}^{\delta}]$, $i, j = 1, \ldots, n$, we obtain the inequality (4.123). $\qquad \square$

The following theorem shows that it is sufficient to verify the inequality (4.124) only for $i, j = 1, \ldots, n$, $i < j$, thus saving half of the computations.

Theorem 50 *A trapezoidal FAPCM-A* $\widetilde{R} = \left\{\widetilde{r}_{ij}\right\}_{i,j=1}^n$, $\widetilde{r}_{ij} = \left(r_{ij}^\alpha, r_{ij}^\beta, r_{ij}^\gamma, r_{ij}^\delta\right)$, *is additively weakly consistent according to Definition 59 if and only if*

$$\max_{k=1,\ldots,n} \left\{r_{ik}^\alpha + r_{kj}^\alpha - 0.5\right\} \leq \min_{k=1,\ldots,n} \left\{r_{ik}^\delta + r_{kj}^\delta - 0.5\right\}, \qquad i,j = 1,\ldots,n, \ i < j.$$
(4.125)

Proof It is sufficient to show that the validity of inequalities (4.125) for $i,j = 1,\ldots,n$, $i < j$, implies automatically their validity for all $i,j = 1,\ldots,n$, i.e. the validity of (4.124). The validity of inequalities (4.124) for $i = j$ is trivial from the definition of trapezoidal FAPCMs-A since

$$\max_{k=1,\ldots,n} \left\{r_{ik}^\alpha + r_{ki}^\alpha - 0.5\right\} = \max_{k=1,\ldots,n} \left\{r_{ik}^\alpha + 1 - r_{ik}^\delta - 0.5\right\} \leq 0.5 \leq$$

$$\min_{k=1,\ldots,n} \left\{r_{ik}^\delta + 1 - r_{ik}^\alpha - 0.5\right\} = \min_{k=1,\ldots,n} \left\{r_{ik}^\delta + r_{ki}^\delta - 0.5\right\}.$$

Further, for $i > j$, by using (4.125) and the additive-reciprocity properties, we obtain

$$\max_{k=1,\ldots,n} \left\{r_{ik}^\alpha + r_{kj}^\alpha - 0.5\right\} = \max_{k=1,\ldots,n} \left\{1 - r_{ki}^\delta + 1 - r_{jk}^\delta - 0.5\right\} =$$

$$1 - \min_{k=1,\ldots,n} \left\{r_{jk}^\delta + r_{ki}^\delta - 0.5\right\} \leq 1 - \max_{k=1,\ldots,n} \left\{r_{jk}^\alpha + r_{ki}^\alpha - 0.5\right\} =$$

$$\min_{k=1,\ldots,n} \left\{1 - 1 + r_{kj}^\delta - 1 + r_{ik}^\delta + 0.5\right\} = \min_{k=1,\ldots,n} \left\{r_{ik}^\delta + r_{kj}^\delta - 0.5\right\}.$$

$\square$

Remark 22 An alternative definition of additive weak consistency to Definition 59 might be formulated as follows.

Let $\widetilde{R} = \left\{\widetilde{r}_{ij}\right\}_{i,j=1}^n$, $\widetilde{r}_{ij} = \left(r_{ij}^\alpha, r_{ij}^\beta, r_{ij}^\gamma, r_{ij}^\delta\right)$, be a trapezoidal FAPCM-A. $\widetilde{R}$ is said to be additively weakly consistent if there exists a vector $\underline{v} = (v_1, \ldots, v_n)^T$, $|v_i - v_j| \leq 1$, $i,j = 1,\ldots,n$, such that

$$r_{ij}^\beta \leq 0.5(v_i - v_j + 1) \leq r_{ij}^\gamma, \qquad i,j = 1,\ldots,n.$$
(4.126)

Notice that this definition is stronger than Definition 59. In fact, every trapezoidal FAPCM-A additively weakly consistent according to this definition is also additively weakly consistent according to Definition 59 since (4.126) automatically implies (4.123).

All theorems regarding FAPCMs-A additively weakly consistent according to Definition 59 formulated above can be easily reformulated for FAPCMs-A additively weakly consistent according to this definition; it is sufficient to consider r_{ij}^β and r_{ij}^γ instead of r_{ij}^α and r_{ij}^δ, respectively, where appropriate.

In the following definition, a stronger version of additive consistency for trapezoidal FAPCMs-A is formulated.

Definition 60 Let $\widetilde{R} = \{\widetilde{r}_{ij}\}_{i,j=1}^n, \widetilde{r}_{ij} = \left(r_{ij}^\alpha, r_{ij}^\beta, r_{ij}^\gamma, r_{ij}^\delta\right)$, be a trapezoidal FAPCM-A. $\widetilde{R}$ is said to be additively consistent if for each triplet $(i, j, k) \subseteq \{1, \ldots, n\}$ the following holds:

$$\forall r_{ij} \in \left[r_{ij}^\alpha, r_{ij}^\delta\right] \exists r_{ik} \in \left[r_{ik}^\alpha, r_{ik}^\delta\right] \wedge \exists r_{kj} \in \left[r_{kj}^\alpha, r_{kj}^\delta\right] : r_{ij} = r_{ik} + r_{kj} - 0.5, \quad (4.127)$$

$$\forall r_{ij} \in \left[r_{ij}^\beta, r_{ij}^\gamma\right] \exists r_{ik} \in \left[r_{ik}^\beta, r_{ik}^\gamma\right] \wedge \exists r_{kj} \in \left[r_{kj}^\beta, r_{kj}^\gamma\right] : r_{ij} = r_{ik} + r_{kj} - 0.5. \quad (4.128)$$

Remark 23 Definition 60 is a natural fuzzy extension of Definition 9 of additive consistency proposed by Tanino (1984). According to this definition, for any possible value $r_{ij} \in \widetilde{r}_{ij}, i, j \in \{1, \ldots, n\}$, there exist possible values $r_{ik} \in \widetilde{r}_{ik}$ and $r_{kj} \in \widetilde{r}_{kj}, k \in \{1, \ldots, n\}$, such that they satisfy the additive-transitivity property (2.28). Analogously, for any possible value $r_{ij} \in Core\,\widetilde{r}_{ij}, i, j \in \{1, \ldots, n\}$, there exist possible values $r_{ik} \in Core\,\widetilde{r}_{ik}$ and $r_{kj} \in Core\,\widetilde{r}_{kj}, k \in \{1, \ldots, n\}$, such that they satisfy (2.28). Clearly, in comparison to the additive weak consistency given by Definition 59, the additive consistency given by Definition 60 is very strong.

Unlike Definitions 56 and 58 of additively consistent interval FAPCMs-A proposed by Liu et al. (2012a) and Xu et al. (2014a), respectively, new Definition 60 is invariant under permutation of objects compared in FAPCMs-A.

Theorem 51 *Definition 60 of additive consistency is invariant under permutation of objects in trapezoidal FAPCMs-A.*

Proof For a trapezoidal FAPCM-A $\widetilde{R} = \{\widetilde{r}_{ij}\}_{i,j=1}^n, \widetilde{r}_{ij} = (r_{ij}^\alpha, r_{ij}^\beta, r_{ij}^\gamma, r_{ij}^\delta)$, additively consistent according to Definition 60, the conditions (4.127) and (4.128) are satisfied for every triplet $(i, j, k) \subseteq \{1, \ldots, n\}$. By permuting the FAPCM-A $\widetilde{R}$ to $\widetilde{R}^\pi = P\widetilde{R}P^T$, the original PC $\widetilde{r}_{ij}$ in the i–th row and in the j–th column of $\widetilde{R}$ moves to the $\pi(i)$–th row and the $\pi(j)$–th column of $\widetilde{R}^\pi$ preserving $\widetilde{r}_{\pi(i)\pi(j)}^\pi = \widetilde{r}_{ij}$. Thus, by permuting $\widetilde{R}$ also the validity of the conditions (4.127) and (4.128) is preserved, i.e.,

$$\forall r_{ij}^\pi \in \left[r_{ji}^{\pi\alpha}, r_{ij}^{\pi\delta}\right] \exists r_{ik}^\pi \in \left[r_{ik}^{\pi\alpha}, r_{ik}^{\pi\delta}\right] \wedge \exists r_{kj}^\pi \in \left[r_{kj}^{\pi\alpha}, r_{kj}^{\pi\delta}\right] : r_{ij}^\pi = r_{ik}^\pi + r_{kj}^\pi - 0.5,$$

$$\forall r_{ij}^\pi \in \left[r_{ji}^{\pi\beta}, r_{ij}^{\pi\gamma}\right] \exists r_{ik}^\pi \in \left[r_{ik}^{\pi\beta}, r_{ik}^{\pi\gamma}\right] \wedge \exists r_{kj}^\pi \in \left[r_{kj}^{\pi\beta}, r_{kj}^{\pi\gamma}\right] : r_{ij}^\pi = r_{ik}^\pi + r_{kj}^\pi - 0.5,$$

for every triplet $(i, j, k) \in \{1, \ldots, n\}$. Thus, $\widetilde{R}^\pi$ is additively consistent according to Definition 60. $\qquad\square$

Further, unlike Definition 57 of additively consistent interval FAPCMs-A proposed by Wang and Li (2012) and the equivalent conditions derived by Wang (2014), Definition 60 does not violate the additive reciprocity of the related PCs.

Theorem 52 *Definition 60 of additive consistency preserves the additive reciprocity of the related PCs in trapezoidal FAPCMs-A in the sense that any fixed value $r_{ij} \in [r_{ij}^\alpha, r_{ij}^\delta], i, j \in \{1, \ldots, n\}$, representing the intensity of preference of object o_i over object o_j is associated with the corresponding value $r_{ji} \in [r_{ji}^\alpha, r_{ji}^\delta]$ representing the intensity of preference of object o_j over object o_i such that $r_{ji} = 1 - r_{ij}$.*

Proof It is sufficient to show that expressions (4.127) and (4.128) do not violate the additive-reciprocity property in the sense that when two particular intensities of preference $r_{ij} \in \tilde{r}_{ij}$ and $r_{ji} \in \tilde{r}_{ji}$ on the pair of objects o_i and o_j are considered at the same time in the expressions (4.127) and (4.128), then they are such that $r_{ji} = 1 - r_{ij}$.

For a triplet $(i, j, k) \subseteq \{1, \ldots, n\}$, $i \neq j \neq k$, no reciprocals appear in the expression $r_{ij} = r_{ik} + r_{kj} - 0.5$ for any $r_{ij} \in [r_{ij}^\alpha, r_{ij}^\delta]$. For $i = j = k$, the expression (4.127) reduces to: $\forall r_{ii} = 0.5 \, \exists r_{ii}^* = 0.5 \wedge \exists r_{ii}^{**} = 0.5$: $0.5 = 0.5 + 0.5 - 0.5$, which again does not violate the additive reciprocity. Further, for $i \neq j = k$, the expression (4.127) is as: $\forall r_{ij} \in [r_{ij}^\alpha, r_{ij}^\delta] \, \exists r_{ij}^* \in [r_{ij}^\alpha, r_{ij}^\delta] \wedge \exists r_{jj} = 0.5$: $r_{ij} = r_{ij}^* + 0.5 - 0.5$. This means that $r_{ij}^* = r_{ij}$, and therefore, the additive reciprocity is not violated. For $i = k \neq j$ the proof is analogous. Finally, for $i = j \neq k$, the expression (4.127) is as:

$$\forall r_{ij} = 0.5 \, \exists r_{ik} \in [r_{ik}^\alpha, r_{ik}^\delta] \wedge \exists r_{ki}^* \in [r_{ki}^\alpha, r_{ki}^\delta]: \quad 0.5 = r_{ik} + r_{ki}^* - 0.5.$$

This means that $r_{ik} = 1 - r_{ki}^*$, and thus, the additive reciprocity is again preserved.

The proof for the expression (4.128) is analogous. $\square$

Remark 24 Similarly to Theorem 48, also Theorem 52 does not simply state that a FAPCM-A $\tilde{R} = \{\tilde{r}_{ij}\}_{i,j=1}^n$ additively consistent according to Definition 60 is additively reciprocal since this property automatically follows from Definition 52 of a FAPCM. Theorem 52 states that Definition 60 is in accordance with the conception of additive reciprocity discussed on p. 160, i.e., only additively reciprocal PCs are involved in Definition 60 of additive consistency. For more details, see Remark 21.

By handling properly the additive-reciprocity property of PCs, Theorem 3 can be easily extended to trapezoidal FAPCMs-A as follows.

Theorem 53 *For a trapezoidal FAPCM-A $\tilde{R} = \{\tilde{r}_{ij}\}_{i,j=1}^n$, $\tilde{r}_{ij} = \left(r_{ij}^\alpha, r_{ij}^\beta, r_{ij}^\gamma, r_{ij}^\delta\right)$, the following statements are equivalent:*

(i) $\tilde{R}$ is additively consistent according to Definition 60.

(ii) For every $i, j, k = 1, \ldots, n$:

$$\forall r_{ij} \in \left[r_{ij}^\alpha, r_{ij}^\delta\right] \, \exists r_{jk} \in \left[r_{jk}^\alpha, r_{jk}^\delta\right] \wedge \exists r_{ki} \in \left[r_{ki}^\alpha, r_{ki}^\delta\right]: \quad r_{ij} + r_{jk} + r_{ki} = r_{ik} + r_{kj} + r_{ji},$$

$$r_{ij} = 1 - r_{ji}, \quad r_{jk} = 1 - r_{kj}, \quad r_{ki} = 1 - r_{ik}. \quad (4.129)$$

$$\forall r_{ij} \in \left[r_{ij}^{\beta}, r_{ij}^{\gamma} \right] \exists r_{jk} \in \left[r_{jk}^{\beta}, r_{jk}^{\gamma} \right] \ \wedge \ \exists r_{ki} \in \left[r_{ki}^{\beta}, r_{ki}^{\gamma} \right] : \ r_{ij} + r_{jk} + r_{ki} = r_{ik} + r_{kj} + r_{ji},$$

$$r_{ij} = 1 - r_{ji}, \ r_{jk} = 1 - r_{kj}, \ r_{ki} = 1 - r_{ik}. \quad (4.130)$$

(iii) For every $i, j, k = 1, \ldots, n$:

$$\forall r_{ij} \in \left[r_{ij}^{\alpha}, r_{ij}^{\delta} \right] \exists r_{jk} \in \left[r_{jk}^{\alpha}, r_{jk}^{\delta} \right] \wedge \exists r_{ki} \in \left[r_{ki}^{\alpha}, r_{ki}^{\delta} \right] : r_{ij} + r_{jk} + r_{ki} = \frac{3}{2}, \quad (4.131)$$

$$\forall r_{ij} \in \left[r_{ij}^{\beta}, r_{ij}^{\gamma} \right] \exists r_{jk} \in \left[r_{jk}^{\beta}, r_{jk}^{\gamma} \right] \wedge \exists r_{ki} \in \left[r_{ki}^{\beta}, r_{ki}^{\gamma} \right] : r_{ij} + r_{jk} + r_{ki} = \frac{3}{2}. \tag{4.132}$$

Proof From the additive-reciprocity property $\tilde{r}_{ij} = 1 - \tilde{r}_{ji}$, $i, j = 1, \ldots, n$, it follows that $\forall r_{ij} \in \left[r_{ij}^{\alpha}, r_{ij}^{\delta} \right] \exists r_{ji} \in \left[r_{ji}^{\alpha}, r_{ji}^{\delta} \right] : r_{ji} = 1 - r_{ij}$, and $\forall r_{ij} \in \left[r_{ij}^{\beta}, r_{ij}^{\gamma} \right] \exists r_{ji} \in \left[r_{ji}^{\beta}, r_{ji}^{\gamma} \right] : r_{ji} = 1 - r_{ij}$.

(a) First, let us show that the statements (i) and (iii) are equivalent. Because of the reciprocity property, (4.127) can be equivalently written as

$$\forall r_{ij} \in \left[r_{ij}^{\alpha}, r_{ij}^{\delta} \right] \exists r_{jk} \in \left[r_{jk}^{\alpha}, r_{jk}^{\delta} \right] \wedge \exists r_{ki} \in \left[r_{ki}^{\alpha}, r_{ki}^{\delta} \right] : r_{ij} = (1 - r_{ki}) + (1 - r_{jk}) + 0.5,$$

which is equivalent to (4.131). Analogously, the equivalence of (4.128) and (4.132) is proved.

(b) Now, let us show that the statements (ii) and (iii) are equivalent. Because of the reciprocity property, (4.129) can be equivalently written as

$$\forall r_{ij} \in \left[r_{ij}^{\alpha}, r_{ij}^{\delta} \right] \exists r_{jk} \in \left[r_{jk}^{\alpha}, r_{jk}^{\delta} \right] \wedge \exists r_{ki} \in \left[r_{ki}^{\alpha}, r_{ki}^{\delta} \right] :$$

$$r_{ij} + r_{jk} + r_{ki} = (1 - r_{ki}) + (1 - r_{jk}) + (1 - r_{ij}),$$

which is equivalent to (4.131). Analogously, the equivalence of (4.130) and (4.132) is proved. $\square$

The following theorems give us useful tools for verifying additive consistency of trapezoidal FAPCMs-A.

Theorem 54 *A trapezoidal FAPCM-A $\tilde{R} = \{\tilde{r}_{ij}\}_{i,j=1}^{n}, \tilde{r}_{ij} = \left(r_{ij}^{\alpha}, r_{ij}^{\beta}, r_{ij}^{\gamma}, r_{ij}^{\delta} \right)$, is additively consistent according to Definition 60 if and only if the inequalities*

$$r_{ij}^{\alpha} \geq r_{ik}^{\alpha} + r_{kj}^{\alpha} - 0.5, \qquad r_{ij}^{\delta} \leq r_{ik}^{\delta} + r_{kj}^{\delta} - 0.5, \tag{4.133}$$

$$r_{ij}^{\beta} \geq r_{ik}^{\beta} + r_{kj}^{\beta} - 0.5, \qquad r_{ij}^{\gamma} \leq r_{ik}^{\gamma} + r_{kj}^{\gamma} - 0.5, \tag{4.134}$$

hold for every $i, j, k = 1, \ldots, n$, $i < j$, $k \neq i, j$.

Proof It is sufficient to demonstrate the equivalence of the expressions (4.133) and (4.127). The demonstration of the equivalence of (4.134) and (4.128) is analogous.

First, let us demonstrate that when the inequalities (4.133) hold for every $i, j, k = 1, \ldots, n, i < j, k \neq i, j$, then they hold for every $i, j, k = 1, \ldots, n$. The inequalities (4.133) are always satisfied for $i, j, k = 1, \ldots, n$ such that $i = j \neq k$, or $i \neq j = k$, or $j \neq k = i$, or $i = j = k$:

$$r_{ik}^\alpha + r_{ki}^\alpha - 0.5 = 0.5 - (r_{ik}^\delta - r_{ik}^\alpha) \leq 0.5 = r_{ii}^\alpha,$$

$$r_{ik}^\delta + r_{ki}^\delta - 0.5 = 0.5 + (r_{ik}^\delta - r_{ik}^\alpha) \geq 0.5 = r_{ii}^\delta,$$

$$r_{ij}^\alpha + r_{jj}^\alpha - 0.5 = r_{ij}^\alpha, \quad r_{ij}^\delta + r_{jj}^\delta - 0.5 = r_{ij}^\delta,$$

$$r_{ii}^\alpha + r_{ij}^\alpha - 0.5 = r_{ij}^\alpha, \quad r_{ii}^\delta + r_{ij}^\delta - 0.5 = r_{ij}^\delta,$$

$$r_{ii}^\alpha + r_{ii}^\alpha - 0.5 = 0.5 = r_{ii}^\alpha, \quad r_{ii}^\delta + r_{ii}^\delta - 0.5 = 0.5 = r_{ii}^\delta.$$

Further, when the inequalities (4.133) are satisfied for $i, j, k = 1, \ldots, n, i < j, k \neq i, j$, then they are satisfied also for $j, i, k = 1, \ldots, n, j > i, k \neq i, j$:

$$r_{jk}^\alpha + r_{ki}^\alpha - 0.5 = 1 - r_{ik}^\delta + 1 - r_{kj}^\delta - 0.5 = 1 - (r_{ik}^\delta + r_{kj}^\delta - 0.5) \leq 1 - r_{ij}^\delta = r_{ji}^\alpha,$$

$$r_{jk}^\delta + r_{ki}^\delta - 0.5 = 1 - r_{ik}^\alpha + 1 - r_{kj}^\alpha - 0.5 = 1 - (r_{ik}^\alpha + r_{kj}^\alpha - 0.5) \geq 1 - r_{ij}^\alpha = r_{ji}^\delta.$$

To finalize the proof, it is sufficient to show that the inequalities (4.133) are equivalent to the condition (4.127) for every $i, j, k = 1, \ldots, n$. First, let $\widetilde{R}$ be a trapezoidal FAPCM-A additively consistent according to Definition 60. Then for $r_{ij} := r_{ij}^\alpha \exists r_{ik} \in \left[r_{ik}^\alpha, r_{ik}^\delta \right] \wedge \exists r_{kj} \in \left[r_{kj}^\alpha, r_{kj}^\delta \right] : r_{ij}^\alpha = r_{ik} + r_{kj} - 0.5$. Since $r_{ik} \geq r_{ik}^\alpha, r_{kj} \geq r_{kj}^\alpha$, then clearly $r_{ij}^\alpha \geq r_{ik}^\alpha + r_{ji}^\alpha - 0.5$. Analogously, for $r_{ij} := r_{ij}^\delta \exists r_{ik} \in \left[r_{ik}^\alpha, r_{ik}^\delta \right] \wedge \exists r_{kj} \in \left[r_{kj}^\alpha, r_{kj}^\delta \right] : r_{ij}^\delta = r_{ik} + r_{kj} - 0.5$. Since $r_{ik} \leq r_{ik}^\delta, r_{kj} \leq r_{kj}^\delta$, then clearly $r_{ij}^\delta \leq r_{ik}^\delta + r_{kj}^\delta - 0.5$.

Second, let (4.133) be valid for a trapezoidal FAPCM-A $\widetilde{R}$. From inequalities (4.133) we get $\forall r_{ij} \in \left[r_{ij}^\alpha, r_{ij}^\delta \right] : r_{ik}^\alpha + r_{kj}^\alpha - 0.5 \leq r_{ij} \leq r_{ik}^\delta + r_{kj}^\delta - 0.5$, and therefore, (4.127) is satisfied.   □

Theorem 55 *A trapezoidal FAPCM-A* $\widetilde{R} = \left\{ \widetilde{r}_{ij} \right\}_{i,j=1}^n, \widetilde{r}_{ij} = \left(r_{ij}^\alpha, r_{ij}^\beta, r_{ij}^\gamma, r_{ij}^\delta \right)$, *is additively consistent according to Definition 60 if and only if*

$$r_{ij}^\alpha \geq \max_{\substack{k=1,\ldots,n \\ k \neq i,j}} \left\{ r_{ik}^\alpha + r_{kj}^\alpha - 0.5 \right\}, \quad r_{ij}^\delta \leq \min_{\substack{k=1,\ldots,n \\ k \neq i,j}} \left\{ r_{ik}^\delta + r_{kj}^\delta - 0.5 \right\}, \quad (4.135)$$

$$r_{ij}^{\beta} \geq \max_{\substack{k=1,\dots,n \\ k \neq i,j}} \left\{ r_{ik}^{\beta} + r_{kj}^{\beta} - 0.5 \right\}, \quad r_{ij}^{\gamma} \leq \min_{\substack{k=1,\dots,n \\ k \neq i,j}} \left\{ r_{ik}^{\gamma} + r_{kj}^{\gamma} - 0.5 \right\}, \quad (4.136)$$

hold for every $i, j = 1, \dots, n, \ i < j$.

Proof The inequatilies (4.135) and (4.136) follow immediately from Theorem 54. □

In the following example, Definition 60 of additive consistency is confronted with Definitions 56, 57, and 58. In particular, it is demonstrated how the drawbacks regarding the dependence of Definitions 56 and 58 on permutation of objects and violation of the additive-reciprocity property in Definition 57 are removed by Definition 60.

Example 55 Let us examine the interval FAPCM-A given by (4.114). In Example 52, it was demonstrated that Definition 56 is not invariant under permutation of objects since the interval FAPCM-A (4.114) is judged as additively consistent while its permutation (4.116) is judged as additively inconsistent.

Similarly, the interval FAPCM-A (4.114) is judged additively consistent also according to Definition 58 since the expression (4.120) is valid for $i = 1, j = 2, k = 3 : [0.6, 0.7] + [0.7, 0.8] = [0.8, 1] + [0.5, 0.5]$. The permuted interval FAPCM-A (4.116) is, however, judged as additively inconsistent since the expression (4.120) is not valid: $[0, 0.2] + [0.6, 0.7] \neq [0.2, 0.3] + [0.5, 0.5]$.

Now let us apply Definition 60 to the interval FAPCM-A (4.114). By using Theorem 54, the interval FAPCM-A (4.114) is judged additively consistent since it satisfies the inequalities (4.133); see Table 4.10. Also the permuted interval FAPCM-A (4.116) satisfies the inequalities (4.133); see Table 4.11. Therefore, it is again judged as additively consistent. Moreover, from Theorem 51 it follows that any permutation of the interval FAPCM-A (4.114) is additively consistent.

In Example 53, it was demonstrated that Definition 57 violates the additive reciprocity of the related PCs. According to Theorem 52, the additive-reciprocity property is preserved in new Definition 60. This basically means that by taking any value from any interval PC in the interval FAPCM-A (4.114), there exist values in the remaining interval PCs such that they form an additively consistent APCM-A. Let us examine the triplet $i = 1, j = 2, k = 3$ of indices and let us consider the value $r_{12} = 0.65 \in [0.6, 0.7]$. Then, according to (4.127), there exist values $r_{13} \in [0.8, 1]$ and $r_{32} \in [0.2, 0.3]$ such that $0.65 = r_{13} + r_{32} - 0.5$. It is, for example, $r_{13} = 0.9, r_{32} = 0.25$. The additive reciprocity is clearly not violated. More interestingly, let us consider the triplet $i = 1, j = 1, k = 2$. Then, according to (4.127), there exist values $r_{12} \in [0.6, 0.7]$ and $r_{21} \in [0.3, 0.4]$ such that $0.5 = r_{12} + r_{21} - 0.5$. This equality is satisfied by any value $r_{12} \in [0.6, 0.7]$ and the corresponding value $r_{21} \in [0.3, 0.4]$ such that $r_{21} = 1 - r_{12}$, which again preserves the additive reciprocity. △

Table 4.10 Inequality conditions (4.133) for the interval FAPCM-A (4.114)

$i < j$:	$r_{ij}^L \geq r_{ik}^L + r_{kj}^L - 0.5$	$r_{ij}^U \leq r_{ik}^U + r_{kj}^U - 0.5$
1, 2:	$0.6 \geq 0.8 + 0.2 - 0.5$	$0.7 \leq 1 + 0.3 - 0.5$
1, 3:	$0.8 \geq 0.6 + 0.7 - 0.5$	$1 \leq 0.7 + 0.8 - 0.5$
2, 3:	$0.7 \geq 0.3 + 0.8 - 0.5$	$0.8 \leq 0.4 + 1 - 0.5$

Table 4.11 Inequality conditions (4.133) for the permuted interval FAPCM-A (4.116)

$i < j$:	$r_{ij}^L \geq r_{ik}^L + r_{kj}^L - 0.5$	$r_{ij}^U \leq r_{ik}^U + r_{kj}^U - 0.5$
1, 2:	$0 \geq 0.2 + 0.3 - 0.5$	$0.2 \leq 0.3 + 0.4 - 0.5$
1, 3:	$0.2 \geq 0 + 0.6 - 0.5$	$0.3 \leq 0.2 + 0.7 - 0.5$
2, 3:	$0.6 \geq 0.8 + 0.2 - 0.5$	$0.7 \leq 1 + 0.3 - 0.5$

In the rest of this section, some interesting properties of additively weakly consistent and additively consistent trapezoidal FAPCMs-A are examined. The following theorem shows the relation between Definition 60 of additive consistency and Definition 59 of additive weak consistency.

Theorem 56 *Let* $\widetilde{R} = \{\widetilde{r}_{ij}\}_{i,j=1}^n$, $\widetilde{r}_{ij} = \left(r_{ij}^\alpha, r_{ij}^\beta, r_{ij}^\gamma, r_{ij}^\delta \right)$, *be a trapezoidal FAPCM-A.*
If $\widetilde{R} = \{\widetilde{r}_{ij}\}_{i,j=1}^n$ *is additively consistent according to Definition 60, then it is also additively weakly consistent according to Definition 59.*

Proof The statement follows immediately from Theorem 55. In particular, the inequality (4.125) is obtained immediately from the inequalities (4.135).   □

Remark 25 According to Theorem 56, when a trapezoidal FAPCM-A is additively consistent according to Definition 60, then it is also automatically additively weakly consistent according to Definition 59. However, this does not hold true the other way around. Clearly, the definition of additive weak consistency is much weaker then the definition of additive consistency; it only requires existence of one crisp additively consistent APCM-A obtainable by combining particular elements from the closures of the supports of the trapezoidal fuzzy numbers in the trapezoidal FAPCM-A. Thus, the set of all trapezoidal FAPCMs-A additively consistent according to Definition 60 is a proper subset of the set of all trapezoidal FAPCMs-A additively weakly consistent according to Definition 59.

Example 56 Let us examine additive consistency and additive weak consistency of the trapezoidal FAPCM-A

$$\widetilde{R} = \begin{pmatrix} 0.5 & (0.5, 0.6, 0.65, 0.7) & (0.7, 0.8, 0.9, 0.95) & (0.85, 0.9, 0.95, 1) \\ (0.3, 0.35, 0.4, 0.5) & 0.5 & (0.6, 0.65, 0.7, 0.7) & (0.6, 0.7, 0.8, 0.9) \\ (0.05, 0.1, 0.2, 0.3) & (0.3, 0.3, 0.35, 0.4) & 0.5 & (0.5, 0.55, 0.6, 0.7) \\ (0, 0.05, 0.1, 0.15) & (0.1, 0.2, 0.3, 0.4) & (0.3, 0.4, 0.45, 0.5) & 0.5 \end{pmatrix}.$$

$$(4.137)$$

Table 4.12 Condition (4.125) for the trapezoidal FAPCM-A (4.137)	$i < j$:	$\max\limits_{k=1,\ldots,4}\left\{r_{ik}^\alpha + r_{kj}^\alpha - 0.5\right\} \leq \min\limits_{k=1,\ldots,4}\left\{r_{ik}^\delta + r_{kj}^\delta - 0.5\right\}$
	1, 2:	max {0.5, 0.5, 0.45, 0.45} ≤ min {0.7, 0.7, 0.8, 0.9}
	1, 3:	max {0.7, 0.6, 0.7, 0.65} ≤ min {0.9, 0.95, 0.9, 1}
	1, 4:	max {0.85, 0.6, 0.7, 0.85} ≤ min {1, 1.1, 1.1, 1}
	2, 3:	max {0.5, 0.6, 0.6, 0.4} ≤ min {0.9, 0.75, 0.75, 0.9}
	2, 4:	max {0.65, 0.6, 0.6, 0.6} ≤ min {1, 0.9, 0.95, 0.9}
	3, 4:	max {0.45, 0.35, 0.5, 0.5} ≤ min {0.8, 0.8, 0.7, 0.7}

Let us, for example, use Theorem 54 to verify the additive consistency of $\widetilde{R}$. We find out that $\widetilde{R}$ is not additively consistent because it violates the inequalities (4.133) and (4.134); for example, $r_{13}^\delta = 0.95 \nleq r_{12}^\delta + r_{23}^\delta - 0.5 = 0.9$.

Even though $\widetilde{R}$ is not additively consistent, it can still be at least additively weakly consistent. Let us verify that by using Theorem 50. According to Table 4.12, the condition (4.125) is satisfied, and thus $\widetilde{R}$ is additively weakly consistent.

A vector satisfying the inequalities (4.123) in Definition 59 is, for example, $\underline{v} = (1.45, 1.25, 0.85, 0.45)^T$. The corresponding APCM-A R^* is in the form

$$R^* = \begin{pmatrix} 0.5 & 0.6 & 0.8 & 1 \\ 0.4 & 0.5 & 0.7 & 0.9 \\ 0.2 & 0.3 & 0.5 & 0.7 \\ 0 & 0.1 & 0.3 & 0.5 \end{pmatrix}. \tag{4.138}$$

△

Theorem 57 *Let $\widetilde{R}$ be a trapezoidal FAPCM-A additively weakly consistent according to Definition 59. The trapezoidal FAPCM-A $\widetilde{R}^*$ constructed from $\widetilde{R}$ by eliminating the l-th row and the l-th column, $l \in \{1, \ldots, n\}$, is again additively weakly consistent.*

Proof For $\widetilde{R}$, the inequalities (4.123) are valid for every $i, j = 1, \ldots, n$. After eliminating the l-th row and the l-th column of $\widetilde{R}$, (4.123) is still valid for every remaining $i, j \in \{1, \ldots, n\} \setminus \{l\}$. Therefore, the new trapezoidal FAPCM-A $\widetilde{R}^*$ is still additively weakly consistent. □

The same holds also for additively consistent trapezoidal FAPCMs-A.

Theorem 58 *Let $\widetilde{R}$ be a trapezoidal FAPCM-A additively weakly consistent according to Definition 60. The trapezoidal FAPCM-A $\widetilde{R}^*$ constructed from $\widetilde{R}$ by eliminating the l-th row and the l-th column, $l \in \{1, \ldots, n\}$, is again additively consistent.*

Proof For $\widetilde{R}$, the inequalities (4.127) and (4.128) are valid for every $i, j, k = 1, \ldots, n$. After eliminating the l-th row and the l-th column of $\widetilde{R}$, (4.127) and (4.128) is still valid for every remaining $i, j, k \in \{1, \ldots, n\} \setminus \{l\}$. Therefore, the new trapezoidal FAPCM-A $\widetilde{R}^*$ is additively consistent. □

Remark 26 Theorems 57 and 58 are useful in situations when the set of objects compared pairwisely is being reduced. According to the theorems, elimination of one or more objects has no impact on the additive or additive weak consistency of fuzzy PCs of the remaining objects.

The following theorems provide some results regarding aggregation of additively and additively weakly consistent trapezoidal FAPCMs-A into one trapezoidal FAPCM-A, which are particularly useful in group decision making.

Theorem 59 *Let* $\widetilde{R}^1 = \left\{ \widetilde{r}_{ij}^1 \right\}_{i,j=1}^n$, $\widetilde{r}_{ij}^1 = \left(r_{ij}^{1\alpha}, r_{ij}^{1\beta}, r_{ij}^{1\gamma}, r_{ij}^{1\delta} \right)$, *and* $\widetilde{R}^2 = \left\{ \widetilde{r}_{ij}^2 \right\}_{i,j=1}^n$, $\widetilde{r}_{ij}^2 = \left(r_{ij}^{2\alpha}, r_{ij}^{2\beta}, r_{ij}^{2\gamma}, r_{ij}^{2\delta} \right)$, *be trapezoidal FAPCMs-A additively weakly consistent according to Definition 59. Then* $\widetilde{R} = \{ \widetilde{r}_{ij} \}_{i,j=1}^n$, $\widetilde{r}_{ij} = \left(r_{ij}^\alpha, r_{ij}^\beta, r_{ij}^\gamma, r_{ij}^\delta \right)$, *such that*

$$r_{ij}^\alpha = \epsilon r_{ij}^{1\alpha} + (1-\epsilon) r_{ij}^{2\alpha}, \quad r_{ij}^\beta = \epsilon r_{ij}^{1\beta} + (1-\epsilon) r_{ij}^{2\beta},$$
$$r_{ij}^\gamma = \epsilon r_{ij}^{1\gamma} + (1-\epsilon) r_{ij}^{2\gamma}, \quad r_{ij}^\delta = \epsilon r_{ij}^{1\delta} + (1-\epsilon) r_{ij}^{2\delta},$$

is an additively weakly consistent trapezoidal FAPCM-A for any $\epsilon \in [0, 1]$.

Proof First, let us show that $\widetilde{R}$ is a trapezoidal FAPCM-A. For $i = 1, \dots, n$ we get

$$r_{ii}^\alpha = \epsilon r_{ii}^{1\alpha} + (1-\epsilon) r_{ii}^{2\alpha} = 0.5\epsilon + 0.5(1-\epsilon) = 0.5,$$
$$r_{ii}^\delta = \epsilon r_{ii}^{1\delta} + (1-\epsilon) r_{ii}^{2\delta} = 0.5\epsilon + 0.5(1-\epsilon) = 0.5.$$

Similarly, $r_{ii}^\beta = 0.5$, $r_{ii}^\gamma = 0.5$, and thus, $\widetilde{r}_{ii} = 0.5$, $i = 1, \dots, n$. Further, for $i \neq j$ we have

$$r_{ij}^\alpha = \epsilon r_{ij}^{1\alpha} + (1-\epsilon) r_{ij}^{2\alpha} = \epsilon(1 - r_{ji}^{1\delta}) + (1-\epsilon)(1 - r_{ji}^{2\delta}) = 1 - \left[\epsilon r_{ji}^{1\delta} + (1-\epsilon) r_{ji}^{2\delta} \right] = 1 - r_{ji}^\delta,$$

$$r_{ij}^\delta = \epsilon r_{ij}^{1\delta} + (1-\epsilon) r_{ij}^{2\delta} = \epsilon(1 - r_{ji}^{1\alpha}) + (1-\epsilon)(1 - r_{ji}^{2\alpha}) = 1 - \left[\epsilon r_{ji}^{1\alpha} + (1-\epsilon) r_{ji}^{2\alpha} \right] = 1 - r_{ji}^\alpha,$$

and analogously we obtain $r_{ij}^\beta = 1 - r_{ji}^\gamma$, $r_{ij}^\gamma = 1 - r_{ji}^\beta$. Therefore, $\widetilde{r}_{ij} = 1 - \widetilde{r}_{ji}$, $i, j = 1, \dots, n$. Finally,

$$r_{ij}^\alpha = \epsilon r_{ij}^{1\alpha} + (1-\epsilon) r_{ij}^{2\alpha} \geq \epsilon \cdot 0 + (1-\epsilon) \cdot 0 = 0,$$
$$r_{ij}^\delta = \epsilon r_{ij}^{1\delta} + (1-\epsilon) r_{ij}^{2\delta} \leq \epsilon + (1-\epsilon) = 1,$$

i.e. $[r_{ij}^\alpha, r_{ij}^\delta] \subseteq [0, 1]$.

Second, let us show that $\widetilde{R}$ is additively weakly consistent. It is sufficient to prove inequalities (4.125). Since (4.125) is valid for FAPCMs-A $\widetilde{R}^1$ and $\widetilde{R}^2$, we obtain

$$\max_{k=1,\ldots,n}\left\{r_{ik}^\alpha + r_{kj}^\alpha - 0.5\right\} = \max_{k=1,\ldots,n}\left\{\epsilon r_{ik}^{1\alpha} + (1-\epsilon)r_{ik}^{2\alpha} + \epsilon r_{kj}^{1\alpha} + (1-\epsilon)r_{kj}^{2\alpha} - 0.5\right\} \le$$

$$\max_{k=1,\ldots,n}\left\{\epsilon[r_{ik}^{1\alpha} + r_{kj}^{1\alpha} - 0.5]\right\} + \max_{k=1,\ldots,n}\left\{(1-\epsilon)[r_{ik}^{2\alpha} + r_{kj}^{2\alpha} - 0.5]\right\} \le$$

$$\min_{k=1,\ldots,n}\left\{\epsilon[r_{ik}^{1\delta} + r_{kj}^{1\delta} - 0.5]\right\} + \min_{k=1,\ldots,n}\left\{(1-\epsilon)[r_{ik}^{2\delta} + r_{kj}^{2\delta} - 0.5]\right\} \le$$

$$\min_{k=1,\ldots,n}\left\{\epsilon r_{ik}^{1\delta} + (1-\epsilon)r_{ik}^{2\delta} + \epsilon r_{kj}^{1\delta} + (1-\epsilon)r_{kj}^{2\delta} - 0.5\right\} = \min_{k=1,\ldots,n}\left\{r_{ik}^\delta + r_{kj}^\delta - 0.5\right\},$$

which proves the theorem. $\qquad\square$

Theorem 59 can be further extended to the aggregation of $p \ge 2$ additively weakly consistent trapezoidal FAPCMs-A as follows.

Theorem 60 Let $\widetilde{R}^\tau = \left\{\widetilde{r}_{ij}^\tau\right\}_{i,j=1}^n$, $\widetilde{r}_{ij}^\tau = \left(r_{ij}^{\tau\alpha}, r_{ij}^{\tau\beta}, r_{ij}^{\tau\gamma}, r_{ij}^{\tau\delta}\right)$, $\tau = 1, \ldots, p$, be trapezoidal FAPCMs-A additively weakly consistent according to Definition 59. Then $\widetilde{R} = \left\{\widetilde{r}_{ij}\right\}_{i,j=1}^n$ such that

$$\widetilde{r}_{ij} = \left(r_{ij}^\alpha, r_{ij}^\beta, r_{ij}^\gamma, r_{ij}^\delta\right) = \left(\sum_{\tau=1}^p \epsilon_\tau r_{ij}^{\tau\alpha}, \sum_{\tau=1}^p \epsilon_\tau r_{ij}^{\tau\beta}, \sum_{\tau=1}^p \epsilon_\tau r_{ij}^{\tau\gamma}, \sum_{\tau=1}^p \epsilon_\tau r_{ij}^{\tau\delta}\right), \quad (4.139)$$

is an additively weakly consistent trapezoidal FAPCM-A for any $\epsilon_\tau \in [0, 1]$, $\tau = 1, \ldots, p$, with $\sum_{\tau=1}^p \epsilon_\tau = 1$.

Proof The proof is analogous to the proof of Theorem 59. $\qquad\square$

Similar theorems are formulated also for additively consistent trapezoidal FAPCMs-A.

Theorem 61 Let $\widetilde{R}^1 = \left\{\widetilde{r}_{ij}^1\right\}_{i,j=1}^n$, $\widetilde{r}_{ij}^1 = \left(r_{ij}^{1\alpha}, r_{ij}^{1\beta}, r_{ij}^{1\gamma}, r_{ij}^{1\delta}\right)$, and $\widetilde{R}^2 = \left\{\widetilde{r}_{ij}^2\right\}_{i,j=1}^n$, $\widetilde{r}_{ij}^2 = \left(r_{ij}^{2\alpha}, r_{ij}^{2\beta}, r_{ij}^{2\gamma}, r_{ij}^{2\delta}\right)$, be trapezoidal FAPCMs-A additively consistent according to Definition 60. Then $\widetilde{R} = \left\{\widetilde{r}_{ij}\right\}_{i,j=1}^n$, $\widetilde{r}_{ij} = \left(r_{ij}^\alpha, r_{ij}^\beta, r_{ij}^\gamma, r_{ij}^\delta\right)$, such that

$$r_{ij}^\alpha = \epsilon r_{ij}^{1\alpha} + (1-\epsilon)r_{ij}^{2\alpha}, \quad r_{ij}^\beta = \epsilon r_{ij}^{1\beta} + (1-\epsilon)r_{ij}^{2\beta},$$

$$r_{ij}^\gamma = \epsilon r_{ij}^{1\gamma} + (1-\epsilon)r_{ij}^{2\gamma}, \quad r_{ij}^\delta = \epsilon r_{ij}^{1\delta} + (1-\epsilon)r_{ij}^{2\delta},$$

is an additively consistent trapezoidal FAPCM-A for any $\epsilon \in [0, 1]$.

Proof We already know from Theorem 59 that $\widetilde{R}$ is a trapezoidal FAPCM-A. Therefore, we only need to show that $\widetilde{R}$ is additively consistent according to Definition 60.

It is sufficient to prove the inequalities (4.133) and (4.134). Since (4.133) are valid for interval FAPCMs-A $\widetilde{R}^1$ and $\widetilde{R}^2$, we obtain

$$
\begin{aligned}
r_{ik}^\alpha + r_{kj}^\alpha - 0.5 &= \left[\epsilon r_{ik}^{1\alpha} + (1-\epsilon)r_{ik}^{2\alpha}\right] + \left[\epsilon r_{kj}^{1\alpha} + (1-\epsilon)r_{kj}^{2\alpha}\right] - 0.5 \\
&= \epsilon(r_{ik}^{1\alpha} + r_{kj}^{1\alpha} - 0.5) + (1-\epsilon)(r_{ik}^{2\alpha} + r_{kj}^{2\alpha} - 0.5) \\
&\le \epsilon r_{ij}^{1\alpha} + (1-\epsilon)r_{ij}^{2\alpha} = r_{ij}^\alpha, \\
r_{ik}^\delta + r_{kj}^\delta - 0.5 &= \left[\epsilon r_{ik}^{1\delta} + (1-\epsilon)r_{ik}^{2\delta}\right] + \left[\epsilon r_{kj}^{1\delta} + (1-\epsilon)r_{kj}^{2\delta}\right] - 0.5 \\
&= \epsilon(r_{ik}^{1\delta} + r_{kj}^{1\delta} - 0.5) + (1-\epsilon)(r_{ik}^{2\delta} + r_{kj}^{2\delta} - 0.5) \\
&\ge \epsilon r_{ij}^{1\delta} + (1-\epsilon)r_{ij}^{2\delta} = r_{ij}^\delta.
\end{aligned}
$$

Analogously, the validity of inequalities (4.134) is proved.   □

Theorem 61 can be further extended to the aggregation of $p \ge 2$ additively consistent interval FAPCMs-A as follows.

Theorem 62 *Let* $\widetilde{R}^\tau = \left\{\widetilde{r}_{ij}^\tau\right\}_{i,j=1}^n$, $\widetilde{r}_{ij}^\tau = \left(r_{ij}^{\tau\alpha}, r_{ij}^{\tau\beta}, r_{ij}^{\tau\gamma}, r_{ij}^{\tau\delta}\right), \tau = 1, \ldots, p$, *be trapezoidal FAPCMs-A additively consistent according to Definition 60. Then* $\widetilde{R} = \left\{\widetilde{r}_{ij}\right\}_{i,j=1}^n$ *such that*

$$
\widetilde{r}_{ij} = \left(r_{ij}^\alpha, r_{ij}^\beta, r_{ij}^\gamma, r_{ij}^\delta\right) = \left(\sum_{\tau=1}^p \epsilon_\tau r_{ij}^{\tau\alpha}, \sum_{\tau=1}^p \epsilon_\tau r_{ij}^{\tau\beta}, \sum_{\tau=1}^p \epsilon_\tau r_{ij}^{\tau\gamma}, \sum_{\tau=1}^p \epsilon_\tau r_{ij}^{\tau\delta}\right),
$$

is an additively consistent trapezoidal FAPCM-A for any $\epsilon_\tau \in [0, 1]$, $\tau = 1, \ldots, p$, *with* $\sum_{\tau=1}^p \epsilon_\tau = 1$.

Proof The proof is analogous to the proof of Theorem 61.   □

Example 57 Let us assume that two DMs m_1 and m_2 compare pairwisely four objects in interval FAPCMs-A $\widetilde{R}^1$ and $\widetilde{R}^2$, respectively, as follows:

$$
\widetilde{R}^1 = \begin{pmatrix} 0.5 & [0.5, 0.6] & [0.6, 0.7] & [0.7, 0.85] \\ [0.4, 0.5] & 0.5 & [0.5, 0.6] & [0.65, 0.8] \\ [0.3, 0.4] & [0.4, 0.5] & 0.5 & [0.6, 0.7] \\ [0.15, 0.3] & [0.2, 0.35] & [0.3, 0.4] & 0.5 \end{pmatrix}, \tag{4.140}
$$

$$
\widetilde{R}^2 = \begin{pmatrix} 0.5 & [0.5, 0.6] & [0.55, 0.7] & [0.8, 1] \\ [0.4, 0.5] & 0.5 & [0.55, 0.65] & [0.7, 0.9] \\ [0.3, 0.45] & [0.35, 0.45] & 0.5 & [0.6, 0.8] \\ [0, 0.2] & [0.1, 0.3] & [0.2, 0.4] & 0.5 \end{pmatrix}. \tag{4.141}
$$

First, let us verify their additive consistency by applying Theorem 55. From Table 4.13 we see that the inequality conditions (4.135) are satisfied for the interval FAPCM-A $\widetilde{R}^1$. Therefore, $\widetilde{R}^1$ is additively consistent according to Definition 60. Additive consistency of the interval FAPCM-A $\widetilde{R}^2$ is verified in an analogous way.

Table 4.13 Inequality conditions (4.135) for the interval FAPCM-A (4.140)

$i < j$:	$r_{ij}^\alpha \geq \max\limits_{k=2,3} \left\{ r_{ik}^\alpha + r_{kj}^\alpha - 0.5 \right\}$	$r_{ij}^\delta \leq \min\limits_{k=2,3} \left\{ r_{ik}^\delta + r_{kj}^\delta - 0.5 \right\}$
1, 2:	$0.5 \geq \max \{0.5, 0.4\}$	$0.6 \leq \min \{0.7, 0.7\}$
1, 3:	$0.6 \geq \max \{0.5, 0.5\}$	$0.7 \leq \min \{0.7, 0.75\}$
1, 4:	$0.7 \geq \max \{0.65, 0.7\}$	$0.85 \leq \min \{0.9, 0.9\}$
2, 3:	$0.5 \geq \max \{0.5, 0.45\}$	$0.6 \leq \min \{0.7, 0.7\}$
2, 4:	$0.65 \geq \max \{0.6, 0.6\}$	$0.8 \leq \min \{0.85, 0.8\}$
3, 4:	$0.6 \geq \max \{0.5, 0.55\}$	$0.7 \leq \min \{0.75, 0.8\}$

Because the interval FAPCMs-A $\widetilde{R}^1$ and $\widetilde{R}^2$ are additively consistent according to Definition 60, then, based on Theorem 56, they are also additively weakly consistent according to Definition 59.

Let us now aggregate the interval FAPCMs-A (4.140) and (4.141) into one interval FAPCM-A $\widetilde{R}$ representing the preferences of both DMs. Let us use the formula (4.139) with the importance of the DM m_1 given as $\epsilon_1 = 0.4$ and the importance of the DM m_2 given as $\epsilon_2 = 0.6$, $\epsilon_1 + \epsilon_2 = 1$. The resulting interval FAPCM-A $\widetilde{R}$ is in the form

$$\widetilde{R} = \begin{pmatrix} 0.5 & [0.5, 0.6] & [0.57, 0.7] & [0.76, 0.94] \\ [0.4, 0.5] & 0.5 & [0.53, 0.63] & [0.68, 0.86] \\ [0.3, 0.43] & [0.37, 0.47] & 0.5 & [0.6, 0.76] \\ [0.06, 0.24] & [0.14, 0.32] & [0.24, 0.4] & 0.5 \end{pmatrix}. \qquad (4.142)$$

According to Theorem 55, the interval FAPCM-A (4.142) is again additively consistent (the reader can again verify that the inequalities (4.135) are satisfied). Further, based on Theorem 56, it is also additively weakly consistent. $\triangle$

4.3.2.2 Deriving Priorities from FAPCMs-A

In this section, the focus is put on methods for obtaining fuzzy priorities of objects from FAPCMs-A. The notation $\widetilde{\underline{v}} = (\widetilde{v}_1, \ldots, \widetilde{v}_n)^T$, $\widetilde{v}_i = (v_i^\alpha, v_i^\beta, v_i^\gamma, v_i^\delta)$, $i = 1, \ldots, n$, will be used hereafter to represent exclusively a fuzzy priority vector associated with a FAPCM-A.

In the literature, various methods have been proposed to obtain interval priorities from interval FAPCMs-A. Most of these methods are based on linear programming models rather than on interval arithmetic. Xu (2007a) and Xu and Chen (2008a) proposed linear programming models for obtaining interval priorities of objects from interval FAPCMs-A additively consistent according to Definition 55. The solution of the models is a set of priority vectors satisfying the inequalities (4.108) and the normalization condition (4.110). Further, Xu and Chen (2008a) proposed a modification of these linear programming models to obtain interval priorities also from additively inconsistent interval FAPCMs-A. The modification is based on introducing a set of deviation variables relaxing the inequalities (4.108). However, as already mentioned in Sect. 2.3.2.2, the normalization condition (4.110) is not compatible with Tanino's characterization (2.32). Similar linear programming models for obtaining interval priorities from interval FAPCMs-A were proposed by Wang and Li (2012). However, they again employed the inappropriate normalization condition (4.110) in their models. Hu et al. (2014) later proposed a modification of the linear programming models introduced by Xu and Chen (2008a) by replacing Tanino's characterization with the characterization (2.46), and Xu et al. (2014b) generalized the models by adding a parameter into the characterization. Wang et al. (2012) introduced linear programming models for obtaining interval priorities using a particular characterization based on logarithms.

The number of papers proposing linear programming models based on Tanino's characterization (2.32) or on alternative characterizations is extensive. Nevertheless, the focus of this book is put only on the methods based on applying fuzzy arithmetic to the fuzzy extension of Tanino's characterization.

We know from Sect. 2.3.2.2 that the only appropriate formula (up to addition of a constant) for obtaining priorities from APCMs-A compatible with Tanino's characterization (2.32) is the formula (2.36). So far, I have not encountered any research paper dealing with the fuzzy extension of this formula to FAPCMs-A by applying fuzzy arithmetic. As far as I am aware, the only approach for obtaining interval priorities from interval FAPCMs-A based on interval arithmetic is the approach proposed by Liu et al. (2012a).

Liu et al. (2012a) proposed to transform an interval FAPCM-A into an interval FMPCM and then to compute interval priorities by using the formulas (4.90). Putting aside the fact that such interval priorities are not invariant under permutation of objects, it is important to realize that priorities obtained in such way do not reflect the preference information in the interval FAPCMs-A by means of differences; they reflect, instead, by means of ratios, the preference information contained in the corresponding interval FMPCMs. It is very important to realize this difference. Nevertheless, Liu et al. (2012a) completely omitted this issue.

The necessity of distinguishing between the two types of priorities mentioned above was emphasized by Krejčí (2017e), who proposed a method based on the constrained fuzzy arithmetic for obtaining "multiplicative" fuzzy priorities from triangular FAPCMs-A. Note that the method for obtaining "multiplicative" fuzzy priorities proposed by Krejčí (2017e) is invariant under permutation of objects and preserves the additive reciprocity of the related PCs as it is based on the formulas (4.97)–(4.100).

In this section, the fuzzy extension of the formula (2.36) to FAPCMs-A is introduced and properties of the new formulas are discussed. Further, it is demonstrated that the new formulas preserve two desirable properties—additive reciprocity of the related PCs and invariance under permutation of objects. The findings summarized in this section are mostly based on the original paper of Krejčí (2016) on obtaining fuzzy priorities from triangular FAPCMs-A.

We already know from Sect. 2.3.2.2 that the usual normalization condition (2.39)—$\sum_{i=1}^{n} v_i = 1$, $v_i \in [0, 1]$, $i = 1, \ldots, n$—is not reachable for the priorities obtainable from APCMs-A. Thus, Fedrizzi and Brunelli (2009) proposed the normalization condition (2.40)—$\min_{i \in \{1,\ldots,n\}} v_i = 0$, $v_i \in [0, 1]$, $i = 1, \ldots, n$—for the priorities obtainable from APCMs-A. However, as demonstrated in Sect. 2.3.2.2, when an APCM-A is not additively consistent, reaching the property (2.40) is not guaranteed anymore. Therefore, the weakened normalization condition (2.42)—$\sum_{i=1}^{n} v_i = 1$—was introduced in Sect. 2.3.2.2 for the priorities obtainable from APCMs-A in order to make an analogy to the normalization condition (2.39) for the priorities obtainable from MPCMs and from APCMs-M. In this section, the normalization condition (2.42) is extended to fuzzy priorities obtainable from FAPCMs-A.

First, let us start with the fuzzy extension of the formulas (2.36) for obtaining non-normalized priorities from APCMs-A. By applying constrained fuzzy arithmetic (3.47) to the formula (2.36), the representing values of the fuzzy priorities $\tilde{v}_i = \left(v_i^{\alpha}, v_i^{\beta}, v_i^{\gamma}, v_i^{\delta} \right)$, $i = 1, \ldots, n$, obtained from a FAPCM-A $\tilde{R} = \{\tilde{r}_{ij}\}_{i,j=1}^{n}$, $\tilde{r}_{ij} = \left(r_{ij}^{\alpha}, r_{ij}^{\beta}, r_{ij}^{\gamma}, r_{ij}^{\delta} \right)$, are given as:

$$v_i^{\alpha} = \min \left\{ \frac{2}{n} \sum_{j=1}^{n} r_{ij}; \begin{array}{l} r_{pq} \in \left[r_{pq}^{\alpha}, r_{pq}^{\delta} \right], \\ r_{pq} = 1 - r_{qp}, \\ p, q = 1, \ldots, n \end{array} \right\}, \tag{4.143}$$

$$v_i^{\beta} = \min \left\{ \frac{2}{n} \sum_{j=1}^{n} r_{ij}; \begin{array}{l} r_{pq} \in \left[r_{pq}^{\beta}, r_{pq}^{\gamma} \right], \\ r_{pq} = 1 - r_{qp}, \\ p, q = 1, \ldots, n \end{array} \right\}, \tag{4.144}$$

$$v_i^{\gamma} = \max \left\{ \frac{2}{n} \sum_{j=1}^{n} r_{ij}; \begin{array}{l} r_{pq} \in \left[r_{pq}^{\beta}, r_{pq}^{\gamma} \right], \\ r_{pq} = 1 - r_{qp}, \\ p, q = 1, \ldots, n \end{array} \right\}, \tag{4.145}$$

$$v_i^{\delta} = \max \left\{ \frac{2}{n} \sum_{j=1}^{n} r_{ij}; \begin{array}{l} r_{pq} \in \left[r_{pq}^{\alpha}, r_{pq}^{\delta} \right], \\ r_{pq} = 1 - r_{qp}, \\ p, q = 1, \ldots, n \end{array} \right\}. \tag{4.146}$$

Because the function optimized in the formulas (4.143)–(4.146) is increasing in all variables, the formulas can be further simplified so that no optimization is needed:

$$v_i^{\alpha} = \frac{2}{n} \sum_{j=1}^{n} r_{ij}^{\alpha}, \quad v_i^{\beta} = \frac{2}{n} \sum_{j=1}^{n} r_{ij}^{\beta}, \quad v_i^{\gamma} = \frac{2}{n} \sum_{j=1}^{n} r_{ij}^{\gamma}, \quad v_i^{\delta} = \frac{2}{n} \sum_{j=1}^{n} r_{ij}^{\delta}. \tag{4.147}$$

Remark 27 It is worth to note that the elimination of the optimization problems in the formulas (4.143)–(4.146) and their replacement by very simple formulas (4.147) was possible to do only because the constraints of the optimization problems have no effect on the optima; the additive-reciprocity condition $r_{ij} = 1 - r_{ji}$ has no influence since only the PCs from the i−th row of the FAPCM-A are present in the optimized function. Thus, in this particular case, the formulas (4.143)–(4.146) based on the constrained fuzzy arithmetic actually give the same results as the formulas (4.147) based on the standard fuzzy arithmetic.

Notice that also the formulas (4.85) for computing fuzzy priorities from FMPCMs proposed by Buckley (1985a) reviewed in Sect. 4.2.3.2 have this simple form. Also in this case the multiplicative-reciprocity condition $m_{ij} = \frac{1}{m_{ji}}$ would have no impact since only the PCs from the i−th row of the FMPCM are present in the optimized function. Thus, the standard fuzzy arithmetic is sufficient here.

Usually, however, the formulas based on the constrained fuzzy arithmetic cannot be simplified to standard fuzzy arithmetic by simply eliminating the constraints and thus avoiding solving an optimization problem; the formulas (4.91)–(4.94) for obtaining normalized fuzzy priorities from FMPCMs serve as an example.

There are interactions between the fuzzy priorities $\widetilde{v}_i, i = 1, \ldots, n$, obtained by formulas (4.147). The property (2.37) valid for the priorities obtained from an APCM-A by the formulas (2.36) is extended to the fuzzy priorities as

$$\forall v_i \in \widetilde{v}_{i(\alpha)} \,\exists v_j \in \widetilde{v}_{j(\alpha)}, j = 1, \ldots, n, j \neq i: \quad v_i + \sum_{\substack{j=1 \\ j \neq i}}^{n} v_j = n, \qquad (4.148)$$

for all $\alpha \in [0, 1]$ and $i = 1, \ldots, n$. This interaction property will be formulated properly and proved later. First, the following proposition is needed.

Proposition 14 *The interaction property* (4.148) *between the trapezoidal fuzzy numbers* $\widetilde{v}_i = \left(v_i^\alpha, v_i^\beta, v_i^\gamma, v_i^\delta \right), i = 1, \ldots, n,$ *is valid if and only if*

$$v_i^\alpha + \sum_{\substack{j=1 \\ j \neq i}}^{n} v_j^\delta \geq n, \quad v_i^\delta + \sum_{\substack{j=1 \\ j \neq i}}^{n} v_j^\alpha \leq n, \quad v_i^\beta + \sum_{\substack{j=1 \\ j \neq i}}^{n} v_j^\gamma \geq n, \quad v_i^\gamma + \sum_{\substack{j=1 \\ j \neq i}}^{n} v_j^\beta \leq n.$$
$$(4.149)$$

Proof First, let us show that (4.148) implies (4.149). For $\alpha = 0$ it follows from (4.148) that for $v_i^\alpha, i \in \{1, \ldots, n\}$, $\exists v_j \in \left[v_j^\alpha, v_j^\delta \right], j = 1, \ldots, n, j \neq i: v_i^\alpha + \sum_{\substack{j=1 \\ j \neq i}}^{n} v_j = n$. Because $v_j^\delta \geq v_j$, then clearly $v_i^\alpha + \sum_{\substack{j=1 \\ j \neq i}}^{n} v_j^\delta \geq n$. Similarly, for $v_i^\delta, i \in \{1, \ldots, n\}$, $\exists v_j \in \left[v_j^\alpha, v_j^\delta \right], j = 1, \ldots, n, j \neq i: v_i^\delta + \sum_{\substack{j=1 \\ j \neq i}}^{n} v_j = n$. Because $v_j^\alpha \leq v_j$, then clearly $v_i^\delta + \sum_{\substack{j=1 \\ j \neq i}}^{n} v_j^\alpha \leq n$. Analogously, for $\alpha = 1$ it follows from (4.148) that for $v_i^\beta, i \in \{1, \ldots, n\}$, $\exists v_j \in \left[v_j^\beta, v_j^\gamma \right], j = 1, \ldots, n, j \neq i: v_i^\beta + \sum_{\substack{j=1 \\ j \neq i}}^{n} v_j = n$.

Because $v_j^\gamma \geq v_j$, then clearly $v_i^\beta + \sum_{\substack{j=1 \\ j\neq i}}^n v_j^\gamma \geq n$. Similarly, for $v_i^\gamma, i \in \{1, \ldots, n\}$,

$\exists v_j \in \left[v_j^\beta, v_j^\gamma\right], j = 1, \ldots, n, j \neq i : v_i^\gamma + \sum_{\substack{j=1 \\ j\neq i}}^n v_j = n$. Because $v_j^\beta \leq v_j$, then

clearly $v_i^\gamma + \sum_{\substack{j=1 \\ j\neq i}}^n v_j^\beta \leq n$.

Now, let us show that (4.149) implies (4.148). From the inequalities $v_i^\delta + \sum_{\substack{j=1 \\ j\neq i}}^n$

$v_j^\alpha \leq n$ and $v_i^\alpha + \sum_{\substack{j=1 \\ j\neq i}}^n v_j^\delta \geq n$, the inequalities $v_i + \sum_{\substack{j=1 \\ j\neq i}}^n v_j^\alpha \leq n$ and $v_i + \sum_{\substack{j=1 \\ j\neq i}}^n$

$v_j^\delta \geq n$ follow $\forall v_i \in \left[v_i^\alpha, v_i^\delta\right]$. Therefore, $\exists v_j \in \left[v_j^\alpha, v_j^\delta\right] : v_i + \sum_{\substack{j=1 \\ j\neq i}}^n v_j = n$, which

implies (4.148) for $\alpha = 0$. Analogously, from the inequalities $v_i^\gamma + \sum_{\substack{j=1 \\ j\neq i}}^n v_j^\beta \leq n$ and

$v_i^\beta + \sum_{\substack{j=1 \\ j\neq i}}^n v_j^\gamma \geq n$, the inequalities $v_i + \sum_{\substack{j=1 \\ j\neq i}}^n v_j^\beta \leq n$ and $v_i + \sum_{\substack{j=1 \\ j\neq i}}^n v_j^\gamma \geq n$ follow

$\forall v_i \in \left[v_i^\beta, v_i^\gamma\right]$. Therefore, $\exists v_j \in \left[v_j^\beta, v_j^\gamma\right] : v_i + \sum_{\substack{j=1 \\ j\neq i}}^n v_j = n$, which implies

(4.148) for $\alpha = 1$.

The proof of the validity of (4.148) for $\alpha \in]0, 1[$ is analogous; it is sufficient to show that the inequalities (4.149) hold also for the α−cuts $\tilde{v}_{i(\alpha)} = \left[v_{i(\alpha)}^L, v_{i(\alpha)}^U\right]$ of the trapezoidal fuzzy numbers $\tilde{v}_i = (v_i^\alpha, v_i^\beta, v_i^\gamma, v_i^\delta)$, i.e.

$$v_{i(\alpha)}^L + \sum_{\substack{j=1 \\ j\neq i}}^n v_{j(\alpha)}^U \geq n, \qquad v_{i(\alpha)}^U + \sum_{\substack{j=1 \\ j\neq i}}^n v_{j(\alpha)}^L \leq n. \qquad (4.150)$$

Then it is enough to take the α−cuts $\tilde{v}_{i(\alpha)} = \left[v_{i(\alpha)}^L, v_{i(\alpha)}^U\right]$ of $\tilde{v}_i, i = 1, \ldots, n$, for $\left[v_i^\alpha, v_i^\delta\right], i = 1, \ldots, n$, in the above part of the proof.

Using the definition (3.8) of α−cuts and formulas (4.149), we have

$$v_{i(\alpha)}^U + \sum_{\substack{j=1 \\ j\neq i}}^n v_{j(\alpha)}^L = \alpha v_i^\gamma + (1 - \alpha)v_i^\delta + \sum_{\substack{j=1 \\ j\neq i}}^n \left[\alpha v_j^\beta + (1 - \alpha)v_j^\alpha\right] =$$

$$\alpha\left[v_i^\gamma + \sum_{\substack{j=1 \\ j\neq i}}^n v_j^\beta\right] + (1 - \alpha)\left[v_i^\delta + \sum_{\substack{j=1 \\ j\neq i}}^n v_j^\alpha\right] \leq \alpha n + (1 - \alpha)n = n,$$

and analogously the inequality $v_{i(\alpha)}^L + \sum_{\substack{j=1 \\ j\neq i}}^n v_{j(\alpha)}^U \geq n$ could be demonstrated. $\square$

Now, by utilizing Proposition 14, we can formulate and prove the following theorem.

Theorem 63 *Let* $\tilde{v}_i = \left(v_i^\alpha, v_i^\beta, v_i^\gamma, v_i^\delta \right), i = 1, \ldots, n$, *be trapezoidal fuzzy priorities obtained from a FAPCM-A by formulas* (4.147). *Then the property* (4.148) *holds for all* $\alpha \in [0, 1]$ *and* $i = 1, \ldots, n$.

Proof By utilizing Proposition 14, it is sufficient to show that the fuzzy priorities obtained by formulas (4.147) satisfy (4.149).

$$
v_i^\alpha + \sum_{\substack{j=1 \\ j \neq i}}^n v_j^\delta = \frac{2}{n} \sum_{k=1}^n r_{ik}^\alpha + \sum_{\substack{j=1 \\ j \neq i}}^n \frac{2}{n} \sum_{k=1}^n r_{jk}^\delta = \frac{2}{n} \left(\sum_{k=1}^n r_{ik}^\alpha + \sum_{\substack{j=1 \\ j \neq i}}^n \sum_{k=1}^n r_{jk}^\delta \right) =
$$

$$
\frac{2}{n} \left(0.5n + (n-1) + \sum_{\substack{j=1 \\ j \neq i}}^n \sum_{\substack{k=1 \\ k \neq i \\ k \neq j}}^n r_{jk}^\delta \right) \geq \frac{2}{n} \left(0.5n + (n-1) + \frac{(n-1)(n-2)}{2} \right) = n
$$

$$
v_i^\delta + \sum_{\substack{j=1 \\ j \neq i}}^n v_j^\alpha = \frac{2}{n} \sum_{k=1}^n r_{ik}^\delta + \sum_{\substack{j=1 \\ j \neq i}}^n \frac{2}{n} \sum_{k=1}^n r_{jk}^\alpha = \frac{2}{n} \left(\sum_{k=1}^n r_{ik}^\delta + \sum_{\substack{j=1 \\ j \neq i}}^n \sum_{k=1}^n r_{jk}^\alpha \right) =
$$

$$
\frac{2}{n} \left(0.5n + (n-1) + \sum_{\substack{j=1 \\ j \neq i}}^n \sum_{\substack{k=1 \\ k \neq i \\ k \neq j}}^n r_{jk}^\alpha \right) \leq \frac{2}{n} \left(0.5n + (n-1) + \frac{(n-1)(n-2)}{2} \right) = n
$$

Validity of the remaining two inequalities is proved in an analogous way.   □

The interaction property (4.148) corresponds to the fact that, for any $i \in \{1, \ldots, n\}$, and for any value $v_i \in \tilde{v}_{i(\alpha)}$, $\alpha \in [0, 1]$, there exist values $v_j \in \tilde{v}_{j(\alpha)}$, $j \neq i$, such that they are all obtained by the formula (2.36) from the same APCM-A R belonging to the FAPCM-A $\tilde{R}$. According to Proposition 3, these priorities are such that $\sum_{i=1}^n v_i = n$. The following example is given to illustrate better this interaction property.

Example 58 Let us consider the FAPCM-A

$$
\tilde{R} = \begin{pmatrix} 0.5 & (0.6, 0.7, 0.8, 0.9) & (0.8, 0.9, 0.9, 1) \\ (0.1, 0.2, 0.3, 0.4) & 0.5 & (0.5, 0.6, 0.7, 0.8) \\ (0, 0.1, 0.1, 0.2) & (0.2, 0.3, 0.4, 0.5) & 0.5 \end{pmatrix}. \tag{4.151}
$$

The fuzzy priorities of objects obtained by formulas (4.147) are

$$\tilde{v}_1 = \left(\frac{19}{15}, \frac{21}{15}, \frac{22}{15}, \frac{24}{15}\right), \quad \tilde{v}_2 = \left(\frac{11}{15}, \frac{13}{15}, \frac{15}{15}, \frac{17}{15}\right), \quad \tilde{v}_3 = \left(\frac{7}{15}, \frac{9}{15}, \frac{10}{15}, \frac{12}{15}\right).$$

(4.152)

Let us fix, for example, the upper boundary value $v_1^\delta = \frac{24}{15}$ of $\tilde{v}_1$, and let us show that there exist priorities $v_2 \in \tilde{v}_2$ and $v_3 \in \tilde{v}_3$ such that all three priorities are obtained from the same APCM-A belonging to the FAPCM-A (4.151) and $\sum_{i=1}^3 v_i = 3$.

In order not to violate the additive reciprocity of the related PCs, v_1^δ must have been obtained from the matrix

$$\begin{pmatrix} 0.5 & 0.9 & 1 \\ 0.1 & 0.5 & - \\ 0 & - & 0.5 \end{pmatrix}$$

(4.153)

since $\frac{2}{3}(0.5 + 0.9 + 1) = \frac{24}{15} = v_1^\delta$. Notice that in order to obtain the priority v_i, $i \in \{1, \ldots, n\}$, by formula (2.36) we do not need to know all the PCs in the APCM-A; the PCs in the i-th row are sufficient. The possible values v_2, v_3 of the fuzzy priorities $\tilde{v}_2$, $\tilde{v}_3$ corresponding to the possible value v_1 are then obtainable from APCMs-A

$$\begin{pmatrix} 0.5 & 0.9 & 1 \\ 0.1 & 0.5 & x \\ 0 & 1-x & 0.5 \end{pmatrix}, \qquad x \in [0.5, 0.8],$$

in order to preserve the additive reciprocity of the related PCs. The sum of the priorities obtained from such matrices is always equal to 3:

$$\sum_{i=1}^3 v_i = \frac{24}{15} + \frac{2}{3}(0.1 + 0.5 + x) + \frac{2}{3}(0 + (1-x) + 0.5) = 3 \qquad \triangle$$

Let us now focus on the fuzzy extension of the weakened normalization condition (2.42) introduced in Sect. 2.3.2.2 for the priorities obtainable from APCMs-A. By applying constrained fuzzy arithmetic (3.47) to the fuzzy extension of the formula (2.43), the normalized fuzzy priorities $\tilde{v}_i = \left(v_i^\alpha, v_i^\beta, v_i^\gamma, v_i^\delta\right)$, $i = 1, \ldots, n$, are derived from a FAPCM-A $\tilde{R} = \{\tilde{r}_{ij}\}_{i,j=1}^n$, $\tilde{r}_{ij} = \left(r_{ij}^\alpha, r_{ij}^\beta, r_{ij}^\gamma, r_{ij}^\delta\right)$, as:

$$v_i^\alpha = \min \left\{ \frac{2}{n} \sum_{j=1}^n r_{ij} - \frac{n-1}{n}; \begin{array}{l} r_{pq} \in \left[r_{pq}^\alpha, r_{pq}^\delta\right], \\ r_{pq} = 1 - r_{qp}, \\ p, q = 1, \ldots, n \end{array} \right\},$$

(4.154)

$$v_i^\beta = \min \left\{ \frac{2}{n} \sum_{j=1}^{n} r_{ij} - \frac{n-1}{n}; \begin{array}{l} r_{pq} \in \left[r_{pq}^\beta, r_{pq}^\gamma \right], \\ r_{pq} = 1 - r_{qp}, \\ p, q = 1, \ldots, n \end{array} \right\}, \tag{4.155}$$

$$v_i^\gamma = \max \left\{ \frac{2}{n} \sum_{j=1}^{n} r_{ij} - \frac{n-1}{n}; \begin{array}{l} r_{pq} \in \left[r_{pq}^\beta, r_{pq}^\gamma \right], \\ r_{pq} = 1 - r_{qp}, \\ p, q = 1, \ldots, n \end{array} \right\}, \tag{4.156}$$

$$v_i^\delta = \max \left\{ \frac{2}{n} \sum_{j=1}^{n} r_{ij} - \frac{n-1}{n}; \begin{array}{l} r_{pq} \in \left[r_{pq}^\alpha, r_{pq}^\delta \right], \\ r_{pq} = 1 - r_{qp}, \\ p, q = 1, \ldots, n \end{array} \right\}. \tag{4.157}$$

Analogously as in the case of the formulas (4.143)–(4.146), also the formulas (4.154)–(4.157) can be simplified so that no optimization is needed:

$$v_i^\alpha = \frac{2}{n} \sum_{j=1}^{n} r_{ij}^\alpha - \frac{n-1}{n}, \qquad v_i^\beta = \frac{2}{n} \sum_{j=1}^{n} r_{ij}^\beta - \frac{n-1}{n},$$
$$v_i^\gamma = \frac{2}{n} \sum_{j=1}^{n} r_{ij}^\gamma - \frac{n-1}{n}, \qquad v_i^\delta = \frac{2}{n} \sum_{j=1}^{n} r_{ij}^\delta - \frac{n-1}{n}. \tag{4.158}$$

The normalization property (2.44) valid for the priorities (2.43) obtained from an APCM-A is extended to the fuzzy priorities (4.158) as

$$\forall v_i \in \tilde{v}_{i(\alpha)} \; \exists v_j \in \tilde{v}_{j(\alpha)}, j = 1, \ldots, n, j \neq i : \quad v_i + \sum_{\substack{j=1 \\ j \neq i}}^{n} v_j = 1, \tag{4.159}$$

for all $\alpha \in [0, 1]$ and $i = 1, \ldots, n$. Similarly to Proposition 14 and Theorem 63, the following properties are valid.

Proposition 15 *The interaction property* (4.159) *between the trapezoidal fuzzy numbers* $\tilde{v}_i = \left(v_i^\alpha, v_i^\beta, v_i^\gamma, v_i^\delta \right), i = 1, \ldots, n,$ *is valid if and only if*

$$v_i^\alpha + \sum_{\substack{j=1 \\ j \neq i}}^{n} v_j^\delta \geq 1, \quad v_i^\delta + \sum_{\substack{j=1 \\ j \neq i}}^{n} v_j^\alpha \leq 1, \quad v_i^\beta + \sum_{\substack{j=1 \\ j \neq i}}^{n} v_j^\gamma \geq 1, \quad v_i^\gamma + \sum_{\substack{j=1 \\ j \neq i}}^{n} v_j^\beta \leq 1,$$
$$\tag{4.160}$$

for all $\alpha \in [0, 1]$ *and* $i = 1, \ldots, n$.

Proof The proof is analogous to the proof of Proposition 14. $\square$

Theorem 64 *Let* $\tilde{v}_i = \left(v_i^\alpha, v_i^\beta, v_i^\gamma, v_i^\delta \right), i = 1, \ldots, n,$ *be fuzzy priorities obtained from a FAPCM-A by formulas (4.158). Then the property (4.159) holds for all* $\alpha \in [0, 1]$ *and* $i = 1, \ldots, n.$

Proof The proof is analogous to the proof of Theorem 63.   □

Remark 28 The interaction property (4.159) is in fact the interaction property (3.14) from Definition 29 of the normalized fuzzy vector. However, the fuzzy priorities obtainable from a FAPCM-A by the formulas (4.158) satisfying the property (4.159) are not constrained to the interval [0, 1] as it is required in Definition 29. Therefore, they are not normalized in the sense of Definition 29. Nevertheless, for the simplicity, they will be called "normalized" here (keeping in mind the slight difference in the definitions).

Proposition 16 *Let* $\tilde{v}_i = \left(v_i^\alpha, v_i^\beta, v_i^\gamma, v_i^\delta \right), i = 1, \ldots, n,$ *be normalized fuzzy priorities obtained by formulas (4.158). Then*

$$ -1 < \tilde{v}_i \le 1, \quad i = 1, \ldots, n. \tag{4.161} $$

Proof It is sufficient to prove inequalities $-1 < v_i^\alpha$ and $v_i^\delta \le 1, i = 1, \ldots, n.$ The proof is analogous to the proof of Proposition 5.   □

Remark 29 As mentioned in Sect. 2.3.2.2, for an APCM-A, any vector derived from the priority vector (2.36) by adding an arbitrary constant, i.e. by the transformation (2.38), is again a priority vector. For the case of FAPCMs-A, the formulas (4.158) for obtaining normalized fuzzy priorities are in fact obtained from the formulas (4.147) by adding constant $-\frac{n-1}{n}$. That is, the shape of the trapezoidal fuzzy numbers and the distances between them remain unchanged by applying the normalization condition (4.159); the whole set of trapezoidal fuzzy numbers is just shifted back on the scale of real numbers by $-\frac{n-1}{n}$.

Example 59 Let us consider the FAPCM-A (4.151). The fuzzy priorities obtained by formulas (4.147) are given as (4.152), and the normalized fuzzy priorities obtained by formulas (4.158) are given as

$$ \tilde{v}_{N1} = \left(\frac{9}{15}, \frac{11}{15}, \frac{12}{15}, \frac{14}{15} \right), \; \tilde{v}_{N2} = \left(\frac{1}{15}, \frac{3}{15}, \frac{5}{15}, \frac{7}{15} \right), \; \tilde{v}_{N3} = \left(\frac{-3}{15}, \frac{-1}{15}, 0, \frac{2}{15} \right). \tag{4.162} $$

The fuzzy priorities (4.152) and the normalized fuzzy priorities (4.162) are depicted in Fig. 4.9. It is evident from the figure that the normalized fuzzy priorities have the same shape as the original non-normalized fuzzy priorities; they are just moved backwards by $-\frac{2}{3}$.

△

Theorem 65 *The method for obtaining the normalized fuzzy priorities of objects from FAPCMs-A by using the formulas (4.158) is invariant under permutation of objects in FAPCMs-A.*

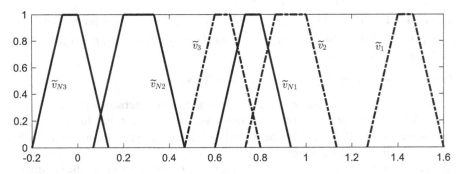

Fig. 4.9 Fuzzy priorities of the FAPCM-A (4.151)

Proof It is sufficient to show that for a given object o_i, $i \in \{1, \ldots, n\}$, its priority $\tilde{v}_i$ obtained by the formulas (4.158) does not change under permutation of objects in a FAPCM-A $\tilde{R}$.

From the invariance of the formula (2.43) reviewed in Sect. 2.3.2.2 it follows that the priority v_i of object o_i determined by the formula (2.43) from the given APCM-A R does not change under any permutation $R^\pi = PRP^T$ of R, it is just permuted accordingly. This means that the priority v_i obtained from R is equal to the corresponding priority $v_{\pi(i)}^\pi$ obtained from R^π.

From this it follows that the priorities v_i^α, v_i^β, v_i^γ, and v_i^δ obtained by the formulas (4.158) from a FAPCM-A $\tilde{R} = \{\tilde{r}_{ij}\}_{i,j=1}^n$, $\tilde{r}_{ij} = (r_{ij}^\alpha, r_{ij}^\beta, r_{ij}^\gamma, r_{ij}^\delta)$, do not change under permutation (they are just permuted accordingly). Therefore, the fuzzy priority $\tilde{v}_i = (v_i^\alpha, v_i^\beta, v_i^\gamma, v_i^\delta)$ does not change for any permutation of objects in a FAPCM-A (it is only permuted accordingly), which concludes the proof.   □

4.3.3 Fuzzy Additive Pairwise Comparison Matrices with Multiplicative Representation

In this Section, the fuzzy extension of the methods related to APCMs-M reviewed in Sect. 2.3.3 is dealt with. In particular, Sect. 4.3.3.1 is dedicated to the extension of the multiplicative-consistency condition (2.48) to FAPCMs-M, and Sect. 4.3.3.2 is focused on methods for obtaining fuzzy priorities of objects from FAPCMs-M.

4.3.3.1 Multiplicative Consistency of FAPCMs-M

In this section, multiplicative consistency of FAPCMs-M is studied. First, definitions of multiplicative consistency for interval FAPCMs-M based on Tanino's characterization (2.54) proposed in the literature are reviewed and some drawbacks of the definitions are pointed out in Sect. "Review of Fuzzy Extensions of Multiplicative

Consistency". Afterwards, in Sect. "New Fuzzy Extension of Multiplicative Consistency", a new definition of multiplicative consistency for FAPCMs-M is proposed.

Review of Fuzzy Extensions of Multiplicative Consistency

In order to examine consistency of interval FAPCMs-M, many definitions of consistency have been proposed in the literature. These definitions are mostly based on interval extension of multiplicative-transitivity property (2.48) and related Tanino's characterization (2.54).

Xu and Chen (2008a) proposed a weak version of multiplicative consistency for interval FAPCMs-M based on Tanino's characterization. Xia and Xu (2011) defined perfect multiplicative consistency for interval FAPCMs-M based on an extension of Tanino's multiplicative-transitivity property and discussed its properties. Wang and Li (2012) proposed another definition of multiplicative consistency for interval FAPCMs-M based on an interval extension of a property equivalent to Tanino's multiplicative-transitivity property. Wu and Chiclana (2014a) defined multiplicative consistency of interval FAPCMs-M based on a direct extension of Tanino's multiplicative-transitivity property to intervals and proposed a consistency index for measuring inconsistency of interval FAPCMs-M. Wu and Chiclana (2014b) proposed another extension of Tanino's multiplicative-transitivity property to define multiplicative consistency of interval FAPCMs-M, and they further extended the definition to intuitionistic FAPCMs-M.

In this section, all these definitions of multiplicatively consistent interval FAPCMs-M are reviewed in detail and some drawbacks are pointed out. In particular, it is shown that some definitions are not invariant under permutation of objects in interval FAPCMs-M and some violate additive reciprocity of the related PCs of objects. Afterwards, it is shown that the drawbacks can be eliminated by employing the constrained fuzzy arithmetic into computations instead of the standard fuzzy arithmetic. The critical review in this section is mostly based on the original paper of Krejčí (2017d) on multiplicative consistency of interval FAPCMs-M.

Xu and Chen (2008) defined a weak version of multiplicative consistency for interval FAPCMs-M. This definition of consistency will be studied in more detail in the following section. In order to distinguish easily this definition of consistency from others, consistency according to this definition will be simply called multiplicative weak consistency.

Definition 61 (Xu and Chen 2008a) Let $\overline{Q} = \left\{\overline{q}_{ij}\right\}_{i,j=1}^{n}$, $\overline{q}_{ij} = \left[q_{ij}^{L}, q_{ij}^{U}\right]$, be an interval FAPCM-M. If there exists a positive vector $\underline{u} = (u_1, \ldots, u_n)^T$ such that

$$q_{ij}^{L} \leq \frac{u_i}{u_i + u_j} \leq q_{ij}^{U}, \quad i, j = 1, \ldots, n, \quad (4.163)$$

then $\overline{Q}$ is called a multiplicatively weakly consistent interval FAPCM-M.

Definition 61 of multiplicative weak consistency for interval FAPCMs-M is clearly based on Proposition 6 for APCMs-M. According to the definition, an interval FAPCM-M $\overline{Q}$ is multiplicatively weakly consistent if there exists a vector $\underline{u} = (u_1, \ldots, u_n)^T$ using which we can construct a multiplicatively consistent APCM-M $Q^* = \left\{ q_{ij}^* \right\}_{i,j=1}^n$ such that $q_{ij}^* \in \left[q_{ij}^L, q_{ij}^U \right]$, $i, j = 1, \ldots, n$.

The requirement of at least one multiplicatively consistent APCM-M obtainable from the interval FAPCM-M is very weak. Therefore, it is quite easy to satisfy the multiplicative-consistency condition in Definition 61 when constructing an interval FAPCM-M. The multiplicative consistency conditions reviewed in the rest of this section are significantly stronger and thus much more difficult to fulfill.

Xia and Xu (2011) defined multiplicative consistency for interval FAPCMs-M as follows.

Definition 62 (Xia and Xu 2011) Let $\overline{Q} = \left\{ \overline{q}_{ij} \right\}_{i,j=1}^n$, $\overline{q}_{ij} = \left[q_{ij}^L, q_{ij}^U \right]$, be an interval FAPCM-M. $\overline{Q}$ is called multiplicatively consistent if the APCMs-M $C = \left\{ c_{ij} \right\}_{i,j=1}^n$, $D = \left\{ d_{ij} \right\}_{i,j=1}^n$ such that

$$c_{ij} = \begin{cases} q_{ij}^L, & i < j \\ 0.5, & i = j \\ q_{ij}^U, & i > j \end{cases}, \quad d_{ij} = \begin{cases} q_{ij}^U, & i < j \\ 0.5, & i = j \\ q_{ij}^L, & i > j \end{cases}, \quad i, j = 1, \ldots, n, \quad (4.164)$$

are multiplicatively consistent according to (2.53).

However, Definition 62 is not invariant under permutation of objects in the interval FAPCM-M. This serious drawback is demonstrated on the following example.

Example 60 (Krejčí 2017d) Let us consider interval FAPCM-M $\overline{Q}$ of three objects o_1, o_2, and o_3 in the form

$$\overline{Q} = \begin{pmatrix} \frac{1}{2} & \left[\frac{1}{2}, \frac{3}{5} \right] & \left[\frac{3}{5}, \frac{6}{7} \right] \\ \left[\frac{2}{5}, \frac{1}{2} \right] & \frac{1}{2} & \left[\frac{3}{5}, \frac{4}{5} \right] \\ \left[\frac{1}{7}, \frac{2}{5} \right] & \left[\frac{1}{5}, \frac{2}{5} \right] & \frac{1}{2} \end{pmatrix}. \quad (4.165)$$

The corresponding APCMs-M C and D given by (4.164) are in the form

$$C = \begin{pmatrix} \frac{1}{2} & \frac{1}{2} & \frac{3}{5} \\ \frac{1}{2} & \frac{1}{2} & \frac{3}{5} \\ \frac{2}{5} & \frac{2}{5} & \frac{1}{2} \end{pmatrix}, \quad D = \begin{pmatrix} \frac{1}{2} & \frac{3}{5} & \frac{6}{7} \\ \frac{2}{5} & \frac{1}{2} & \frac{4}{5} \\ \frac{1}{7} & \frac{1}{5} & \frac{1}{2} \end{pmatrix}.$$

Both APCMs-M C and D satisfy the property (2.53), which means that they are multiplicatively consistent according to Definition 11. Therefore, according to Definition 62, the interval FAPCM-M $\overline{Q}$ is multiplicatively consistent.

Now, let us permute the interval FAPCM-M $\overline{Q}$ to the interval FAPCM-M $\overline{Q}^{\pi} = P\overline{Q}P^{T}$ by using the permutation matrix (4.115):

$$
\overline{Q}^{\pi} = \begin{pmatrix}
\dfrac{1}{2} & \left[\dfrac{1}{7}, \dfrac{2}{5}\right] & \left[\dfrac{1}{5}, \dfrac{2}{5}\right] \\[2ex]
\left[\dfrac{3}{5}, \dfrac{6}{7}\right] & \dfrac{1}{2} & \left[\dfrac{1}{2}, \dfrac{3}{5}\right] \\[2ex]
\left[\dfrac{3}{5}, \dfrac{4}{5}\right] & \left[\dfrac{2}{5}, \dfrac{1}{2}\right] & \dfrac{1}{2}
\end{pmatrix}. \tag{4.166}
$$

The corresponding APCMs-M C^{π} and D^{π} given by (4.164) are in the form

$$
C^{\pi} = \begin{pmatrix}
\dfrac{1}{2} & \dfrac{1}{7} & \dfrac{1}{5} \\[1.5ex]
\dfrac{6}{7} & \dfrac{1}{2} & \dfrac{1}{2} \\[1.5ex]
\dfrac{4}{5} & \dfrac{1}{2} & \dfrac{1}{2}
\end{pmatrix}, \quad
D^{\pi} = \begin{pmatrix}
\dfrac{1}{2} & \dfrac{2}{5} & \dfrac{2}{5} \\[1.5ex]
\dfrac{3}{5} & \dfrac{1}{2} & \dfrac{3}{5} \\[1.5ex]
\dfrac{3}{5} & \dfrac{2}{5} & \dfrac{1}{2}
\end{pmatrix},
$$

and they clearly do not satisfy the property (2.53). Therefore, according to Definition 62, the interval FAPCM-M $\overline{Q}^{\pi}$ is not multiplicatively consistent.

Obviously, the PCs of objects o_1, o_2, and o_3 in interval FAPCMs-M $\overline{Q}$ and $\overline{Q}^{\pi}$ are the same, they vary only in the order in which they are associated with the rows and the columns of the interval FAPCM-M. Therefore, also the conclusion about the multiplicative consistency should be the same for both interval FAPCMs-M. However, as demonstrated, the interval FAPCM-M $\overline{Q}$ is judged as multiplicatively consistent while the interval FAPCM-M $\overline{Q}^{\pi}$ results to be inconsistent according to Definition 62. $\triangle$

Wang and Li (2012) proposed the following definition of multiplicative consistency for interval FAPCMs-M.

Definition 63 (Wang and Li 2012) Let $\overline{Q} = \{\overline{q}_{ij}\}_{i,j=1}^{n}$, $\overline{q}_{ij} = \left[q_{ij}^{L}, q_{ij}^{U}\right]$, be an interval FAPCM-M. $\overline{Q}$ is called multiplicatively consistent if the multiplicative-transitivity condition

$$
\frac{\overline{q}_{ji}}{\overline{q}_{ij}} \frac{\overline{q}_{kj}}{\overline{q}_{jk}} \frac{\overline{q}_{ik}}{\overline{q}_{ki}} = \frac{\overline{q}_{jk}}{\overline{q}_{kj}} \frac{\overline{q}_{ij}}{\overline{q}_{ji}} \frac{\overline{q}_{ki}}{\overline{q}_{ik}}, \quad i, j, k = 1, \ldots, n, \tag{4.167}
$$

is satisfied.

Wang and Li (2012) applied the standard interval arithmetic to extend the multiplicative consistency of APCMs-M to interval FAPCMs-M in Definition 63. Therefore, the equation (4.167) is equivalent to the equation

$$\frac{q_{ji}^L q_{kj}^L q_{ik}^L}{q_{ij}^U q_{jk}^U q_{ki}^U} = \frac{q_{jk}^L q_{ij}^L q_{ki}^L}{q_{kj}^U q_{ji}^U q_{ik}^U}, \qquad i,j,k = 1, \ldots, n. \tag{4.168}$$

Beside claiming that $\overline{q}_{ij} + \overline{q}_{ji} = 1$ cannot be reached, as reviewed in Sect. "Review of Fuzzy Extensions of Additive Consistency", Wang and Li (2012) pointed out that "due to the possibility of $\frac{\overline{a}}{\overline{a}} \neq 1$ for intervals, (4.167) is not equivalent to

$$\frac{\overline{q}_{ik}}{\overline{q}_{ki}} \frac{\overline{q}_{kj}}{\overline{q}_{jk}} = \frac{\overline{q}_{ij}}{\overline{q}_{ji}}, \qquad i,j,k = 1, \ldots, n, \tag{4.169}$$

as in the case of regular APCMs-M" (Wang and Li 2012, p. 183). However, it is necessary to clarify here that, in contrast to Wang and Li's assertion, it is possible to ensure easily the validity of $\frac{\overline{a}}{\overline{a}} = 1$. At the end of this section, it will be demonstrated that the validity of equation $\frac{\overline{a}}{\overline{a}} = 1$ can be easily achieved by applying appropriately the constrained fuzzy arithmetic instead of the standard fuzzy arithmetic. Furthermore, it will be shown that Definition 63 of multiplicative consistency is inappropriate since it violates the additive reciprocity of the related PCs. Before doing that, let us finalize the literature review.

Wu and Chiclana (2014a) and Wu and Chiclana (2014b) defined multiplicative transitivity for interval FAPCMs-M. However, since the multiplicative-transitivity property is normally used to define the multiplicative consistency (and this is what is done in this book), the expression "multiplicative consistency" will be used in their definition instead of "multiplicative transitivity" in order to keep the same terminology.

Definition 64 (Wu and Chiclana 2014a) Let $\overline{Q} = \{\overline{q}_{ij}\}_{i,j=1}^n$, $\overline{q}_{ij} = \left[q_{ij}^L, q_{ij}^U \right]$, be an interval FAPCM-M. $\overline{Q}$ is called multiplicatively consistent if

$$\frac{\overline{q}_{ki}}{\overline{q}_{ik}} = \frac{\overline{q}_{kj}}{\overline{q}_{jk}} \frac{\overline{q}_{ji}}{\overline{q}_{ij}}, \qquad i < j < k, \ i,j,k = 1, \ldots, n. \tag{4.170}$$

Wu and Chiclana (2014a) defined multiplicatively consistent interval FAPCMs-M by extending the multiplicative-transitivity property (2.48). For the extension they used the standard interval arithmetic. Therefore, Wu and Chiclana (2014a) derived that the equation (4.170) is equivalent to the equations

$$\begin{aligned} \frac{1}{q_{ik}^L} - 1 &= \left(\frac{1}{q_{ij}^L} - 1 \right) \left(\frac{1}{q_{jk}^L} - 1 \right), \\ \frac{1}{q_{ik}^U} - 1 &= \left(\frac{1}{q_{ij}^U} - 1 \right) \left(\frac{1}{q_{jk}^U} - 1 \right), \end{aligned} \qquad i < j < k, \ i,j,k = 1, \ldots, n, \tag{4.171}$$

which can be further written as

$$q_{ik}^L = \frac{q_{ij}^L q_{jk}^L}{q_{ij}^L q_{jk}^L + (1 - q_{ij}^L)(1 - q_{jk}^L)},$$
$$q_{ik}^U = \frac{q_{ij}^U q_{jk}^U}{q_{ij}^U q_{jk}^U + (1 - q_{ij}^U)(1 - q_{jk}^U)}, \quad i < j < k, \ i, j, k = 1, \dots, n. \quad (4.172)$$

However, it will be demonstrated on the following example that Definition 64, similarly to Definition 62, is not invariant under permutation of objects in interval FAPCMs-M.

Example 61 Let us consider again the interval FAPCM-M $\overline{Q}$ of three objects o_1, o_2, and o_3 in the form (4.165). In order to evaluate whether $\overline{Q}$ is multiplicatively consistent according to Definition 64, we only need to verify whether the equations (4.172) are satisfied for $i = 1, j = 2, k = 3$. Since

$$\frac{q_{12}^L q_{23}^L}{q_{12}^L q_{23}^L + (1 - q_{12}^L)(1 - q_{23}^L)} = \frac{\frac{1}{2}\frac{3}{5}}{\frac{1}{2}\frac{3}{5} + \frac{1}{2}\frac{2}{5}} = \frac{3}{5} = q_{13}^L$$

and

$$\frac{q_{12}^U q_{23}^U}{q_{12}^U q_{23}^U + (1 - q_{12}^U)(1 - q_{23}^U)} = \frac{\frac{3}{5}\frac{4}{5}}{\frac{3}{5}\frac{4}{5} + \frac{2}{5}\frac{1}{5}} = \frac{6}{7} = q_{13}^U,$$

we can conclude that the interval FAPCM-M (4.165) is multiplicatively consistent according to Definition 64.

Now, let us consider the permuted interval FAPCM-M $\overline{Q}^\pi$ given by (4.166). By verifying equations (4.172) for $i = 1, j = 2, k = 3$, we obtain

$$\frac{q_{12}^{*L} q_{23}^{*L}}{q_{12}^{*L} q_{23}^{*L} + (1 - q_{12}^{*L})(1 - q_{23}^{*L})} = \frac{\frac{1}{7}\frac{1}{2}}{\frac{1}{7}\frac{1}{2} + \frac{6}{7}\frac{1}{2}} = \frac{1}{7} \neq q_{13}^{*L}$$

and

$$\frac{q_{12}^{*U} q_{23}^{*U}}{q_{12}^{*U} q_{23}^{*U} + (1 - q_{12}^{*U})(1 - q_{23}^{*U})} = \frac{\frac{2}{5}\frac{3}{5}}{\frac{2}{5}\frac{3}{5} + \frac{3}{5}\frac{2}{5}} = \frac{1}{2} \neq q_{13}^{*U},$$

which means that the permuted interval FAPCM-M (4.166) is not multiplicatively consistent according to Definition 64. Therefore, it results that Definition 64 of multiplicative consistency is not invariant under permutation of objects in interval FAPCMs-M. △

Definition 65 (Wu and Chiclana 2014b) Let $\overline{Q} = \{\overline{q}_{ij}\}_{i,j=1}^n$, $\overline{q}_{ij} = \left[q_{ij}^L, q_{ij}^U \right]$, be an interval FAPCM-M. $\overline{Q}$ is called multiplicatively consistent if

$$q_{ij}^L q_{jk}^L q_{ki}^L = q_{ik}^L q_{kj}^L q_{ji}^L,$$
$$q_{ij}^U q_{jk}^U q_{ki}^U = q_{ik}^U q_{kj}^U q_{ji}^U, \quad i, j, k = 1, \dots, n. \quad (4.173)$$

The equations (4.173) are equivalent to the equation

$$\bar{q}_{ij}\bar{q}_{jk}\bar{q}_{ki} = \bar{q}_{ik}\bar{q}_{kj}\bar{q}_{ji}, \qquad i,j,k = 1,\ldots,n, \tag{4.174}$$

when using standard fuzzy arithmetic (and in particular the formula (3.28) for the multiplication of trapezoidal fuzzy numbers). Therefore, Definition 65 suffers again from the same drawbacks as Definition 63; the additive reciprocity of the related PCs is violated. For a detailed discussion on the issue of additive reciprocity, see p. 159.

The problem is caused by the fact that both Wang and Li (2012) and Wu and Chiclana (2014b), similarly to other researchers whose work has been reviewed in this section, applied the standard interval arithmetic to the computations with intervals. This means that the multiplicative-consistency condition (4.167) is equivalent to the equation (4.168) and the multiplicative-consistency condition (4.173) is equivalent to the equation (4.174). Similarly to the additive-consistency conditions (4.117), (4.121), and (4.122) described in Sect. "Review of Fuzzy Extensions of Additive Consistency", none of the multiplicative-consistency conditions (4.167) and (4.173) preserves the additive reciprocity of the related PCs. This drawback is demonstrated on the following example.

Example 62 Let us examine Definitions 63 and 65 of multiplicative consistency on the interval FAPCM-M $\bar{Q}$ given by (4.165). The expressions (4.173) basically mean that we construct matrices $Q^L = \left\{q_{ij}^L\right\}_{i=1}^n$ and $Q^U = \left\{q_{ij}^U\right\}_{i=1}^n$ from the interval FAPCM-M (4.165) as

$$Q^L = \begin{pmatrix} \dfrac{1}{2} & \dfrac{1}{2} & \dfrac{3}{5} \\[2mm] \dfrac{2}{5} & \dfrac{1}{2} & \dfrac{3}{5} \\[2mm] \dfrac{1}{7} & \dfrac{1}{5} & \dfrac{1}{2} \end{pmatrix}, \qquad Q^U = \begin{pmatrix} \dfrac{1}{2} & \dfrac{3}{5} & \dfrac{6}{7} \\[2mm] \dfrac{2}{5} & \dfrac{1}{2} & \dfrac{4}{5} \\[2mm] \dfrac{2}{5} & \dfrac{2}{5} & \dfrac{1}{2} \end{pmatrix},$$

and we verify their multiplicative consistency by utilizing the property (2.50). However, we can easily see that neither Q^L nor Q^U is additively reciprocal, which means that both Q^L and Q^U are not even APCMs-M according to Definition 10. Therefore, it is nonsensical to verify their "multiplicative consistency".

An analogous drawback appears also when using Definition 63. In addition, two values of each intensity of preference $\bar{q}_{ij}$, $i,j = 1, 2, 3$, $i \neq j$, appear in the expression (4.168) at the same time. For example, the intensities $\frac{1}{2}$ and $\frac{3}{5}$ of preference of object o_1 over object o_2 are considered at the same time, which is nonsensical. $\triangle$

For reader's convenience the properties regarding the reciprocity of the related PCs and invariance under permutations of all definitions of multiplicative consistency reviewed in this section are listed in Table 4.14. To summarize, only Definition 61

Table 4.14 Properties of the definitions of multiplicative consistency for FAPCMs-M

Definition	Authors	Invariance under permutation	Multiplicative reciprocity
Definition 61	Xu and Chen (2008a)	✓	✓
Definition 62	Xia and Xu (2011)	✗	✓
Definition 63	Wang and Li (2012)	✓	✗
Definition 64	Wu and Chiclana (2014a)	✗	✓
Definition 65	Wu and Chiclana (2014b)	✓	✗

of multiplicative weak consistency introduced by Xu and Chen (2008a) for interval FAPCMs-M keeps both properties; all other definitions violate one of them.

As already mentioned in the discussion following Definition 63, Wang and Li (2012) claim that it is not possible to obtain the equality $\frac{\overline{a}}{\overline{a}} = 1$. Obviously, for $\overline{a} = [a^L, a^U]$, we obtain

$$\frac{\overline{a}}{\overline{a}} = \frac{[a^L, a^U]}{[a^L, a^U]} = \left[\frac{a^L}{a^U}, \frac{a^U}{a^L} \right] \neq 1, \qquad \text{unless } a^L = a^U.$$

Applying the standard fuzzy arithmetic in this case is not appropriate; the constrained fuzzy arithmetic needs to be applied whenever there are any interactions among fuzzy numbers. The presence of interactions in the expression $\frac{\overline{a}}{\overline{a}}$ is clear since the intervals in the expression represent the same variable. According to Krejčí (2017d), $\overline{d} = \frac{\overline{a}}{\overline{a}}$ should be computed by using the constrained fuzzy arithmetic (3.42) as $\overline{d} = [d^L, d^U]$:

$$d^L = \min \left\{ \frac{a_1}{a_2}; a_1 \in [a^L, a^U], a_2 \in [a^L, a^U], a_1 = a_2 \right\} =$$
$$= \min \left\{ \frac{a_1}{a_1}; a_1 \in [a^L, a^U] \right\} = 1,$$
$$d^U = \max \left\{ \frac{a_1}{a_2}; a_1 \in [a^L, a^U], a_2 \in [a^L, a^U], a_1 = a_2 \right\} =$$
$$= \max \left\{ \frac{a_1}{a_1}; a_1 \in [a^L, a^U] \right\} = 1.$$

Keeping in mind the importance of the additive-reciprocity property of PCs in interval FAPCMs-M, multiplicative consistency needs to be defined accordingly so that it does not violate the additive reciprocity. In the following section, two definitions of multiplicatively consistent trapezoidal FAPCMs-M keeping the additive reciprocity of the related PCs and invariant under permutation of objects are proposed.

New Fuzzy Extension of Multiplicative Consistency

In this section, Definition 61 of multiplicative weak consistency given by Xu and Chen (2008a) is extended to trapezoidal FAPCMs-M and another definition of multiplicative consistency much stronger than Definition 61 is proposed. Tools for verifying both the multiplicative weak consistency and the multiplicative consistency are provided and some properties of multiplicatively weakly consistent and multiplicatively consistent trapezoidal FAPCMs-M are derived. Both definitions preserve two desired properties—invariance under permutation of objects and additive reciprocity of the related PCs. The findings summarized in this section are mostly based on the original paper of Krejčí (2017d) on the multiplicative consistency for interval FAPCMs-M.

Definition 66 Let $\widetilde{Q} = \{\widetilde{q}_{ij}\}_{i,j=1}^{n}$, $\widetilde{q}_{ij} = (q_{ij}^{\alpha}, q_{ij}^{\beta}, q_{ij}^{\gamma}, q_{ij}^{\delta})$, be a trapezoidal FAPCM-M. $\widetilde{Q}$ is said to be multiplicatively weakly consistent if there exists a positive vector $\underline{u} = (u_1, \ldots, u_n)^T$ such that

$$q_{ij}^{\alpha} \leq \frac{u_i}{u_i + u_j} \leq q_{ij}^{\delta}, \qquad i, j = 1, \ldots, n. \qquad (4.175)$$

Note that when Definition 66 is applied to interval FAPCMs-M, it is identical to Definition 61 proposed by Xu and Chen (2008a).

Proposition 17 *Let $\widetilde{Q} = \{\widetilde{q}_{ij}\}_{i,j=1}^{n}$, $\widetilde{q}_{ij} = (q_{ij}^{\alpha}, q_{ij}^{\beta}, q_{ij}^{\gamma}, q_{ij}^{\delta})$, be a trapezoidal FAPCM-M. $\widetilde{Q} = \{\widetilde{q}_{ij}\}_{i,j=1}^{n}$ is multiplicatively weakly consistent according to Definition 66 if and only if there exist elements $q_{ij}^{*} \in \left[q_{ij}^{\alpha}, q_{ij}^{\delta}\right]$, $i, j = 1, \ldots, n$, such that $Q^{*} = \left\{q_{ij}^{*}\right\}_{i,j=1}^{n}$ is an APCM-M multiplicatively consistent according to Definition 11.*

Proof First, let $\widetilde{Q} = \{\widetilde{q}_{ij}\}_{i,j=1}^{n}$, $\widetilde{q}_{ij} = (q_{ij}^{\alpha}, q_{ij}^{\beta}, q_{ij}^{\gamma}, q_{ij}^{\delta})$, be a trapezoidal FAPCM-M multiplicatively weakly consistent according to Definition 66. Let us denote $q_{ij}^{*} := \frac{u_i}{u_i + u_j}$. From (4.175) it follows that $q_{ij}^{*} \in \left[q_{ij}^{\alpha}, q_{ij}^{\delta}\right]$, $i, j = 1, \ldots, n$. Further, we have $q_{ii}^{*} = \frac{u_i}{u_i + u_i} = 0.5$ and $q_{ji}^{*} = \frac{u_j}{u_j + u_i} = 1 - \frac{u_i}{u_i + u_j} = 1 - q_{ij}^{*}$, $i, j = 1, \ldots, n$. From $[q_{ij}^{\alpha}, q_{ij}^{\delta}] \subseteq]0, 1[$, $i, j = 1, \ldots, n$, it follows that $q_{ij}^{*} \in]0, 1[$, $i, j = 1, \ldots, n$. Therefore, $Q^{*} = \left\{q_{ij}^{*}\right\}_{i,j=1}^{n}$ is a FAPCM-M.

Finally,

$$\frac{q_{ik}^{*}}{q_{ki}^{*}} \frac{q_{kj}^{*}}{q_{jk}^{*}} = \frac{\frac{u_i}{u_i+u_k}}{\frac{u_k}{u_k+u_i}} \frac{\frac{u_k}{u_k+u_j}}{\frac{u_j}{u_j+u_k}} = \frac{u_i}{u_j} = \frac{\frac{u_i}{u_i+u_j}}{\frac{u_j}{u_j+u_i}} = \frac{q_{ij}^{*}}{q_{ji}^{*}}, \qquad i, j, k = 1, \ldots, n,$$

which means that $Q^* = \left\{q_{ij}^*\right\}_{i,j=1}^n$ is multiplicatively consistent according to (2.48).

In the opposite direction, let $\tilde{Q} = \{\tilde{q}_{ij}\}_{i,j=1}^n$, $\tilde{q}_{ij} = (q_{ij}^\alpha, q_{ij}^\beta, q_{ij}^\gamma, q_{ij}^\delta)$, be a trapezoidal FAPCM-M and let $Q^* = \left\{q_{ij}^*\right\}_{i,j=1}^n$, $q_{ij}^* \in \left[q_{ij}^\alpha, q_{ij}^\delta\right]$, $i,j = 1, \ldots, n$, be an APCM-M multiplicatively consistent according to (2.48). Then, from Proposition 6 it follows that there exists a vector $\underline{u} = (u_1, \ldots, u_n)^T$, $u_i > 0$, $i,j = 1, \ldots, n$, such that $q_{ij}^* = \frac{u_i}{u_i + u_j}$, $i,j = 1, \ldots, n$. Because, $q_{ij}^* \in \left[q_{ij}^\alpha, q_{ij}^\delta\right]$, $i,j = 1, \ldots, n$, then (4.175) holds. □

Remark 30 According to Proposition 17 and its proof, a trapezoidal FAPCM-M $\tilde{Q} = \{\tilde{q}_{ij}\}_{i,j=1}^n$, $\tilde{q}_{ij} = (q_{ij}^\alpha, q_{ij}^\beta, q_{ij}^\gamma, q_{ij}^\delta)$, is multiplicatively weakly consistent if and only if there exists a multiplicatively consistent APCM-M $Q^* = \left\{q_{ij}^*\right\}_{i,j=1}^n$ such that $q_{ij}^* \in \left[q_{ij}^\alpha, q_{ij}^\delta\right]$. This consistency condition is quite easy to reach. That is why the multiplicative consistency according to Definition 66 is called weak. Later in this section, a much stronger definition of multiplicative consistency for trapezoidal FAPCMs-M will be given.

Definition 66 of multiplicative weak consistency satisfies two desirable properties—invariance under permutation of objects and additive reciprocity of the related PCs in trapezoidal FAPCMs-M.

Theorem 66 *Definition 66 of multiplicative weak consistency is invariant under permutation of objects in FAPCMs-M.*

Proof For a multiplicatively weakly consistent trapezoidal FAPCM-M $\tilde{Q} = \{\tilde{q}_{ij}\}_{i,j=1}^n$, $\tilde{q}_{ij} = \left(q_{ij}^\alpha, q_{ij}^\beta, q_{ij}^\gamma, q_{ij}^\delta\right)$, there exists a positive priority vector $\underline{u} = (u_1, \ldots, u_n)^T$, such that the inequality $q_{ij}^\alpha \leq \frac{u_i}{u_i + u_j} \leq q_{ij}^\delta$ is required to hold for every single PC $\tilde{q}_{ij}$. By permuting the FAPCM-M $\tilde{Q}$ to $\tilde{Q}^\pi = P\tilde{Q}P^T$, the original PC $\tilde{q}_{ij}$ in the i-th row and in the j-th column of $\tilde{Q}$ is moved to the $\pi(i)$-th row and to the $\pi(j)$-th column of the permuted trapezoidal FAPCM-M $\tilde{Q}^\pi$ as $\tilde{q}_{\pi(i)\pi(j)}^\pi$, but still keeping $\tilde{q}_{ij} = \tilde{q}_{\pi(i)\pi(j)}^\pi$, $i,j = 1, \ldots, n$. Thus, there exists a vector $\underline{u}^\pi = (u_1^\pi, \ldots, u_n^\pi)^T$ obtained by permuting the vector $\underline{u}$, i.e. $\underline{u}^\pi = P\underline{u}$, with the components satisfying the inequalities $q_{ij}^{\pi\alpha} \leq \frac{u_i}{u_i + u_j} \leq q_{ij}^{\pi\delta}$ for every $i,j = 1, \ldots, n$. □

Theorem 67 *Definition 66 of multiplicative weak consistency does not violate the additive reciprocity of the related PCs in trapezoidal FAPCMs-M in the sense that any fixed value $q_{ij} \in [q_{ij}^\alpha, q_{ij}^\delta]$, $i,j \in \{1, \ldots, n\}$, representing the intensity of preference of object o_i over object o_j is associated with the corresponding values $q_{ji} \in [q_{ji}^\alpha, q_{ji}^\delta]$ representing the intensity of preference of object o_j over object o_i such that $q_{ji} = 1 - q_{ij}$.*

Proof The existence of the positive priority vector $\underline{u} = (u_1, \ldots, u_n)^T$ satisfying the inequalities (4.175) means that there exists an APCM-M $Q = \{q_{ij}\}_{i,j=1}^n$, $q_{ij} \in$

$[q_{ij}^{\alpha}, q_{ij}^{\delta}]$, such that $q_{ij} = \frac{u_i}{u_i + u_j}$, $i, j = 1, \ldots, n$. Q is additively reciprocal from the definition, i.e., every PC q_{ij} is associated with the PC q_{ji} such that $q_{ji} = 1 - q_{ij}$. $\quad\square$

Remark 31 Note that Theorem 67 does not simply state that a FAPCM-M $\widetilde{Q} = \{\widetilde{q}_{ij}\}_{i,j=1}^{n}$ multiplicatively weakly consistent according to Definition 66 is additively reciprocal, i.e. $\widetilde{q}_{ji} = 1 - \widetilde{q}_{ij}$, $i, j = 1, \ldots, n$. Theorem 67 states that only additively reciprocal PCs are involved in Definition 66 of multiplicative weak consistency, which is in accordance with the conception of additive reciprocity discussed on p. 160. For more detailed discussion see Remark 21.

The following theorems provide useful tools for verifying multiplicative weak consistency of trapezoidal FAPCMs-M.

Theorem 68 *A trapezoidal FAPCM-M* $\widetilde{Q} = \{\widetilde{q}_{ij}\}_{i,j=1}^{n}$, $\widetilde{q}_{ij} = \left(q_{ij}^{\alpha}, q_{ij}^{\beta}, q_{ij}^{\gamma}, q_{ij}^{\delta}\right)$, *is multiplicatively weakly consistent according to Definition 66 if and only if*

$$\max_{k=1,\ldots,n} \left\{ \frac{q_{ik}^{\alpha} q_{kj}^{\alpha}}{q_{ik}^{\alpha} q_{kj}^{\alpha} + (1 - q_{ik}^{\alpha})(1 - q_{kj}^{\alpha})} \right\} \leq \min_{k=1,\ldots,n} \left\{ \frac{q_{ik}^{\delta} q_{kj}^{\delta}}{q_{ik}^{\delta} q_{kj}^{\delta} + (1 - q_{ik}^{\delta})(1 - q_{kj}^{\delta})} \right\},$$
$$i, j = 1, \ldots, n. \tag{4.176}$$

Proof From the inequalities (4.175) it follows that $q_{ik}^{\alpha} \leq \frac{u_i}{u_i + u_k} \leq q_{ik}^{\delta}$ and $q_{kj}^{\alpha} \leq \frac{u_k}{u_k + u_j} \leq q_{kj}^{\delta}$. Thus, $\forall k \in \{1, \ldots, n\}$ the following inequalities hold:

$$q_{ik}^{\alpha} q_{kj}^{\alpha} \leq \frac{u_i}{u_i + u_k} \frac{u_k}{u_k + u_j} \leq q_{ik}^{\delta} q_{kj}^{\delta},$$

$$1 - q_{ik}^{\alpha} \geq \frac{u_k}{u_i + u_k} \geq 1 - q_{ik}^{\delta}, \quad 1 - q_{kj}^{\alpha} \geq \frac{u_j}{u_k + u_j} \geq 1 - q_{kj}^{\delta},$$

$$(1 - q_{ik}^{\alpha})(1 - q_{kj}^{\alpha}) \geq \frac{u_k}{u_i + u_k} \frac{u_j}{u_k + u_j} \geq (1 - q_{ik}^{\delta})(1 - q_{kj}^{\delta}).$$

Putting all together we obtain $\forall k \in \{1, \ldots, n\}$

$$\frac{q_{ik}^{\alpha} q_{kj}^{\alpha}}{q_{ik}^{\alpha} q_{kj}^{\alpha} + (1 - q_{ik}^{\alpha})(1 - q_{kj}^{\alpha})} \leq \frac{u_i}{u_i + u_j} \leq \frac{q_{ik}^{\delta} q_{kj}^{\delta}}{q_{ik}^{\delta} q_{kj}^{\delta} + (1 - q_{ik}^{\delta})(1 - q_{kj}^{\delta})},$$

and thus (4.176) holds.

In the opposite direction, let (4.176) hold. Then, $\forall i, j, k \in \{1, \ldots, n\}$:

$$q_{ij}^{\alpha} \leq \max_{k=1,\ldots,n} \left\{ \frac{q_{ik}^{\alpha} q_{kj}^{\alpha}}{q_{ik}^{\alpha} q_{kj}^{\alpha} + (1 - q_{ik}^{\alpha})(1 - q_{kj}^{\alpha})} \right\} \leq \min_{k=1,\ldots,n} \left\{ \frac{q_{ik}^{\delta} q_{kj}^{\delta}}{q_{ik}^{\delta} q_{kj}^{\delta} + (1 - q_{ik}^{\delta})(1 - q_{kj}^{\delta})} \right\} \leq q_{ij}^{\delta}.$$

Thus, $\forall i, j, k \in \{1, \ldots, n\}$:

$$\exists q_{ij}^* \in \left[\max_{k=1,\ldots,n} \left\{ \frac{q_{ik}^\alpha q_{kj}^\alpha}{q_{ik}^\alpha q_{kj}^\alpha + (1 - q_{ik}^\alpha)(1 - q_{kj}^\alpha)} \right\}, \min_{k=1,\ldots,n} \left\{ \frac{q_{ik}^\delta q_{kj}^\delta}{q_{ik}^\delta q_{kj}^\delta + (1 - q_{ik}^\delta)(1 - q_{kj}^\delta)} \right\} \right]$$

$$\wedge \exists q_{ik}^* \in [q_{ik}^\alpha, q_{ik}^\delta] \wedge \exists q_{kj}^* \in [q_{kj}^\alpha, q_{kj}^\delta] : q_{ij}^* = q_{ik}^* + q_{kj}^* - 0.5.$$

This means that $Q^* = \left\{ q_{ij}^* \right\}_{i,j=1}^n$ is an APCM-M. Thus, according to Proposition 6, there exists a positive vector $\underline{u} = (u_1, \ldots, u_n)^T$ such that $q_{ij}^* = \frac{u_i}{u_i + u_j}$. Since $q_{ij}^* \in [q_{ij}^\alpha, q_{ij}^\delta], i, j = 1, \ldots, n$, we obtain the inequality (4.175). $\qquad \square$

The following theorem shows that it is sufficient to verify the inequality (4.176) only for $i, j = 1, \ldots, n$, $i < j$, thus saving half of the computations.

Theorem 69 *A trapezoidal FAPCM-M $\tilde{Q} = \{\tilde{q}_{ij}\}_{i,j=1}^n, \tilde{q}_{ij} = \left(q_{ij}^\alpha, q_{ij}^\beta, q_{ij}^\gamma, q_{ij}^\delta \right),$ is multiplicatively weakly consistent according to Definition 66 if and only if*

$$\max_{k=1,\ldots,n} \left\{ \frac{q_{ik}^\alpha q_{kj}^\alpha}{q_{ik}^\alpha q_{kj}^\alpha + (1 - q_{ik}^\alpha)(1 - q_{kj}^\alpha)} \right\} \leq \min_{k=1,\ldots,n} \left\{ \frac{q_{ik}^\delta q_{kj}^\delta}{q_{ik}^\delta q_{kj}^\delta + (1 - q_{ik}^\delta)(1 - q_{kj}^\delta)} \right\},$$
$$i, j = 1, \ldots, n, \ i < j. \tag{4.177}$$

Proof It is sufficient to show that the validity of inequalities (4.177) for $i, j = 1, \ldots, n$, $i < j$, implies automatically their validity for all $i, j = 1, \ldots, n$, i.e. the validity of (4.176). The validity of inequalities (4.176) for $i = j$ is trivial from the definition of trapezoidal FAPCMs-M since

$$\max_{k=1,\ldots,n} \left\{ \frac{q_{ik}^\alpha q_{ki}^\alpha}{q_{ik}^\alpha q_{ki}^\alpha + (1 - q_{ik}^\alpha)(1 - q_{ki}^\alpha)} \right\} = \max_{k=1,\ldots,n} \left\{ \frac{q_{ik}^\alpha q_{ki}^\alpha}{q_{ik}^\alpha q_{ki}^\alpha + q_{ki}^\delta q_{ik}^\delta} \right\} \leq 0.5 \leq$$

$$\min_{k=1,\ldots,n} \left\{ \frac{q_{ik}^\delta q_{ki}^\delta}{q_{ik}^\delta q_{ki}^\delta + q_{ki}^\alpha q_{ik}^\alpha} \right\} = \min_{k=1,\ldots,n} \left\{ \frac{q_{ik}^\delta q_{ki}^\delta}{q_{ik}^\delta q_{ki}^\delta + (1 - q_{ik}^\delta)(1 - q_{ki}^\delta)} \right\}.$$

Further, for $i > j$, by using (4.177) and the additive-reciprocity properties, we obtain

$$\max_{k=1,\ldots,n} \left\{ \frac{q_{ik}^\alpha q_{kj}^\alpha}{q_{ik}^\alpha q_{kj}^\alpha + (1 - q_{ik}^\alpha)(1 - q_{kj}^\alpha)} \right\} = \max_{k=1,\ldots,n} \left\{ 1 - \frac{q_{ki}^\delta q_{jk}^\delta}{q_{ki}^\delta q_{jk}^\delta + (1 - q_{ki}^\delta)(1 - q_{jk}^\delta)} \right\} =$$

$$1 - \min_{k=1,\dots,n} \left\{ \frac{q_{jk}^{\delta} q_{ki}^{\delta}}{q_{jk}^{\delta} q_{ki}^{\delta} + (1 - q_{jk}^{\delta})(1 - q_{ki}^{\delta})} \right\} \leq 1 - \max_{k=1,\dots,n} \left\{ \frac{q_{jk}^{\alpha} q_{ki}^{\alpha}}{q_{jk}^{\alpha} q_{ki}^{\alpha} + (1 - q_{jk}^{\alpha})(1 - q_{ki}^{\alpha})} \right\} =$$

$$\min_{k=1,\dots,n} \left\{ 1 - \frac{q_{jk}^{\alpha} q_{ki}^{\alpha}}{q_{jk}^{\alpha} q_{ki}^{\alpha} + (1 - q_{jk}^{\alpha})(1 - q_{ki}^{\alpha})} \right\} = \min_{k=1,\dots,n} \left\{ \frac{q_{ik}^{\delta} q_{kj}^{\delta}}{q_{ik}^{\delta} q_{kj}^{\delta} + (1 - q_{ik}^{\delta})(1 - q_{kj}^{\delta})} \right\}.$$

□

Remark 32 An alternative definition of multiplicative weak consistency to Definition 66 might be formulated as follows.

Let $\widetilde{Q} = \{\widetilde{q}_{ij}\}_{i,j=1}^{n}$, $\widetilde{q}_{ij} = \left(q_{ij}^{\alpha}, q_{ij}^{\beta}, q_{ij}^{\gamma}, q_{ij}^{\delta} \right)$, be a trapezoidal FAPCM-M. $\widetilde{Q}$ is said to be multiplicatively weakly consistent if there exists a positive vector $\underline{u} = (u_1, \dots, u_n)^T$ such that

$$q_{ij}^{\beta} \leq \frac{u_i}{u_i + u_j} \leq q_{ij}^{\gamma}, \qquad i, j = 1, \dots, n. \tag{4.178}$$

Notice that this definition is stronger than Definition 66. In fact, every trapezoidal FAPCM-M multiplicatively weakly consistent according to this definition is also multiplicatively weakly consistent according to Definition 66 since (4.178) automatically implies (4.175). Furthermore, when this definition is applied to interval FAPCMs-M, it is again identical to Definition 61 proposed by Xu and Chen (2008a).

All theorems regarding FAPCMs-M multiplicatively weakly consistent according to Definition 66 formulated above can be easily reformulated for FAPCMs-M multiplicatively weakly consistent according to this definition; it is sufficient to consider q_{ij}^{β} and q_{ij}^{γ} instead of q_{ij}^{α} and q_{ij}^{δ}, respectively, where appropriate.

In the following definition, a stronger version of multiplicative consistency for trapezoidal FAPCMs-M is formulated.

Definition 67 Let $\widetilde{Q} = \{\widetilde{q}_{ij}\}_{i,j=1}^{n}$, $\widetilde{q}_{ij} = \left(q_{ij}^{\alpha}, q_{ij}^{\beta}, q_{ij}^{\gamma}, q_{ij}^{\delta} \right)$, be a trapezoidal FAPCM-M. $\widetilde{Q}$ is said to be multiplicatively consistent if for each triplet $(i, j, k) \subseteq \{1, \dots, n\}$ the following holds:

$$\forall q_{ij} \in \left[q_{ij}^{\alpha}, q_{ij}^{\delta} \right] \exists q_{ik} \in \left[q_{ik}^{\alpha}, q_{ik}^{\delta} \right] \ \wedge \ \exists q_{kj} \in \left[q_{kj}^{\alpha}, q_{kj}^{\delta} \right] : \ q_{ij} = \frac{q_{ik} q_{kj}}{q_{ik} q_{kj} + (1 - q_{ik})(1 - q_{kj})}, \tag{4.179}$$

$$\forall q_{ij} \in \left[q_{ij}^{\beta}, q_{ij}^{\gamma} \right] \exists q_{ik} \in \left[q_{ik}^{\beta}, q_{ik}^{\gamma} \right] \ \wedge \ \exists q_{kj} \in \left[q_{kj}^{\beta}, q_{kj}^{\gamma} \right] : \ q_{ij} = \frac{q_{ik} q_{kj}}{q_{ik} q_{kj} + (1 - q_{ik})(1 - q_{kj})}. \tag{4.180}$$

Remark 33 Definition 67 provides a natural extension of multiplicative consistency from APCMs-M to trapezoidal FAPCMs-M. According to this definition, for any possible value $q_{ij} \in \widetilde{q}_{ij}$, $i, j \in \{1, \ldots, n\}$, there exist possible values $q_{ik} \in \widetilde{q}_{ik}$ and $q_{kj} \in \widetilde{q}_{kj}$, $k \in \{1, \ldots, n\}$, such that they satisfy (2.53), which is equivalent to the multiplicative-transitivity property (2.48) for APCMs-M. Analogously, for any possible value $q_{ij} \in Core\ \widetilde{q}_{ij}$, $i, j \in \{1, \ldots, n\}$, there exist possible values $q_{ik} \in Core\ \widetilde{q}_{ik}$ and $q_{kj} \in Core\ \widetilde{q}_{kj}$, $k \in \{1, \ldots, n\}$, such that they satisfy (2.53). Clearly, in comparison to the multiplicative weak consistency given by Definition 66, the multiplicative consistency given by Definition 67 is very strong.

Note that any of the properties (2.48)–(2.52) could have been used in Definition 67 instead of the property (2.53) as they are all equivalent (see Theorem 4).

Unlike Definitions 62 and 64 of multiplicatively consistent interval FAPCMs-M proposed by Xia and Xu (2011) and by Wu and Chiclana (2014a), respectively, new Definition 67 is invariant under permutation of objects compared in FAPCMs-M.

Theorem 70 *Definition 67 of multiplicative consistency is invariant under permutation of objects in FAPCMs-M.*

Proof For a multiplicatively consistent trapezoidal FAPCM-M $\widetilde{Q} = \{\widetilde{q}_{ij}\}_{i,j=1}^{n}$, $\widetilde{q}_{ij} = (q_{ij}^{\alpha}, q_{ij}^{\beta}, q_{ij}^{\gamma}, q_{ij}^{\delta})$, the conditions (4.179) and (4.180) are satisfied for every triplet $(i, j, k) \subseteq \{1, \ldots, n\}$. By permuting the FAPCM-M $\widetilde{Q}$ to $\widetilde{Q}^{\pi} = P\widetilde{Q}P^{T}$, the original PC $\widetilde{q}_{ij}$ in the i−th row and in the j−th column of $\widetilde{Q}$ moves to the $\pi(i)$−th row and the $\pi(j)$−th column of $\widetilde{Q}^{\pi}$ preserving $\widetilde{q}_{\pi(i)\pi(j)}^{\pi} = \widetilde{q}_{ij}$. Thus, by permuting $\widetilde{Q}$, also the validity of the conditions (4.179) and (4.180) is preserved, i.e.

$$\forall q_{ij}^{\pi} \in \left[q_{ij}^{\pi\alpha}, q_{ij}^{\pi\delta}\right] \exists q_{ik}^{\pi} \in \left[q_{ik}^{\pi\alpha}, q_{ik}^{\pi\delta}\right] \wedge \exists q_{kj}^{\pi} \in \left[q_{kj}^{\pi\alpha}, q_{kj}^{\pi\delta}\right] : \quad q_{ij}^{\pi} = \frac{q_{ik}^{\pi} q_{kj}^{\pi}}{q_{ik}^{\pi} q_{kj}^{\pi} + (1 - q_{ik}^{\pi})(1 - q_{kj}^{\pi})},$$

$$\forall q_{ij}^{\pi} \in \left[q_{ij}^{\pi\beta}, q_{ij}^{\pi\gamma}\right] \exists q_{ik}^{\pi} \in \left[q_{ik}^{\pi\beta}, q_{ik}^{\pi\gamma}\right] \wedge \exists q_{kj}^{\pi} \in \left[q_{kj}^{\pi\beta}, q_{kj}^{\pi\gamma}\right] : \quad q_{ij}^{\pi} = \frac{q_{ik}^{\pi} q_{kj}^{\pi}}{q_{ik}^{\pi} q_{kj}^{\pi} + (1 - q_{ik}^{\pi})(1 - q_{kj}^{\pi})},$$

for every triplet $(i, j, k) \subseteq \{1, \ldots, n\}$. Thus, $\widetilde{Q}^{\pi}$ is multiplicatively consistent according to Definition 67. $\qquad\qquad\square$

Further, unlike Definitions 63 and 65 of multiplicatively consistent interval FAPCMs-M proposed in Wang and Li (2012) and Wu and Chiclana (2014b), respectively, Definition 67 does not violate the additive reciprocity of the related PCs.

Theorem 71 *Definition 67 of multiplicative consistency does not violate the additive reciprocity of the related PCs in trapezoidal FAPCMs-M in the sense that any fixed value $q_{ij} \in [q_{ij}^{\alpha}, q_{ij}^{\delta}]$, $i, j \in \{1, \ldots, n\}$, representing the intensity of preference of object o_i over object o_j is associated with the corresponding value $q_{ji} \in [q_{ji}^{\alpha}, q_{ji}^{\delta}]$ representing the intensity of preference of object o_j over object o_i such that $q_{ji} = 1 - q_{ij}$.*

Proof It is sufficient to show that expressions (4.179) and (4.180) do not violate the additive-reciprocity property in the sense that when two particular intensities of preference $q_{ij} \in \widetilde{q}_{ij}$ and $q_{ji} \in \widetilde{q}_{ji}$ on the pair of objects o_i and o_j are considered at the same time in the expressions (4.179) and (4.180), then they are such that $q_{ji} = 1 - q_{ij}$.

For a triplet $(i, j, k) \subseteq \{1, \ldots, n\}$, $i \neq j \neq k$, no reciprocals appear in expression

$$q_{ij} = \frac{q_{ik}q_{kj}}{q_{ik}q_{kj} + (1 - q_{ik})(1 - q_{kj})}$$

for any $q_{ij} \in \left[q_{ij}^{\alpha}, q_{ij}^{\delta}\right]$. For $i = j = k$, expression (4.179) reduces to: $\forall q_{ii} = 0.5 \; \exists q_{ii}^{*} = 0.5 \land \exists q_{ii}^{**} = 0.5: 0.5 = \frac{0.5 \cdot 0.5}{0.5 \cdot 0.5 + 0.5 \cdot 0.5}$, which again does not violate the additive reciprocity. Further, for $i \neq j = k$, expression (4.179) is as:

$$\forall q_{ij} \in \left[q_{ij}^{\alpha}, q_{ij}^{\delta}\right] \exists q_{ij}^{*} \in \left[q_{ij}^{\alpha}, q_{ij}^{\delta}\right] \land \exists q_{jj} = 0.5: q_{ij} = \frac{q_{ij}^{*} \cdot 0.5}{q_{ij}^{*} \cdot 0.5 + (1 - q_{ij}^{*}) \cdot 0.5}.$$

This means that $q_{ij}^{*} = q_{ij}$ and, therefore, the additive reciprocity is not violated. For $i = k \neq j$ the proof is analogous. Finally, for $i = j \neq k$, expression (4.179) is as

$$\forall q_{ii} = 0.5 \; \exists q_{ik} \in \left[q_{ik}^{\alpha}, q_{ik}^{\delta}\right] \land \exists q_{ki}^{*} \in \left[q_{ki}^{\alpha}, q_{ki}^{\delta}\right]: 0.5 = \frac{q_{ik}q_{ki}^{*}}{q_{ik}q_{ki}^{*} + (1 - q_{ik})(1 - q_{ki}^{*})}.$$

This means that $q_{ki}^{*} = 1 - q_{ik}$ and, therefore, the additive reciprocity is preserved. The proof for the expression (4.180) is analogous.   □

Remark 34 Similarly to Theorem 67, also Theorem 71 does not simply state that a FAPCM-M $\widetilde{Q} = \{\widetilde{q}_{ij}\}_{i,j=1}^{n}$ multiplicatively consistent according to Definition 67 is additively reciprocal since this property automatically follows from Definition 52 of a FAPCM. Theorem 67 states that only additively reciprocal PCs are involved in Definition 67 of multiplicative consistency, which is in accordance with the conception of additive reciprocity discussed on p. 160. For more detailed discussion, see Remark 21.

By handling properly the additive-reciprocity property of PCs, Theorem 4 can be easily extended to interval FAPCMs-M as follows.

Theorem 72 *For a trapezoidal FAPCM-M* $\widetilde{Q} = \{\widetilde{q}_{ij}\}_{i,j=1}^{n}$, $\widetilde{q}_{ij} = \left(q_{ij}^{\alpha}, q_{ij}^{\beta}, q_{ij}^{\gamma}, q_{ij}^{\delta}\right)$, *the following statements are equivalent:*

(i) $\widetilde{Q}$ *is multiplicatively consistent according to Definition 67.*
(ii) *For every* $i, j, k = 1, \ldots, n$:

$$\forall q_{ij} \in \left[q_{ij}^{\alpha}, q_{ij}^{\delta}\right] \exists q_{jk} \in \left[q_{jk}^{\alpha}, q_{jk}^{\delta}\right] \land \exists q_{ki} \in \left[q_{ki}^{\alpha}, q_{ki}^{\delta}\right]: q_{ij}q_{jk}q_{ki} = q_{ik}q_{kj}q_{ji},$$

$$q_{ij} = 1 - q_{ji}, \; q_{jk} = 1 - q_{kj}, \; q_{ki} = 1 - q_{ik},$$

$$(4.181)$$

$$\forall q_{ij} \in \left[q_{ij}^{\beta}, q_{ij}^{\gamma} \right] \exists q_{jk} \in \left[q_{jk}^{\beta}, q_{jk}^{\gamma} \right] \wedge \exists q_{ki} \in \left[q_{ki}^{\beta}, q_{ki}^{\gamma} \right] : q_{ij} q_{jk} q_{ki} = q_{ik} q_{kj} q_{ji},$$

$$q_{ij} = 1 - q_{ji}, \; q_{jk} = 1 - q_{kj}, \; q_{ki} = 1 - q_{ik}.$$

(4.182)

(iii) For every $i, j, k = 1, \ldots, n$:

$$\forall q_{ij} \in \left[q_{ij}^{\alpha}, q_{ij}^{\delta} \right] \exists q_{jk} \in \left[q_{jk}^{\alpha}, q_{jk}^{\delta} \right] \wedge \exists q_{ki} \in \left[q_{ki}^{\alpha}, q_{ki}^{\delta} \right] : \frac{q_{ij}}{q_{ji}} \frac{q_{jk}}{q_{kj}} \frac{q_{ki}}{q_{ik}} = 1,$$

$$q_{ij} = 1 - q_{ji}, \; q_{jk} = 1 - q_{kj}, \; q_{ki} = 1 - q_{ik},$$

(4.183)

$$\forall q_{ij} \in \left[q_{ij}^{\beta}, q_{ij}^{\gamma} \right] \exists q_{jk} \in \left[q_{jk}^{\beta}, q_{jk}^{\gamma} \right] \wedge \exists q_{ki} \in \left[q_{ki}^{\beta}, q_{ki}^{\gamma} \right] : \frac{q_{ij}}{q_{ji}} \frac{q_{jk}}{q_{kj}} \frac{q_{ki}}{q_{ik}} = 1,$$

$$q_{ij} = 1 - q_{ji}, \; q_{jk} = 1 - q_{kj}, \; q_{ki} = 1 - q_{ik}.$$

(4.184)

(iv) For every $i, j, k = 1, \ldots, n$:

$$\forall q_{ij} \in \left[q_{ij}^{\alpha}, q_{ij}^{\delta} \right] \exists q_{jk} \in \left[q_{jk}^{\alpha}, q_{jk}^{\delta} \right] \wedge \exists q_{ki} \in \left[q_{ki}^{\alpha}, q_{ki}^{\delta} \right] : \frac{q_{ij}}{q_{ji}} \frac{q_{jk}}{q_{kj}} \frac{q_{ki}}{q_{ik}} = \frac{q_{ik}}{q_{ki}} \frac{q_{kj}}{q_{jk}} \frac{q_{ji}}{q_{ij}},$$

$$q_{ij} = 1 - q_{ji}, \; q_{jk} = 1 - q_{kj}, \; q_{ki} = 1 - q_{ik},$$

(4.185)

$$\forall q_{ij} \in \left[q_{ij}^{\beta}, q_{ij}^{\gamma} \right] \exists q_{jk} \in \left[q_{jk}^{\beta}, q_{jk}^{\gamma} \right] \wedge \exists q_{ki} \in \left[q_{ki}^{\beta}, q_{ki}^{\gamma} \right] : \frac{q_{ij}}{q_{ji}} \frac{q_{jk}}{q_{kj}} \frac{q_{ki}}{q_{ik}} = \frac{q_{ik}}{q_{ki}} \frac{q_{kj}}{q_{jk}} \frac{q_{ji}}{q_{ij}},$$

$$q_{ij} = 1 - q_{ji}, \; q_{jk} = 1 - q_{kj}, \; q_{ki} = 1 - q_{ik}.$$

(4.186)

(v) For every $i, j, k = 1, \ldots, n$:

$$\forall q_{ij} \in \left[q_{ij}^{\alpha}, q_{ij}^{\delta} \right] \exists q_{jk} \in \left[q_{jk}^{\alpha}, q_{jk}^{\delta} \right] \wedge \exists q_{ki} \in \left[q_{ki}^{\alpha}, q_{k\bar{i}}^{\delta} \right] : \frac{q_{ij}}{q_{ji}} = \frac{q_{ik}}{q_{ki}} \frac{q_{kj}}{q_{jk}},$$

$$q_{ij} = 1 - q_{ji}, \; q_{jk} = 1 - q_{kj}, \; q_{ki} = 1 - q_{ik},$$

(4.187)

$$\forall q_{ij} \in \left[q_{ij}^{\beta}, q_{ij}^{\gamma} \right] \exists q_{jk} \in \left[q_{jk}^{\beta}, q_{jk}^{\gamma} \right] \wedge \exists q_{ki} \in \left[q_{ki}^{\beta}, q_{k\bar{i}}^{\gamma} \right] : \frac{q_{ij}}{q_{ji}} = \frac{q_{ik}}{q_{ki}} \frac{q_{kj}}{q_{jk}},$$

$$q_{ij} = 1 - q_{ji}, \; q_{jk} = 1 - q_{kj}, \; q_{ki} = 1 - q_{ik}.$$

(4.188)

Proof From the additive-reciprocity property $\tilde{q}_{ij} = 1 - \tilde{q}_{ji}, i,j = 1, \ldots, n$, it follows that $\forall q_{ij} \in \left[q_{ij}^{\alpha}, q_{ij}^{\delta}\right] \exists q_{ji} \in \left[q_{ji}^{\alpha}, q_{ji}^{\delta}\right] : q_{ji} = 1 - q_{ij}$ and $\forall q_{ij} \in \left[q_{ij}^{\beta}, q_{ij}^{\gamma}\right] \exists q_{ji} \in \left[q_{ji}^{\beta}, q_{ji}^{\gamma}\right] : q_{ji} = 1 - q_{ij}$.

Obviously, the statements (ii), (iii) and (v) are equivalent since the expressions in the statements are obtained by a simple reordering. Therefore, it is sufficient to prove the equivalence of statements (i) and (ii), and the equivalence of statements (iii) and (iv).

(a) First, let us show that the statements (i) and (ii) are equivalent. Because of the reciprocity property, (4.181) can be equivalently written as

$$\forall q_{ij} \in \left[q_{ij}^{\alpha}, q_{ij}^{\delta}\right] \exists q_{ik} \in [q_{ik}^{\alpha}, q_{ik}^{\delta}] \wedge \exists q_{kj} \in \left[q_{kj}^{\alpha}, q_{kj}^{\delta}\right] :$$
$$q_{ij}(1 - q_{kj})(1 - q_{ik}) = q_{ik}q_{kj}(1 - q_{ij}). \tag{4.189}$$

The expression $q_{ij}(1 - q_{kj})(1 - q_{ik}) = q_{ik}q_{kj}(1 - q_{ij})$ can be further rewritten as

$$q_{ij} = \frac{q_{ik}q_{kj}}{q_{ik}q_{kj} + (1 - q_{ik})(1 - q_{kj})}.$$

This means that the statement (4.181) is equivalent to (4.179). Analogously, the equivalence of (4.182) and (4.180) is proved.

(b) Now, let us show that the statements (iii) and (iv) are equivalent. The expression $\frac{q_{ij}}{q_{ji}} \frac{q_{jk}}{q_{kj}} \frac{q_{ki}}{q_{ik}} = \frac{q_{ik}}{q_{ki}} \frac{q_{kj}}{q_{jk}} \frac{q_{ji}}{q_{ij}}$ in the statement (4.185) can be equivalently written as $\left(\frac{q_{ij}}{q_{ji}} \frac{q_{jk}}{q_{kj}} \frac{q_{ki}}{q_{ik}}\right)^2 = 1$, which is equivalent to $\frac{q_{ij}}{q_{ji}} \frac{q_{jk}}{q_{kj}} \frac{q_{ki}}{q_{ik}} = 1$. Therefore, the statement (4.185) is equivalent to the statement (4.183). Analogously, the equivalence of (4.186) and (4.184) is proved.   □

The following theorems give us useful tools for verifying the multiplicative consistency of trapezoidal FAPCMs-M.

Theorem 73 *A trapezoidal FAPCM-M* $\tilde{Q} = \{\tilde{q}_{ij}\}_{i,j=1}^{n}, \tilde{q}_{ij} = \left(q_{ij}^{\alpha}, q_{ij}^{\beta}, q_{ij}^{\gamma}, q_{ij}^{\delta}\right)$, *is multiplicatively consistent according to Definition 67 if and only if the inequalities*

$$q_{ij}^{\alpha} \geq \frac{q_{ik}^{\alpha}q_{kj}^{\alpha}}{q_{ik}^{\alpha}q_{kj}^{\alpha} + (1 - q_{ik}^{\alpha})(1 - q_{kj}^{\alpha})}, \quad q_{ij}^{\delta} \leq \frac{q_{ik}^{\delta}q_{kj}^{\delta}}{q_{ik}^{\delta}q_{kj}^{\delta} + (1 - q_{ik}^{\delta})(1 - q_{kj}^{\delta})}, \tag{4.190}$$

$$q_{ij}^{\beta} \geq \frac{q_{ik}^{\beta}q_{kj}^{\beta}}{q_{ik}^{\beta}q_{kj}^{\beta} + (1 - q_{ik}^{\beta})(1 - q_{kj}^{\beta})}, \quad q_{ij}^{\gamma} \leq \frac{q_{ik}^{\gamma}q_{kj}^{\gamma}}{q_{ik}^{\gamma}q_{kj}^{\gamma} + (1 - q_{ik}^{\gamma})(1 - q_{kj}^{\gamma})}, \tag{4.191}$$

hold for every $i,j,k = 1, \ldots, n, \ i < j, \ k \neq i,j$.

Proof It is sufficient to demonstrate the equivalence of the expression (4.190) and (4.179). The demonstration of the equivalence of (4.191) and (4.180) is analogous.

First, let us demonstrate that when the inequalities (4.190) hold for every $i, j, k = 1, \ldots, n$, $i < j$, $k \neq i, j$, then they hold for every $i, j, k = 1, \ldots, n$. The inequalities (4.190) are always satisfied for $i, j, k = 1, \ldots, n$ such that

(i) $i = j \neq k$:

$$\frac{q_{ik}^{\alpha} q_{ki}^{\alpha}}{q_{ik}^{\alpha} q_{ki}^{\alpha} + (1 - q_{ik}^{\alpha})(1 - q_{ki}^{\alpha})} = \frac{q_{ik}^{\alpha} q_{ki}^{\alpha}}{q_{ik}^{\alpha} q_{ki}^{\alpha} + q_{ik}^{\delta} q_{ki}^{\delta}} \leq \frac{q_{ik}^{\alpha} q_{ki}^{\alpha}}{2 q_{ik}^{\alpha} q_{ki}^{\alpha}} = 0.5 = q_{ii}^{\alpha}$$

$$\frac{q_{ik}^{\delta} q_{ki}^{\delta}}{q_{ik}^{\delta} q_{ki}^{\delta} + (1 - q_{ik}^{\delta})(1 - q_{ki}^{\delta})} = \frac{q_{ik}^{\delta} q_{ki}^{\delta}}{q_{ik}^{\delta} q_{ki}^{\delta} + q_{ik}^{\alpha} q_{ki}^{\alpha}} \geq \frac{q_{ik}^{\delta} q_{ki}^{\delta}}{2 q_{ik}^{\delta} q_{ki}^{\delta}} = 0.5 = q_{ii}^{\delta},$$

(ii) $i \neq j = k$:

$$\frac{q_{ij}^{\alpha} q_{jj}^{\alpha}}{q_{ij}^{\alpha} q_{jj}^{\alpha} + (1 - q_{ij}^{\alpha})(1 - q_{jj}^{\alpha})} = \frac{0.5 q_{ij}^{\alpha}}{0.5 q_{ij}^{\alpha} + 0.5(1 - q_{ij}^{\alpha})} = q_{ij}^{\alpha},$$

$$\frac{q_{ij}^{\delta} q_{jj}^{\delta}}{q_{ij}^{\delta} q_{jj}^{\delta} + (1 - q_{ij}^{\delta})(1 - q_{jj}^{\delta})} = \frac{0.5 q_{ij}^{\delta}}{0.5 q_{ij}^{\delta} + 0.5(1 - q_{ij}^{\delta})} = q_{ij}^{\delta},$$

(iii) $j \neq k = i$:

$$\frac{q_{ii}^{\alpha} q_{ij}^{\alpha}}{q_{ii}^{\alpha} q_{ij}^{\alpha} + (1 - q_{ii}^{\alpha})(1 - q_{ij}^{\alpha})} = \frac{0.5 q_{ij}^{\alpha}}{0.5 q_{ij}^{\alpha} + 0.5(1 - q_{ij}^{\alpha})} = q_{ij}^{\alpha},$$

$$\frac{q_{ij}^{\delta} q_{jj}^{\delta}}{q_{ii}^{\delta} q_{ij}^{\delta} + (1 - q_{ii}^{\delta})(1 - q_{ij}^{\delta})} = \frac{0.5 q_{ij}^{\delta}}{0.5 q_{ij}^{\delta} + 0.5(1 - q_{ij}^{\delta})} = q_{ij}^{\delta},$$

(iv) $i = j = k$:

$$\frac{q_{ii}^{\alpha} q_{ii}^{\alpha}}{q_{ii}^{\alpha} q_{ii}^{\alpha} + (1 - q_{ii}^{\alpha})(1 - q_{ii}^{\alpha})} = \frac{0.5 \cdot 0.5}{0.5 \cdot 0.5 + 0.5 \cdot 0.5} = 0.5 = q_{ij}^{\alpha},$$

$$\frac{q_{ii}^{\delta} q_{ii}^{\delta}}{q_{ii}^{\delta} q_{ii}^{\delta} + (1 - q_{ii}^{\delta})(1 - q_{ii}^{\delta})} = \frac{0.5 \cdot 0.5}{0.5 \cdot 0.5 + 0.5 \cdot 0.5} = 0.5 = q_{ij}^{\delta}.$$

Further, when the inequalities (4.190) are satisfied for $i, j, k = 1, \ldots, n$, $i < j$, $k \neq i, j$, then they are satisfied also for $j, i, k = 1, \ldots, n$, $j > i$, $k \neq i, j$:

$$\frac{q_{jk}^{\alpha}q_{ki}^{\alpha}}{q_{jk}^{\alpha}q_{ki}^{\alpha}+(1-q_{jk}^{\alpha})(1-q_{ki}^{\alpha})} = \frac{q_{jk}^{\alpha}q_{ki}^{\alpha}}{q_{jk}^{\alpha}q_{ki}^{\alpha}+q_{jk}^{\delta}q_{ki}^{\delta}} = 1 - \frac{q_{ik}^{\delta}q_{kj}^{\delta}}{q_{jk}^{\alpha}q_{ki}^{\alpha}+q_{ik}^{\delta}q_{kj}^{\delta}}$$

$$= 1 - \frac{q_{ik}^{\delta}q_{kj}^{\delta}}{q_{ik}^{\delta}q_{kj}^{\delta}+(1-q_{ik}^{\delta})(1-q_{kj}^{\delta})} \le 1 - q_{ij}^{\delta} = q_{ji}^{\alpha},$$

$$\frac{q_{jk}^{\delta}q_{ki}^{\delta}}{q_{jk}^{\delta}q_{ki}^{\delta}+(1-q_{jk}^{\delta})(1-q_{ki}^{\delta})} = \frac{q_{jk}^{\delta}q_{ki}^{\delta}}{q_{jk}^{\delta}q_{ki}^{\delta}+q_{jk}^{\alpha}q_{ki}^{\alpha}} = 1 - \frac{q_{ik}^{\alpha}q_{kj}^{\alpha}}{q_{jk}^{\delta}q_{ki}^{\delta}+q_{ik}^{\alpha}q_{kj}^{\alpha}}$$

$$= 1 - \frac{q_{ik}^{\alpha}q_{kj}^{\alpha}}{q_{ik}^{\alpha}q_{kj}^{\alpha}+(1-q_{ik}^{\alpha})(1-q_{kj}^{\alpha})} \ge 1 - q_{ij}^{\alpha} = q_{ji}^{\delta}.$$

To finalize the proof, it is sufficient to show that the inequalities (4.190) are equivalent to the condition (4.179) for every $i, j, k = 1, \ldots, n$.

First, let $\widetilde{Q}$ be a trapezoidal FAPCM-M multiplicatively consistent according to Definition 67. Then for $q_{ij} := q_{ij}^{\alpha} \exists q_{ik} \in \left[q_{ik}^{\alpha}, q_{ik}^{\delta}\right] \wedge \exists q_{kj} \in \left[q_{kj}^{\alpha}, q_{kj}^{\delta}\right] : q_{ij}^{\alpha} = \frac{q_{ik}q_{kj}}{q_{ik}q_{kj}+(1-q_{ik})(1-q_{kj})}$. Since $q_{ik} \ge q_{ik}^{\alpha}, q_{kj} \ge q_{kj}^{\alpha}$, and since $\frac{q_{ik}q_{kj}}{q_{ik}q_{kj}+(1-q_{ik})(1-q_{kj})}$ is increasing in both variables on interval $]0, 1[$, then clearly (4.190) is valid. Analogously, for $q_{ij} := q_{ij}^{\delta} \exists q_{ik} \in \left[q_{ik}^{\alpha}, q_{ik}^{\delta}\right] \wedge \exists q_{kj} \in \left[q_{kj}^{\alpha}, q_{kj}^{\delta}\right] : q_{ij}^{\delta} = \frac{q_{ik}q_{kj}}{q_{ik}q_{kj}+(1-q_{ik})(1-q_{kj})}$. Since $q_{ik} \le q_{ik}^{\delta}, q_{kj} \le q_{kj}^{\delta}$, then clearly (4.191) is valid.

Second, let the expression (4.190) be valid for a trapezoidal FAPCM-M $\widetilde{Q}$. Because $\frac{q_{ik}q_{kj}}{q_{ik}q_{kj}+(1-q_{ik})(1-q_{kj})}$ is increasing in both variables q_{ik} and q_{kj} on intervals $[q_{ik}^{\alpha}, q_{ik}^{\delta}]$ and $[q_{kj}^{\alpha}, q_{kj}^{\delta}]$, respectively, then from the inequalities (4.190) we get

$$\forall q_{ij} \in \left[q_{ij}^{\alpha}, q_{ij}^{\delta}\right] : \frac{q_{ik}^{\alpha}q_{kj}^{\alpha}}{q_{ik}^{\alpha}q_{kj}^{\alpha}+(1-q_{ik}^{\alpha})(1-q_{kj}^{\alpha})} \le q_{ij} \le \frac{q_{ik}^{\delta}q_{kj}^{\delta}}{q_{ik}^{\delta}q_{kj}^{\delta}+(1-q_{ik}^{\delta})(1-q_{kj}^{\delta})}$$

and, therefore, (4.179) is satisfied.   □

Theorem 74 *A trapezoidal FAPCM-M* $\widetilde{Q} = \{\widetilde{q}_{ij}\}_{i,j=1}^{n}, \widetilde{q}_{ij} = \left(q_{ij}^{\alpha}, q_{ij}^{\beta}, q_{ij}^{\gamma}, q_{ij}^{\delta}\right)$, *is multiplicatively consistent according to Definition 67 if and only if*

$$q_{ij}^{\alpha} \ge \max_{\substack{k=1,\ldots,n \\ k \neq i,j}}\left\{\frac{q_{ik}^{\alpha}q_{kj}^{\alpha}}{q_{ik}^{\alpha}q_{kj}^{\alpha}+(1-q_{ik}^{\alpha})(1-q_{kj}^{\alpha})}\right\}, \quad q_{ij}^{\delta} \le \min_{\substack{k=1,\ldots,n \\ k \neq i,j}}\left\{\frac{q_{ik}^{\delta}q_{kj}^{\delta}}{q_{ik}^{\delta}q_{kj}^{\delta}+(1-q_{ik}^{\delta})(1-q_{kj}^{\delta})}\right\},$$

$$(4.192)$$

$$q_{ij}^{\beta} \ge \max_{\substack{k=1,\ldots,n \\ k \neq i,j}}\left\{\frac{q_{ik}^{\beta}q_{kj}^{\beta}}{q_{ik}^{\beta}q_{kj}^{\beta}+(1-q_{ik}^{\beta})(1-q_{kj}^{\beta})}\right\}, \quad q_{ij}^{\gamma} \le \min_{\substack{k=1,\ldots,n \\ k \neq i,j}}\left\{\frac{q_{ik}^{\gamma}q_{kj}^{\gamma}}{q_{ik}^{\gamma}q_{kj}^{\gamma}+(1-q_{ik}^{\gamma})(1-q_{kj}^{\gamma})}\right\},$$

$$(4.193)$$

hold for every $i, j = 1, \ldots, n, i < j$.

Proof The inequalities (4.192) and (4.193) follow immediately from Theorem 73. □

Theorem 75 *A trapezoidal FAPCM-M $\widetilde{Q} = \left\{\widetilde{q}_{ij}\right\}_{i,j=1}^{n}$, $\widetilde{q}_{ij} = \left(q_{ij}^{\alpha}, q_{ij}^{\beta}, q_{ij}^{\gamma}, q_{ij}^{\delta}\right)$, is multiplicatively consistent according to Definition 67 if and only if one of the following conditions holds for every $i, j, k = 1, \ldots, n$, $i < j$, $k \neq i, j$:*

(i)

$$q_{ij}^{\alpha} q_{jk}^{\delta} q_{ki}^{\delta} \geq q_{ik}^{\alpha} q_{kj}^{\alpha} q_{ji}^{\delta}, \qquad q_{ij}^{\delta} q_{jk}^{\alpha} q_{ki}^{\alpha} \leq q_{ik}^{\delta} q_{kj}^{\delta} q_{ji}^{\alpha}, \tag{4.194}$$

$$q_{ij}^{\beta} q_{jk}^{\gamma} q_{ki}^{\gamma} \geq q_{ik}^{\beta} q_{kj}^{\beta} q_{ji}^{\gamma}, \qquad q_{ij}^{\gamma} q_{jk}^{\beta} q_{ki}^{\beta} \leq q_{ik}^{\gamma} q_{kj}^{\gamma} q_{ji}^{\beta}, \tag{4.195}$$

(ii)

$$\frac{q_{ij}^{\alpha}}{q_{ji}^{\delta}} \frac{q_{jk}^{\delta}}{q_{kj}^{\alpha}} \frac{q_{ki}^{\delta}}{q_{ik}^{\alpha}} \geq 1, \qquad \frac{q_{ij}^{\delta}}{q_{ji}^{\alpha}} \frac{q_{jk}^{\alpha}}{q_{kj}^{\delta}} \frac{q_{ki}^{\alpha}}{q_{ik}^{\delta}} \leq 1, \tag{4.196}$$

$$\frac{q_{ij}^{\beta}}{q_{ji}^{\gamma}} \frac{q_{jk}^{\gamma}}{q_{kj}^{\beta}} \frac{q_{ki}^{\gamma}}{q_{ik}^{\beta}} \geq 1, \qquad \frac{q_{ij}^{\gamma}}{q_{ji}^{\beta}} \frac{q_{jk}^{\beta}}{q_{kj}^{\gamma}} \frac{q_{ki}^{\beta}}{q_{ik}^{\gamma}} \leq, 1 \tag{4.197}$$

(iii)

$$\frac{q_{ij}^{\alpha}}{q_{ji}^{\delta}} \frac{q_{jk}^{\delta}}{q_{kj}^{\alpha}} \frac{q_{ki}^{\delta}}{q_{ik}^{\alpha}} \geq \frac{q_{ik}^{\alpha}}{q_{ki}^{\delta}} \frac{q_{kj}^{\delta}}{q_{jk}^{\alpha}} \frac{q_{ji}^{\delta}}{q_{ij}^{\alpha}}, \qquad \frac{q_{ij}^{\delta}}{q_{ji}^{\alpha}} \frac{q_{jk}^{\alpha}}{q_{kj}^{\delta}} \frac{q_{ki}^{\alpha}}{q_{ik}^{\delta}} \leq \frac{q_{ik}^{\delta}}{q_{ki}^{\alpha}} \frac{q_{kj}^{\alpha}}{q_{jk}^{\delta}} \frac{q_{ji}^{\alpha}}{q_{ij}^{\delta}}, \tag{4.198}$$

$$\frac{q_{ij}^{\beta}}{q_{ji}^{\gamma}} \frac{q_{jk}^{\gamma}}{q_{kj}^{\beta}} \frac{q_{ki}^{\gamma}}{q_{ik}^{\beta}} \geq \frac{q_{ik}^{\beta}}{q_{ki}^{\gamma}} \frac{q_{kj}^{\gamma}}{q_{jk}^{\beta}} \frac{q_{ji}^{\gamma}}{q_{ij}^{\beta}}, \qquad \frac{q_{ij}^{\gamma}}{q_{ji}^{\beta}} \frac{q_{jk}^{\beta}}{q_{kj}^{\gamma}} \frac{q_{ki}^{\beta}}{q_{ik}^{\gamma}} \leq \frac{q_{ik}^{\gamma}}{q_{ki}^{\beta}} \frac{q_{kj}^{\beta}}{q_{jk}^{\gamma}} \frac{q_{ji}^{\beta}}{q_{ij}^{\gamma}}, \tag{4.199}$$

(iv)

$$\frac{q_{ij}^{\alpha}}{q_{ji}^{\delta}} \geq \frac{q_{ik}^{\alpha}}{q_{ki}^{\delta}} \frac{q_{kj}^{\alpha}}{q_{jk}^{\delta}}, \qquad \frac{q_{ij}^{\delta}}{q_{ji}^{\alpha}} \leq \frac{q_{ik}^{\delta}}{q_{ki}^{\alpha}} \frac{q_{kj}^{\delta}}{q_{jk}^{\alpha}}, \tag{4.200}$$

$$\frac{q_{ij}^{\beta}}{q_{ji}^{\gamma}} \geq \frac{q_{ik}^{\beta}}{q_{ki}^{\gamma}} \frac{q_{kj}^{\beta}}{q_{jk}^{\gamma}}, \qquad \frac{q_{ij}^{\gamma}}{q_{ji}^{\beta}} \leq \frac{q_{ik}^{\gamma}}{q_{ki}^{\beta}} \frac{q_{kj}^{\gamma}}{q_{jk}^{\beta}}. \tag{4.201}$$

Proof The inequalities (i)–(iv) are obtained directly from the inequalities (4.190) and (4.191) by applying the additive-reciprocity properties $q_{pq}^{\alpha} = 1 - q_{qp}^{\delta}$, $q_{pq}^{\beta} = 1 - q_{qp}^{\gamma}$, $q_{pq}^{\gamma} = 1 - q_{qp}^{\beta}$, $q_{pq}^{\delta} = 1 - q_{qp}^{\alpha}$, $p, q = i, j, k$. $\qquad\square$

Remark 35 Notice the similarities between the inequality conditions (i)–(iv) in Theorem 75 and the inequalities (4.190), (4.191) in Theorem 73, and the definitions of multiplicatively consistent interval FAPCMs-M reviewed in Sect. "Review of Fuzzy Extensions of Multiplicative Consistency". In particular, the inequalities (4.194) are in some sense similar to the multiplicative-transitivity property (4.173) in Wu and Chiclana's Definition 65, condition (4.198) has a form similar to the condition (4.168), which was derived from the condition (4.167), and the inequalities (4.190) are similar to the condition (4.172) in Wu and Chiclana's Definition 64.

However, there is a significant difference between the definitions reviewed in Sect. "Review of Fuzzy Extensions of Multiplicative Consistency" and the new Definition 67 and the inequality properties in Theorem 75. Definitions 62–65 were obtained by extending the expressions from Theorem 4 that are mutually equivalent thanks to the additive reciprocity of the related PCs in APCMs-M. However, because the interval extension of the expressions from Theorem 4 was done by applying the standard interval arithmetic, the additive reciprocity of the related PCs is not preserved anymore in interval FAPCMs-M. This drawback leads to the fact that Definitions 62–65 are not mutually equivalent. Remember that Wang and Li (2012) even stated that "(4.169) is not equivalent to (4.167) as in the case of regular APCMs-M" (Wang and Li 2012, p. 183). This does not hold true for new Definition 67. The constrained fuzzy arithmetic was applied properly to the fuzzy extension of the condition (2.53) in order to preserve the additive reciprocity of the related PCs in trapezoidal FAPCMs-M. As a result, it was possible to extend also the conditions from Theorem 4 so that their fuzzy extensions are mutually equivalent, see Theorem 72.

In the following example, Definition 67 of multiplicative consistency is confronted with Definitions 62–66. In particular, it is demonstrated how the drawbacks regarding the dependence of Definitions 62 and 64 on permutation of objects and violation of the additive-reciprocity property in Definitions 63 and 65 are removed by Definition 67.

Example 63 Let us examine the interval FAPCM-M given by (4.165). In Example 60, it was demonstrated that Definition 62 is not invariant under permutation of objects since the interval FAPCM-M (4.165) is judged as multiplictively consistent while its permutation (4.166) is judged as multiplicatively inconsistent. Analogously, in Example 61, it was demonstrated that Definition 64 is not invariant under permutation of objects since the interval FAPCM-M (4.165) is judged as multiplictively consistent while its permutation (4.166) is judged as multiplicatively inconsistent according to the definition.

Now, let us apply Definition 67 to the interval FAPCM-M (4.165). By using Theorem 73, the interval FAPCM-A (4.165) is judged multiplicatively consistent since it satisfies the inequalities (4.190); see Table 4.15. Also the permuted interval FAPCM-A (4.166) satisfies the inequalities (4.190); see Table 4.16. Therefore, it is again judged as multiplicatively consistent. Moreover, from Theorem 70 it follows that any permutation of the interval FAPCM-M (4.165) is multiplicatively consistent.

In Example 62, it was shown that Definitions 63 and 65 violate additive reciprocity of the related PCs in the interval FAPCM-M (4.165). According to Theorem 71, the additive-reciprocity property is preserved in new Definition 67. This basically means that by taking any value from any interval PC in the interval FAPCM-M (4.165), there exist values in the remaining interval PCs such that they form a multiplicatively consistent APCM-M. Let us examine the triplet $i = 1, j = 2, k = 3$ of indices and let us consider the value $q_{12} = \frac{3}{5} \in [\frac{1}{2}, \frac{3}{5}]$. Then, according to (4.179), there exist values $q_{13} \in [\frac{3}{5}, \frac{6}{7}]$ and $q_{32} \in [\frac{1}{5}, \frac{2}{5}]$ such that $\frac{3}{5} = \frac{q_{13}q_{32}}{q_{13}q_{32}+(1-q_{13})(1-q_{32})}$. It is, for example,

Table 4.15 Inequality conditions (4.190) for the interval FAPCM-M (4.165)

$i<j$:	$q_{ij}^L \geq \dfrac{q_{ik}^L q_{kj}^L}{q_{ik}^L q_{kj}^L+(1-q_{ik}^L)(1-q_{kj}^L)}$	$q_{ij}^U \leq \dfrac{q_{ik}^U q_{kj}^U}{q_{ik}^U q_{kj}^U+(1-q_{ik}^U)(1-q_{kj}^U)}$
1, 2:	$\dfrac{1}{2} \geq \dfrac{\frac{3}{5}\frac{1}{5}}{\frac{3}{5}\frac{1}{5}+\frac{2}{5}\frac{4}{5}}$	$\dfrac{3}{5} \leq \dfrac{\frac{6}{7}\frac{2}{5}}{\frac{6}{7}\frac{2}{5}+\frac{1}{7}\frac{3}{5}}$
1, 3:	$\dfrac{3}{5} \geq \dfrac{\frac{1}{2}\frac{3}{5}}{\frac{1}{2}\frac{3}{5}+\frac{1}{2}\frac{2}{5}}$	$\dfrac{6}{7} \leq \dfrac{\frac{3}{5}\frac{4}{5}}{\frac{3}{5}\frac{4}{5}+\frac{2}{5}\frac{1}{5}}$
2, 3:	$\dfrac{3}{5} \geq \dfrac{\frac{2}{5}\frac{3}{5}}{\frac{2}{5}\frac{3}{5}+\frac{3}{5}\frac{2}{5}}$	$\dfrac{4}{5} \leq \dfrac{\frac{1}{2}\frac{6}{7}}{\frac{1}{2}\frac{6}{7}+\frac{1}{2}\frac{1}{7}}$

Table 4.16 Inequality conditions (4.190) for the permuted interval FAPCM-M (4.166)

$i<j$:	$q_{ij}^L \geq \dfrac{q_{ik}^L q_{kj}^L}{q_{ik}^L q_{kj}^L+(1-q_{ik}^L)(1-q_{kj}^L)}$	$q_{ij}^U \leq \dfrac{q_{ik}^U q_{kj}^U}{q_{ik}^U q_{kj}^U+(1-q_{ik}^U)(1-q_{kj}^U)}$
1, 2:	$\dfrac{1}{7} \geq \dfrac{\frac{1}{5}\frac{2}{5}}{\frac{1}{5}\frac{2}{5}+\frac{4}{5}\frac{3}{5}}$	$\dfrac{2}{5} \leq \dfrac{\frac{2}{5}\frac{1}{2}}{\frac{2}{5}\frac{1}{2}+\frac{3}{5}\frac{1}{2}}$
1, 3:	$\dfrac{1}{5} \geq \dfrac{\frac{1}{7}\frac{1}{2}}{\frac{1}{7}\frac{1}{2}+\frac{6}{7}\frac{1}{2}}$	$\dfrac{2}{5} \leq \dfrac{\frac{2}{5}\frac{3}{5}}{\frac{2}{5}\frac{3}{5}+\frac{3}{5}\frac{2}{5}}$
2, 3:	$\dfrac{1}{2} \geq \dfrac{\frac{3}{5}\frac{1}{5}}{\frac{3}{5}\frac{1}{5}+\frac{2}{5}\frac{4}{5}}$	$\dfrac{3}{5} \leq \dfrac{\frac{6}{7}\frac{2}{5}}{\frac{6}{7}\frac{2}{5}+\frac{1}{7}\frac{3}{5}}$

$q_{13} = \frac{4}{5}$, $q_{32} = \frac{3}{11}$. The additive reciprocity is clearly not violated. More interestingly, let us consider the triplet $i=1, j=1, k=2$. Then, according to (4.179), there exist values $q_{12} \in [\frac{1}{2}, \frac{3}{5}]$ and $q_{21} \in [\frac{2}{5}, \frac{1}{2}]$ such that $\frac{1}{2} = \frac{q_{12}q_{21}}{q_{12}q_{21}+(1-q_{12})(1-q_{21})}$. This equality is satisfied by any value $q_{12} \in [\frac{1}{2}, \frac{3}{5}]$ and the corresponding value $q_{21} \in [\frac{1}{2}, \frac{2}{5}]$ such that $q_{21} = 1 - q_{12}$, which again preserves the additive reciprocity.

Notice that the interval FAPCM-M (4.165) is also multiplicatively weakly consistent according to Definition 66. For example, the priority vector $\underline{u} = (0.5217, 0.3478, 0.1304)^T$ satisfies the condition (4.175). △

In the following theorem, the relation between the multiplicative consistency and the multiplicative weak consistency given by Definitions 67 and 66, respectively, is formulated.

Theorem 76 *Let* $\widetilde{Q} = \{\widetilde{q}_{ij}\}_{i,j=1}^n$, $\widetilde{q}_{ij} = \left(q_{ij}^\alpha, q_{ij}^\beta, q_{ij}^\gamma, q_{ij}^\delta\right)$, *be a trapezoidal FAPCM-M. If* $\widetilde{Q}$ *is multiplicatively consistent according to Definition 67, then it is also multiplicatively weakly consistent according to Definition 66.*

Proof The statement follows immediately from Theorem 74. In particular, the inequality (4.176) is obtained immediately from the inequalities (4.192). □

Remark 36 According to Theorem 76, when a trapezoidal FAPCM-M is multiplicatively consistent according to Definition 67, then it is also automatically multiplicatively weakly consistent according Definition 66. However, this does not hold the other way around. Clearly, the multiplicative weak consistency is much weaker than the multiplicative consistency; it only requires existence of one crisp multiplicatively consistent FAPCM-M obtainable by combining particular elements from the closures of the supports of trapezoidal FAPCM-M. Thus, the set of all trapezoidal FAPCMs-M multiplicatively consistent according to Definition 67 is a proper subset of the set of all trapezoidal FAPCMs-M multiplicatively weakly consistent according to Definition 66.

Example 64 Let us examine multiplicative consistency and multiplicative weak consistency of the trapezoidal FAPCM-M

$$
\widetilde{Q} = \begin{pmatrix}
\frac{1}{2} & \left(\frac{8}{15}, \frac{9}{15}, \frac{10}{15}, \frac{11}{15}\right) & \left(\frac{16}{20}, \frac{16}{20}, \frac{17}{20}, \frac{18}{20}\right) & \left(\frac{16}{20}, \frac{17}{20}, \frac{18}{20}, \frac{19}{20}\right) \\
\left(\frac{4}{15}, \frac{5}{15}, \frac{6}{15}, \frac{7}{15}\right) & \frac{1}{2} & \left(\frac{13}{20}, \frac{14}{20}, \frac{15}{20}, \frac{15}{20}\right) & \left(\frac{15}{20}, \frac{16}{20}, \frac{17}{20}, \frac{17}{20}\right) \\
\left(\frac{2}{20}, \frac{3}{20}, \frac{4}{20}, \frac{4}{20}\right) & \left(\frac{5}{20}, \frac{6}{20}, \frac{6}{20}, \frac{7}{20}\right) & \frac{1}{2} & \left(\frac{9}{20}, \frac{10}{20}, \frac{12}{20}, \frac{13}{20}\right) \\
\left(\frac{1}{20}, \frac{2}{20}, \frac{3}{20}, \frac{4}{20}\right) & \left(\frac{3}{20}, \frac{3}{20}, \frac{4}{20}, \frac{5}{20}\right) & \left(\frac{7}{20}, \frac{8}{20}, \frac{10}{20}, \frac{11}{20}\right) & \frac{1}{2}
\end{pmatrix}.
$$

(4.202)

By verifying the inequalities (4.190) and (4.191), we find out that $\widetilde{Q}$ is not multiplicatively consistent; e.g.

$$
\frac{q_{12}^\delta q_{24}^\delta}{q_{12}^\delta q_{24}^\delta + (1 - q_{12}^\delta)(1 - q_{24}^\delta)} = \frac{\frac{11}{15} \frac{17}{20}}{\frac{11}{15} \frac{17}{20} + \frac{4}{15} \frac{3}{20}} = \frac{187}{199} \not\gtreqless q_{14}^\delta = \frac{19}{20}.
$$

However, $\widetilde{Q}$ can still be at least multiplicatively weakly consistent. Let us verify that by using Theorem 69. According to Table 4.17, the property (4.177) is satisfied, and thus $\widetilde{Q}$ is multiplicatively weakly consistent.

Table 4.17 Condition (4.177) for the trapezoidal FAPCM-M (4.202)

$i < j$:	$\max\limits_{k=1,\dots,4}\left\{\dfrac{q_{ik}^\alpha q_{kj}^\alpha}{q_{ik}^\alpha q_{kj}^\alpha + (1 - q_{ik}^\alpha)(1 - q_{kj}^\alpha)}\right\} \leq \min\limits_{k=1,\dots,4}\left\{\dfrac{q_{ik}^\delta q_{kj}^\delta}{q_{ik}^\delta q_{kj}^\delta + (1 - q_{ik}^\delta)(1 - q_{kj}^\delta)}\right\}$
1, 2:	$\max\left\{\frac{8}{15}, \frac{8}{15}, \frac{4}{7}, \frac{12}{29}\right\} \leq \min\left\{\frac{11}{15}, \frac{11}{15}, \frac{63}{76}, \frac{19}{20}\right\}$
1, 3:	$\max\left\{\frac{16}{20}, \frac{104}{153}, \frac{16}{20}, \frac{28}{41}\right\} \leq \min\left\{\frac{18}{20}, \frac{35}{37}, \frac{18}{20}, \frac{209}{218}\right\}$
1, 4:	$\max\left\{\frac{16}{20}, \frac{24}{31}, \frac{36}{47}, \frac{16}{20}\right\} \leq \min\left\{\frac{19}{20}, \frac{187}{199}, \frac{117}{124}, \frac{19}{20}\right\}$
2, 3:	$\max\left\{\frac{16}{27}, \frac{13}{20}, \frac{13}{20}, \frac{21}{34}\right\} \leq \min\left\{\frac{63}{71}, \frac{15}{20}, \frac{15}{20}, \frac{187}{214}\right\}$
2, 4:	$\max\left\{\frac{16}{27}, \frac{15}{20}, \frac{117}{194}, \frac{15}{20}\right\} \leq \min\left\{\frac{133}{141}, \frac{19}{20}, \frac{39}{46}, \frac{19}{20}\right\}$
3, 4:	$\max\left\{\frac{8}{27}, \frac{1}{2}, \frac{9}{20}, \frac{9}{20}\right\} \leq \min\left\{\frac{19}{23}, \frac{119}{158}, \frac{13}{20}, \frac{13}{20}\right\}$

A vector satisfying the inequalities (4.175) in Definition 66 is, for example, $\underline{u} = \left(\frac{9}{16}, \frac{9}{32}, \frac{3}{32}, \frac{1}{16}\right)^T$ with the corresponding APCM-M Q^* in the form

$$Q^* = \begin{pmatrix} 1 & \frac{2}{3} & \frac{6}{7} & \frac{9}{10} \\ \frac{1}{2} & 1 & \frac{3}{4} & \frac{9}{11} \\ \frac{1}{3} & \frac{1}{2} & 1 & \frac{3}{5} \\ \frac{1}{7} & \frac{1}{4} & \frac{1}{2} & 1 \\ \frac{1}{10} & \frac{2}{11} & \frac{2}{5} & \frac{9}{11} \end{pmatrix}. \tag{4.203}$$

$\triangle$

In the rest of this section, some interesting properties of multiplicatively weakly consistent and multiplicatively consistent trapezoidal FAPCMs-M are examined.

Theorem 77 *Let $\widetilde{Q}$ be a trapezoidal FAPCM-M multiplicatively weakly consistent according to Definition 66. The interval FAPCM-M $\widetilde{Q}^*$ constructed from $\widetilde{Q}$ by eliminating the l-th row and the l-th column, $l \in \{1, \ldots, n\}$, is again multiplicatively weakly consistent.*

Proof For $\widetilde{Q}$, (4.175) is valid for every $i, j = 1, \ldots, n$. After eliminating the l-th row and the l-th column of $\widetilde{Q}$, (4.175) is still valid for every remaining $i, j \in \{1, \ldots, n\} \setminus \{l\}$. Therefore, the new trapezoidal FAPCM-M $\widetilde{Q}^*$ is still multiplicatively weakly consistent. $\qquad\square$

The same holds also for multiplicatively consistent trapezoidal FAPCMs-M.

Theorem 78 *Let $\widetilde{Q}$ be a trapezoidal FAPCM-M multiplicatively consistent according to Definition 67. The trapezoidal FAPCM-M $\widetilde{Q}^*$ constructed from $\widetilde{Q}$ by eliminating the l-th row and the l-th column, $l \in \{1, \ldots, n\}$, is again multiplicatively consistent.*

Proof For $\widetilde{Q}$, (4.179) is valid for every $i, j, k = 1, \ldots, n$. After eliminating the l-th row and the l-th column of $\widetilde{Q}$, (4.179) is still valid for every remaining $i, j, k \in \{1, \ldots, n\} \setminus \{l\}$. Therefore, the new interval FAPCM-M $\widetilde{Q}^*$ is multiplicatively consistent. $\qquad\square$

Remark 37 Theorems 77 and 78 are useful in situations when the set of objects compared pairwisely is being reduced. According to the theorems, elimination of one or more objects has no impact on the multiplicative or multiplicative weak consistency of fuzzy PCs of remaining objects.

The following theorems provide some results regarding aggregation of multi-plicatively and multiplicatively weakly consistent trapezoidal FAPCMs-M into one trapezoidal FAPCM-M, which are particularly useful in group decision making.

Theorem 79 *Let* $\widetilde{Q}^1 = \left\{\widetilde{q}_{ij}^1\right\}_{i,j=1}^n$, $\widetilde{q}_{ij}^1 = \left(q_{ij}^{1\alpha}, q_{ij}^{1\beta}, q_{ij}^{1\gamma}, q_{ij}^{1\delta}\right)$, *and* $\widetilde{Q}^2 = \left\{\widetilde{q}_{ij}^2\right\}_{i,j=1}^n$, $\widetilde{q}_{ij}^2 = \left(q_{ij}^{2\alpha}, q_{ij}^{2\beta}, q_{ij}^{2\gamma}, q_{ij}^{2\delta}\right)$, *be trapezoidal FAPCMs-M multiplicatively weakly consistent according to Definition 66. Then* $\widetilde{Q} = \left\{\widetilde{q}_{ij}\right\}_{i,j=1}^n$, $\widetilde{q}_{ij} = \left(q_{ij}^\alpha, q_{ij}^\beta, q_{ij}^\gamma, q_{ij}^\delta\right)$, *such that*

$$q_{ij}^\alpha = \frac{\left(\frac{q_{ij}^{1\alpha}}{q_{ji}^{1\delta}}\right)^\epsilon \left(\frac{q_{ij}^{2\alpha}}{q_{ji}^{2\delta}}\right)^{1-\epsilon}}{\left(\frac{q_{ij}^{1\alpha}}{q_{ji}^{1\delta}}\right)^\epsilon \left(\frac{q_{ij}^{2\alpha}}{q_{ji}^{2\delta}}\right)^{1-\epsilon} + 1}, \quad q_{ij}^\beta = \frac{\left(\frac{q_{ij}^{1\beta}}{q_{ji}^{1\gamma}}\right)^\epsilon \left(\frac{q_{ij}^{2\beta}}{q_{ji}^{2\gamma}}\right)^{1-\epsilon}}{\left(\frac{q_{ij}^{1\beta}}{q_{ji}^{1\gamma}}\right)^\epsilon \left(\frac{q_{ij}^{2\beta}}{q_{ji}^{2\gamma}}\right)^{1-\epsilon} + 1},$$

$$q_{ij}^\gamma = \frac{\left(\frac{q_{ij}^{1\gamma}}{q_{ji}^{1\beta}}\right)^\epsilon \left(\frac{q_{ij}^{2\gamma}}{q_{ji}^{2\beta}}\right)^{1-\epsilon}}{\left(\frac{q_{ij}^{1\gamma}}{q_{ji}^{1\beta}}\right)^\epsilon \left(\frac{q_{ij}^{2\gamma}}{q_{ji}^{2\beta}}\right)^{1-\epsilon} + 1}, \quad q_{ij}^\delta = \frac{\left(\frac{q_{ij}^{1\delta}}{q_{ji}^{1\alpha}}\right)^\epsilon \left(\frac{q_{ij}^{2\delta}}{q_{ji}^{2\alpha}}\right)^{1-\epsilon}}{\left(\frac{q_{ij}^{1\delta}}{q_{ji}^{1\alpha}}\right)^\epsilon \left(\frac{q_{ij}^{2\delta}}{q_{ji}^{2\alpha}}\right)^{1-\epsilon} + 1},$$

is a multiplicatively weakly consistent trapezoidal FAPCM-M for any $\epsilon \in [0, 1]$.

Proof First, let us show that $\widetilde{Q}$ is a trapezoidal FAPCM-M. For $i = 1, \ldots, n$, we get

$$q_{ii}^\alpha = \frac{\left(\frac{q_{ii}^{1\alpha}}{q_{ii}^{1\delta}}\right)^\epsilon \left(\frac{q_{ii}^{2\alpha}}{q_{ii}^{2\delta}}\right)^{1-\epsilon}}{\left(\frac{q_{ii}^{1\alpha}}{q_{ii}^{1\delta}}\right)^\epsilon \left(\frac{q_{ii}^{2\alpha}}{q_{ii}^{2\delta}}\right)^{1-\epsilon} + 1} = \frac{1^\epsilon 1^{1-\epsilon}}{1^\epsilon 1^{1-\epsilon} + 1} = 0.5.$$

Similarly, $q_{ii}^\beta = 0.5$, $q_{ii}^\gamma = 0.5$, $q_{ii}^\delta = 0.5$, and thus, $\widetilde{q}_{ii} = 0.5$, $i = 1, \ldots, n$. Further, for $i \neq j$, we have

$$q_{ij}^\alpha = \frac{\left(\frac{q_{ij}^{1\alpha}}{q_{ji}^{1\delta}}\right)^\epsilon \left(\frac{q_{ij}^{2\alpha}}{q_{ji}^{2\delta}}\right)^{1-\epsilon}}{\left(\frac{q_{ij}^{1\alpha}}{q_{ji}^{1\delta}}\right)^\epsilon \left(\frac{q_{ij}^{2\alpha}}{q_{ji}^{2\delta}}\right)^{1-\epsilon} + 1} = \frac{(q_{ij}^{1\alpha})^\epsilon (q_{ij}^{2\alpha})^{1-\epsilon}}{(q_{ij}^{1\alpha})^\epsilon (q_{ij}^{2\alpha})^{1-\epsilon} + (q_{ji}^{1\delta})^\epsilon (q_{ji}^{2\delta})^{1-\epsilon}}$$

$$= \frac{1}{\left(\frac{q_{ji}^{1\delta}}{q_{ij}^{1\alpha}}\right)^\epsilon \left(\frac{q_{ji}^{2\delta}}{q_{ij}^{2\alpha}}\right)^{1-\epsilon} + 1} = 1 - \frac{\left(\frac{q_{ji}^{1\delta}}{q_{ij}^{1\alpha}}\right)^\epsilon \left(\frac{q_{ji}^{2\delta}}{q_{ij}^{2\alpha}}\right)^{1-\epsilon}}{\left(\frac{q_{ji}^{1\delta}}{q_{ij}^{1\alpha}}\right)^\epsilon \left(\frac{q_{ji}^{2\delta}}{q_{ij}^{2\alpha}}\right)^{1-\epsilon} + 1} = 1 - q_{ji}^\delta,$$

and, analogously, the equalities $q_{ij}^\beta = 1 - q_{ji}^\gamma$, $q_{ij}^\gamma = 1 - q_{ji}^\beta$, $q_{ij}^\delta = 1 - q_{ji}^\alpha$ are derived. Therefore, $\widetilde{q}_{ij} = 1 - \widetilde{q}_{ji}$, $i, j = 1, \ldots, n$. Finally, because $\forall i, j = 1, \ldots, n : \widetilde{q}_{ij}^1 \subseteq$]0, 1[, $\widetilde{q}_{ij}^2 \subseteq$]0, 1[, then also $\widetilde{q}_{ij} \subseteq$]0, 1[.

Second, let us show that $\widetilde{Q}$ is multiplicatively weakly consistent. It is sufficient to prove inequalities (4.175). Since (4.175) is valid for $\widetilde{Q}^1$ and $\widetilde{Q}^2$, there exist non-negative vectors $\underline{u}^1 = \left(u_1^1, \ldots, u_n^1\right)^T$ and $\underline{u}^2 = \left(u_1^2, \ldots, u_n^2\right)^T$ such that $q_{ij}^{1\alpha} \leq \frac{u_i^1}{u_i^1 + u_j^1} \leq q_{ij}^{1\delta}$, $q_{ij}^{2\alpha} \leq \frac{u_i^2}{u_i^2 + u_j^2} \leq q_{ij}^{2\delta}$, $i, j, k = 1, \ldots, n$. From this it follows $\forall i, j, k = 1, \ldots, n$:

$$
q_{ij}^{\alpha} = \frac{\left(\frac{q_{ij}^{1\alpha}}{q_{ji}^{1\delta}}\right)^{\epsilon} \left(\frac{q_{ij}^{2\alpha}}{q_{ji}^{2\delta}}\right)^{1-\epsilon}}{\left(\frac{q_{ij}^{1\alpha}}{q_{ji}^{1\delta}}\right)^{\epsilon} \left(\frac{q_{ij}^{2\alpha}}{q_{ji}^{2\delta}}\right)^{1-\epsilon} + 1} = \frac{\left(\frac{q_{ij}^{1\alpha}}{1-q_{ij}^{1\alpha}}\right)^{\epsilon} \left(\frac{q_{ij}^{2\alpha}}{1-q_{ij}^{2\alpha}}\right)^{1-\epsilon}}{\left(\frac{q_{ij}^{1\alpha}}{1-q_{ij}^{1\alpha}}\right)^{\epsilon} \left(\frac{q_{ij}^{2\alpha}}{1-q_{ij}^{2\alpha}}\right)^{1-\epsilon} + 1} \leq
$$

$$
\frac{\left(\frac{\frac{u_i^1}{u_i^1+u_j^1}}{1-\frac{u_i^1}{u_i^1+u_j^1}}\right)^{\epsilon} \left(\frac{\frac{u_i^2}{u_i^2+u_j^2}}{1-\frac{u_i^2}{u_i^2+u_j^2}}\right)^{1-\epsilon}}{\left(\frac{\frac{u_i^1}{u_i^1+u_j^1}}{1-\frac{u_i^1}{u_i^1+u_j^1}}\right)^{\epsilon} \left(\frac{\frac{u_i^2}{u_i^2+u_j^2}}{1-\frac{u_i^2}{u_i^2+u_j^2}}\right)^{1-\epsilon} + 1} = \frac{\left(\frac{u_i^1}{u_j^1}\right)^{\epsilon} \left(\frac{u_i^2}{u_j^2}\right)^{1-\epsilon}}{\left(\frac{u_i^1}{u_j^1}\right)^{\epsilon} \left(\frac{u_i^2}{u_j^2}\right)^{1-\epsilon} + 1} =
$$

$$
\frac{(u_i^1)^{\epsilon}(u_i^2)^{1-\epsilon}}{(u_i^1)^{\epsilon}(u_i^2)^{1-\epsilon} + (u_j^1)^{\epsilon}(u_j^2)^{1-\epsilon}} = \cdots \leq q_{ij}^{\delta}.
$$

Thus, by denoting $u_i := (u_i^1)^{\epsilon}(u_i^2)^{1-\epsilon}$, $i = 1, \ldots, n$, we get a non-negative vector $\underline{u} = (u_1, \ldots, u_n)^T$ satisfying the inequalities (4.175), which means that $\widetilde{Q}$ is multiplicatively weakly consistent. $\qquad\square$

Theorem 79 can be further extended to the aggregation of $p \geq 2$ multiplicatively weakly consistent trapezoidal FAPCMs-M as follows.

Theorem 80 *Let* $\widetilde{Q}^{\tau} = \left\{\widetilde{q}_{ij}^{\tau}\right\}_{i,j=1}^{n}, \widetilde{q}_{ij}^{\tau} = \left(q_{ij}^{\tau\alpha}, q_{ij}^{\tau\beta}, q_{ij}^{\tau\gamma}, q_{ij}^{\tau\delta}\right), \tau = 1, \ldots, p,$ *be trapezoidal FAPCMs-M multiplicatively weakly consistent according to Definition 66. Then* $\widetilde{Q} = \{\widetilde{q}_{ij}\}_{i,j=1}^{n}$ *such that* $\widetilde{q}_{ij} = \left(q_{ij}^{\alpha}, q_{ij}^{\beta}, q_{ij}^{\gamma}, q_{ij}^{\delta}\right) =$

$$
\left(\frac{\prod_{\tau=1}^{p} \left(\frac{q_{ij}^{\tau\alpha}}{q_{ji}^{\tau\delta}}\right)^{\epsilon_{\tau}}}{\prod_{\tau=1}^{p} \left(\frac{q_{ij}^{\tau\alpha}}{q_{ji}^{\tau\delta}}\right)^{\epsilon_{\tau}} + 1}, \frac{\prod_{\tau=1}^{p} \left(\frac{q_{ij}^{\tau\beta}}{q_{ji}^{\tau\gamma}}\right)^{\epsilon_{\tau}}}{\prod_{\tau=1}^{p} \left(\frac{q_{ij}^{\tau\beta}}{q_{ji}^{\tau\gamma}}\right)^{\epsilon_{\tau}} + 1}, \frac{\prod_{\tau=1}^{p} \left(\frac{q_{ij}^{\tau\gamma}}{q_{ji}^{\tau\beta}}\right)^{\epsilon_{\tau}}}{\prod_{\tau=1}^{p} \left(\frac{q_{ij}^{\tau\gamma}}{q_{ji}^{\tau\beta}}\right)^{\epsilon_{\tau}} + 1}, \frac{\prod_{\tau=1}^{p} \left(\frac{q_{ij}^{\tau\delta}}{q_{ji}^{\tau\alpha}}\right)^{\epsilon_{\tau}}}{\prod_{\tau=1}^{p} \left(\frac{q_{ij}^{\tau\delta}}{q_{ji}^{\tau\alpha}}\right)^{\epsilon_{\tau}} + 1} \right)
$$

is a multiplicatively weakly consistent trapezoidal FAPCM-M for any $\epsilon_{\tau} \in [0, 1]$, $\tau = 1, \ldots, p$, *with* $\sum_{\tau=1}^{p} \epsilon_{\tau} = 1$.

Proof The proof is analogous to the proof of Theorem 79. $\qquad\square$

Similar theorems are formulated also for multiplicatively consistent trapezoidal FAPCMs-M.

Theorem 81 Let $\widetilde{Q}^1 = \left\{\widetilde{q}_{ij}^1\right\}_{i,j=1}^n$, $\widetilde{q}_{ij}^1 = \left(q_{ij}^{1\alpha}, q_{ij}^{1\beta}, q_{ij}^{1\gamma}, q_{ij}^{1\delta}\right)$, and $\widetilde{Q}^2 = \left\{\widetilde{q}_{ij}^2\right\}_{i,j=1}^n$, $\widetilde{q}_{ij}^2 = \left(q_{ij}^{2\alpha}, q_{ij}^{2\beta}, q_{ij}^{2\gamma}, q_{ij}^{2\delta}\right)$, be trapezoidal FAPCMs-M multiplicatively consistent according to Definition 67. Then $\widetilde{Q} = \left\{\widetilde{q}_{ij}\right\}_{i,j=1}^n$, $\widetilde{q}_{ij} = \left(q_{ij}^\alpha, q_{ij}^\beta, q_{ij}^\gamma, q_{ij}^\delta\right)$, such that

$$q_{ij}^\alpha = \frac{\left(\frac{q_{ij}^{1\alpha}}{q_{ji}^{1\delta}}\right)^\epsilon \left(\frac{q_{ij}^{2\alpha}}{q_{ji}^{2\delta}}\right)^{1-\epsilon}}{\left(\frac{q_{ij}^{1\alpha}}{q_{ji}^{1\delta}}\right)^\epsilon \left(\frac{q_{ij}^{2\alpha}}{q_{ji}^{2\delta}}\right)^{1-\epsilon} + 1}, \quad q_{ij}^\beta = \frac{\left(\frac{q_{ij}^{1\beta}}{q_{ji}^{1\gamma}}\right)^\epsilon \left(\frac{q_{ij}^{2\beta}}{q_{ji}^{2\gamma}}\right)^{1-\epsilon}}{\left(\frac{q_{ij}^{1\beta}}{q_{ji}^{1\gamma}}\right)^\epsilon \left(\frac{q_{ij}^{2\beta}}{q_{ji}^{2\gamma}}\right)^{1-\epsilon} + 1},$$

$$q_{ij}^\gamma = \frac{\left(\frac{q_{ij}^{1\gamma}}{q_{ji}^{1\beta}}\right)^\epsilon \left(\frac{q_{ij}^{2\gamma}}{q_{ji}^{2\beta}}\right)^{1-\epsilon}}{\left(\frac{q_{ij}^{1\gamma}}{q_{ji}^{1\beta}}\right)^\epsilon \left(\frac{q_{ij}^{2\gamma}}{q_{ji}^{2\beta}}\right)^{1-\epsilon} + 1}, \quad q_{ij}^\delta = \frac{\left(\frac{q_{ij}^{1\delta}}{q_{ji}^{1\alpha}}\right)^\epsilon \left(\frac{q_{ij}^{2\delta}}{q_{ji}^{2\alpha}}\right)^{1-\epsilon}}{\left(\frac{q_{ij}^{1\delta}}{q_{ji}^{1\alpha}}\right)^\epsilon \left(\frac{q_{ij}^{2\delta}}{q_{ji}^{2\alpha}}\right)^{1-\epsilon} + 1},$$

is a multiplicatively consistent trapezoidal FAPCM-M for any $\epsilon \in [0, 1]$.

Proof From the first part of the proof of Theorem 79 we know that $\widetilde{Q}$ is a trapezoidal FAPCM-M. Therefore, it is sufficient to show that $\widetilde{Q}$ satisfies the inequalities (4.190) and (4.191). Only the inequalities (4.190) will be proven here; the proof of the inequalities (4.191) is analogous.

$\widetilde{Q}^1$ and $\widetilde{Q}^2$ satisfy the inequalities (4.190), i.e.

$$q_{ij}^{1\alpha} \geq \frac{q_{ik}^{1\alpha} q_{kj}^{1\alpha}}{q_{ik}^{1\alpha} q_{kj}^{1\alpha} + (1 - q_{ik}^{1\alpha})(1 - q_{kj}^{1\alpha})}, \qquad q_{ij}^{1\delta} \leq \frac{q_{ik}^{1\delta} q_{kj}^{1\delta}}{q_{ik}^{1\delta} q_{kj}^{1\delta} + (1 - q_{ik}^{1\delta})(1 - q_{kj}^{1\delta})},$$
$$\tag{4.204}$$

$$q_{ij}^{2\alpha} \geq \frac{q_{ik}^{2\alpha} q_{kj}^{2\alpha}}{q_{ik}^{2\alpha} q_{kj}^{2\alpha} + (1 - q_{ik}^{2\alpha})(1 - q_{kj}^{2\alpha})}, \qquad q_{ij}^{2\delta} \leq \frac{q_{ik}^{2\delta} q_{kj}^{2\delta}}{1 q_{ik}^{2\delta} q_{kj}^{2\delta} + (1 - q_{ik}^{2\delta})(1 - q_{kj}^{2\delta})}.$$
$$\tag{4.205}$$

By applying the inequalities (4.204) and (4.205), we obtain

$$q_{ij}^\alpha = \frac{\left(\frac{q_{ij}^{1\alpha}}{q_{ji}^{1\delta}}\right)^\epsilon \left(\frac{q_{ij}^{2\alpha}}{q_{ji}^{2\delta}}\right)^{1-\epsilon}}{\left(\frac{q_{ij}^{1\alpha}}{q_{ji}^{1\delta}}\right)^\epsilon \left(\frac{q_{ij}^{2\alpha}}{q_{ji}^{2\delta}}\right)^{1-\epsilon} + 1} \geq \frac{\left(\frac{\frac{q_{ik}^{1\alpha} q_{kj}^{1\alpha}}{q_{ik}^{1\alpha} q_{kj}^{1\alpha} + q_{ki}^{1\delta} q_{jk}^{1\delta}}}{1 - \frac{q_{ik}^{1\alpha} q_{kj}^{1\alpha}}{q_{ik}^{1\alpha} q_{kj}^{1\alpha} + q_{ki}^{1\delta} q_{jk}^{1\delta}}}\right)^\epsilon \left(\frac{\frac{q_{ik}^{2\alpha} q_{kj}^{2\alpha}}{q_{ik}^{2\alpha} q_{kj}^{2\alpha} + q_{ki}^{2\delta} q_{jk}^{2\delta}}}{\frac{q_{ik}^{2\alpha} q_{kj}^{2\alpha}}{q_{ik}^{2\alpha} q_{kj}^{2\alpha} + q_{ki}^{2\delta} q_{jk}^{2\delta}}}\right)^{1-\epsilon}}{\left(\frac{q_{ik}^{1\alpha} q_{kj}^{1\alpha}}{q_{jk}^{1\delta} q_{ki}^{1\delta}}\right)^\epsilon \left(\frac{q_{ik}^{2\alpha} q_{kj}^{2\alpha}}{q_{jk}^{2\delta} q_{ki}^{2\delta}}\right)^{1-\epsilon} + 1}$$

$$= \frac{\left[\frac{\left(q_{ik}^{1\alpha}\right)^{\epsilon}\left(q_{ik}^{2\alpha}\right)^{1-\epsilon}}{\left(q_{ki}^{1\delta}\right)^{\epsilon}\left(q_{ki}^{2\delta}\right)^{1-\epsilon}}\right]\left[\frac{\left(q_{kj}^{1\alpha}\right)^{\epsilon}\left(q_{kj}^{2\alpha}\right)^{1-\epsilon}}{\left(q_{jk}^{1\delta}\right)^{\epsilon}\left(q_{jk}^{2\delta}\right)^{1-\epsilon}}\right]}{\left[\frac{\left(q_{ik}^{1\alpha}\right)^{\epsilon}\left(q_{ik}^{2\alpha}\right)^{1-\epsilon}}{\left(q_{ki}^{1\delta}\right)^{\epsilon}\left(q_{ki}^{2\delta}\right)^{1-\epsilon}}+1\right]\left[\frac{\left(q_{kj}^{1\alpha}\right)^{\epsilon}\left(q_{kj}^{2\alpha}\right)^{1-\epsilon}}{\left(q_{jk}^{1\delta}\right)^{\epsilon}\left(q_{jk}^{2\delta}\right)^{1-\epsilon}}+1\right]}$$

$$= \frac{\left[\frac{\left(q_{ik}^{1\alpha}\right)^{\epsilon}\left(q_{ik}^{2\alpha}\right)^{1-\epsilon}}{\left(q_{ki}^{1\delta}\right)^{\epsilon}\left(q_{ki}^{2\delta}\right)^{1-\epsilon}}\right]\left[\frac{\left(q_{kj}^{1\alpha}\right)^{\epsilon}\left(q_{kj}^{2\alpha}\right)^{1-\epsilon}}{\left(q_{jk}^{1\delta}\right)^{\epsilon}\left(q_{jk}^{2\delta}\right)^{1-\epsilon}}\right]}{\left[\frac{\left(q_{ik}^{1\alpha}\right)^{\epsilon}\left(q_{ik}^{2\alpha}\right)^{1-\epsilon}}{\left(q_{ki}^{1\delta}\right)^{\epsilon}\left(q_{ki}^{2\delta}\right)^{1-\epsilon}}+1\right]\left[\frac{\left(q_{kj}^{1\alpha}\right)^{\epsilon}\left(q_{kj}^{2\alpha}\right)^{1-\epsilon}}{\left(q_{jk}^{1\delta}\right)^{\epsilon}\left(q_{jk}^{2\delta}\right)^{1-\epsilon}}+1\right]}+\frac{1}{\left[\frac{\left(q_{ik}^{1\alpha}\right)^{\epsilon}\left(q_{ik}^{2\alpha}\right)^{1-\epsilon}}{\left(q_{ki}^{1\delta}\right)^{\epsilon}\left(q_{ki}^{2\delta}\right)^{1-\epsilon}}+1\right]\left[\frac{\left(q_{kj}^{1\alpha}\right)^{\epsilon}\left(q_{kj}^{2\alpha}\right)^{1-\epsilon}}{\left(q_{jk}^{1\delta}\right)^{\epsilon}\left(q_{jk}^{2\delta}\right)^{1-\epsilon}}+1\right]}}$$

$$= \frac{q_{ik}^{\alpha}q_{kj}^{\alpha}}{q_{ik}^{\alpha}q_{kj}^{\alpha}+\frac{(q_{ki}^{1\delta})^{\epsilon}(q_{ki}^{2\delta})^{1-\epsilon}}{(q_{ki}^{1\delta})^{\epsilon}(q_{ki}^{2\delta})^{1-\epsilon}+(q_{ik}^{2\alpha})^{1-\epsilon}}\frac{(q_{jk}^{1\delta})^{\epsilon}(q_{jk}^{2\delta})^{1-\epsilon}}{(q_{jk}^{1\delta})^{\epsilon}(q_{jk}^{2\delta})^{1-\epsilon}+(q_{kj}^{1\alpha})^{\epsilon}(q_{kj}^{2\alpha})^{1-\epsilon}}}$$

$$= \frac{q_{ik}^{\alpha}q_{kj}^{\alpha}}{q_{ik}^{\alpha}q_{kj}^{\alpha}+\frac{\left(\frac{q_{ik}^{1\delta}}{q_{ik}^{1\alpha}}\right)^{\epsilon}\left(\frac{q_{ik}^{2\delta}}{q_{ik}^{2\alpha}}\right)^{1-\epsilon}}{\left(\frac{q_{ik}^{1\delta}}{q_{ik}^{1\alpha}}\right)^{\epsilon}\left(\frac{q_{ik}^{2\delta}}{q_{ik}^{2\alpha}}\right)^{1-\epsilon}+1}\frac{\left(\frac{q_{jk}^{1\delta}}{q_{kj}^{1\alpha}}\right)^{\epsilon}\left(\frac{q_{jk}^{2\delta}}{q_{kj}^{2\alpha}}\right)^{1-\epsilon}}{\left(\frac{q_{jk}^{1\delta}}{q_{kj}^{1\alpha}}\right)^{\epsilon}\left(\frac{q_{jk}^{2\delta}}{q_{kj}^{2\alpha}}\right)^{1-\epsilon}+1}} = \frac{q_{ik}^{\alpha}q_{kj}^{\alpha}}{q_{ik}^{\alpha}q_{kj}^{\alpha}+(1-q_{ik}^{\alpha})(1-q_{kj}^{\alpha})}.$$

Analogously, the inequality $q_{ij}^{\delta} \leq \frac{q_{ik}^{\delta}q_{kj}^{\delta}}{q_{ik}^{\delta}q_{kj}^{\delta}+(1-q_{ik}^{\delta}(1-q_{kj}^{\delta})}$ is proved. $\qquad\square$

Theorem 81 can be further extended to the aggregation of $p \geq 2$ multiplicatively consistent interval FAPCMs-M as follows.

Theorem 82 *Let* $\widetilde{Q}^{\tau} = \left\{\widetilde{q}_{ij}^{\tau}\right\}_{i,j=1}^{n}$, $\widetilde{q}_{ij}^{\tau} = \left(q_{ij}^{\tau\alpha}, q_{ij}^{\tau\beta}, q_{ij}^{\tau\gamma}, q_{ij}^{\tau\delta}\right)$, $\tau = 1, \ldots, p$, *be trapezoidal FAPCMs-M multiplicatively consistent according to Definition 67. Then* $\widetilde{Q} = \left\{\widetilde{q}_{ij}\right\}_{i,j=1}^{n}$ *such that* $\widetilde{q}_{ij} = \left(q_{ij}^{\alpha}, q_{ij}^{\beta}, q_{ij}^{\gamma}, q_{ij}^{\delta}\right) =$

$$\left(\frac{\prod_{\tau=1}^{p}\left(\frac{q_{ij}^{\tau\alpha}}{q_{ji}^{\tau\delta}}\right)^{\epsilon_{\tau}}}{\prod_{\tau=1}^{p}\left(\frac{q_{ij}^{\tau\alpha}}{q_{ji}^{\tau\delta}}\right)^{\epsilon_{\tau}}+1}, \frac{\prod_{\tau=1}^{p}\left(\frac{q_{ij}^{\tau\beta}}{q_{ji}^{\tau\gamma}}\right)^{\epsilon_{\tau}}}{\prod_{\tau=1}^{p}\left(\frac{q_{ij}^{\tau\beta}}{q_{ji}^{\tau\gamma}}\right)^{\epsilon_{\tau}}+1}, \frac{\prod_{\tau=1}^{p}\left(\frac{q_{ij}^{\tau\gamma}}{q_{ji}^{\tau\beta}}\right)^{\epsilon_{\tau}}}{\prod_{\tau=1}^{p}\left(\frac{q_{ij}^{\tau\gamma}}{q_{ji}^{\tau\beta}}\right)^{\epsilon_{\tau}}+1}, \frac{\prod_{\tau=1}^{p}\left(\frac{q_{ij}^{\tau\delta}}{q_{ji}^{\tau\alpha}}\right)^{\epsilon_{\tau}}}{\prod_{\tau=1}^{p}\left(\frac{q_{ij}^{\tau\delta}}{q_{ji}^{\tau\alpha}}\right)^{\epsilon_{\tau}}+1}\right)$$

is a multiplicatively consistent trapezoidal FAPCM-M for any $\epsilon_{\tau} \in [0, 1]$, $\tau = 1, \ldots, p$, *with* $\sum_{\tau=1}^{p}\epsilon_{\tau} = 1$.

Proof The proof is analogous to the proof of Theorem 81. $\qquad\square$

4.3.3.2 Deriving Priorities from FAPCMs-M

In this section, the focus is put on methods for obtaining fuzzy priorities of objects from FAPCMs-M. The notation $\underline{\widetilde{u}} = (\widetilde{u}_1, \ldots, \widetilde{u}_n)^T$, $\widetilde{u}_i = (u_i^{\alpha}, u_i^{\beta}, u_i^{\gamma}, u_i^{\delta})$, $i = 1, \ldots, n$, will be used hereafter to represent exclusively a fuzzy priority vector associated with a FAPCM-M.

Various methods have been proposed to derive interval priorities of objects from interval FAPCMs-M. These methods are mostly based on linear programming models rather than on interval arithmetic. Xu and Chen (2008a), for example, proposed linear programming models for obtaining interval priorities of objects from interval FAPCMs-M. The models are based on satisfying the inequalities (4.163) in the case when the interval FAPCM-M is multiplicatively consistent according to Definition 61 or on satisfying a relaxed version of the inequalities (4.163) with additional deviation variables in the case when the interval FAPCM-M is not multiplicatively consistent. Genç et al. (2010) showed that in the case of multiplicative consistency, the interval priority vector can be calculated from the multiplicatively consistent interval FAPCM-M directly without the need to solve the linear programming models. Very similar linear programming models for obtaining interval priorities from interval FAPCMs-M based on Tanino's characterization were also proposed by Wang and Li (2012).

As far as I am aware, the only approach for obtaining interval priorities from interval FAPCMs-M not based on linear programming models is the approach presented by Xia and Xu (2011). Xia and Xu (2011) derived formulas for obtaining interval priorities from interval FAPCMs-M based on the extension of the formula (2.62). This approach is reviewed at the beginning of this section and it is shown that this approach is not invariant under permutation of objects. Afterwards, new formulas for obtaining fuzzy priorities from FAPCMs-M are proposed and their properties are discussed. In particular, it is proved that the new formulas preserve the two desired properties - invariance under permutation and additive reciprocity of the related PCs.

Xia and Xu (2011) proposed an extension of the formula (2.62) to interval FAPCMs-M. For an interval FAPCM-M $\overline{Q} = \{\overline{q}_{ij}\}_{i,j=1}^{n}$, $\overline{q}_{ij} = [q_{ij}^{L}, q_{ij}^{U}]$, they constructed two APCMs-M $C = \{c_{ij}\}_{i,j=1}^{n}$ and $D = \{d_{ij}\}_{i,j=1}^{n}$ by applying (4.164). Afterwards, they derived priorities $u_i(C)$ and $u_i(D)$, $i = 1, \ldots, n$, of objects from these APCMs-M C and D, respectively, by using the formula (2.62). The interval priorities $\overline{u}_i = [u_i^{L}, u_i^{U}]$, $i = 1, \ldots, n$, are then determined as

$$u_i^{L} = \min\{u_i(C), u_i(D)\}, \qquad u_i^{U} = \max\{u_i(C), u_i(D)\}. \qquad (4.206)$$

However, this method, similarly to Definition 62 of multiplicative consistency for interval FAPCMs-M proposed by Xia and Xu (2011) and reviewed already in Sect. "Review of Fuzzy Extensions of Multiplicative Consistency", is not invariant under permutation of objects. This drawback is illustrated on the following example.

Example 65 Let us consider the interval FAPCM-M $\overline{Q}$ of three objects o_1, o_2, and o_3 given by (4.165). The interval priorities of the objects obtained by the formulas (4.206) are in the form

$$\overline{u}_1 = [1.1447, 2.0801], \quad \overline{u}_2 = [1.1447, 1.3867], \quad \overline{u}_3 = [0.3467, 0.7631].$$

Now, let us consider the corresponding permuted interval FAPCM-M $\overline{Q}^{\pi}$ given as (4.166). The interval priorities of objects obtained from the permuted interval FAPCM-M $\overline{Q}^{\pi}$ by the formulas (4.206) are in the form

$$\overline{u}_{\pi(1)}^{\pi} = [1.3104, 1.8171], \quad \overline{u}_{\pi(2)}^{\pi} = [1.0000, 1.5874], \quad \overline{u}_{\pi(3)}^{\pi} = [0.3467, 0.7631].$$

As we can see, $\overline{u}_1 \neq \overline{u}_{\pi(1)}^{\pi}$ and $\overline{u}_2 \neq \overline{u}_{\pi(2)}^{\pi}$. $\triangle$

Since the method for deriving interval priorities from interval FAPCMs-M proposed by Xia and Xu (2011) is not invariant under permutation of objects, it is not suitable for deriving interval priorities. It is indispensable to obtain the interval priorities from interval FAPCMs-M in such a way that they do not change under the permutation of objects in the interval FAPCMs-M.

The formula (2.62) for obtaining non-normalized priorities from APCMs-M has to be again extended to FAPCMs-M by properly applying constrained fuzzy arithmetic. For a trapezoidal FAPCM-M $\tilde{Q} = \{\tilde{q}_{ij}\}_{i,j=1}^{n}$, $\tilde{q}_{ij} = (q_{ij}^{\alpha}, q_{ij}^{\beta}, q_{ij}^{\gamma}, q_{ij}^{\delta})$, the non-normalized fuzzy priorities $\tilde{u}_i = (u_i^{\alpha}, u_i^{\beta}, u_i^{\gamma}, u_i^{\delta})$, $i = 1, \ldots, n$, should be obtained by applying (3.47) as:

$$u_i^{\alpha} = \min \left\{ \sqrt[n]{\prod_{j=1}^{n} \frac{q_{ij}}{q_{ji}}}; \begin{array}{l} q_{ij} \in \left[q_{ij}^{\alpha}, q_{ij}^{\delta} \right], \\ q_{ji} = 1 - q_{ij}, \\ j = 1, \ldots, n \end{array} \right\}, \tag{4.207}$$

$$u_i^{\beta} = \min \left\{ \sqrt[n]{\prod_{j=1}^{n} \frac{q_{ij}}{q_{ji}}}; \begin{array}{l} q_{ij} \in \left[q_{ij}^{\beta}, q_{ij}^{\gamma} \right], \\ q_{ji} = 1 - q_{ij}, \\ j = 1, \ldots, n \end{array} \right\}, \tag{4.208}$$

$$u_i^{\gamma} = \max \left\{ \sqrt[n]{\prod_{j=1}^{n} \frac{q_{ij}}{q_{ji}}}; \begin{array}{l} q_{ij} \in \left[q_{ij}^{\beta}, q_{ij}^{\gamma} \right], \\ q_{ji} = 1 - q_{ij}, \\ j = 1, \ldots, n \end{array} \right\}, \tag{4.209}$$

$$u_i^{\delta} = \max \left\{ \sqrt[n]{\prod_{j=1}^{n} \frac{q_{ij}}{q_{ji}}}; \begin{array}{l} q_{ij} \in \left[q_{ij}^{\alpha}, q_{ij}^{\delta} \right], \\ q_{ji} = 1 - q_{ij}, \\ j = 1, \ldots, n \end{array} \right\}. \tag{4.210}$$

Unlike in the formulas (4.143)–(4.146), the additive-reciprocity constraints in the formulas (4.207)–(4.210) are indispensable since with every PC q_{ij} also the reciprocal PC q_{ji} appears in the optimized function. Nevertheless, also in this case the optima of the optimization problems (4.207)–(4.210) can be determined easily:

$$u_i^{\alpha} = \sqrt[n]{\prod_{j=1}^{n} \frac{q_{ij}^{\alpha}}{q_{ji}^{\delta}}} = \sqrt[n]{\prod_{j=1}^{n} \frac{q_{ij}^{\alpha}}{1 - q_{ij}^{\alpha}}}, \tag{4.211}$$

$$u_i^\beta = \sqrt[n]{\prod_{j=1}^{n} \frac{q_{ij}^\beta}{q_{ji}^\gamma}} = \sqrt[n]{\prod_{j=1}^{n} \frac{q_{ij}^\beta}{1 - q_{ij}^\beta}}, \tag{4.212}$$

$$u_i^\gamma = \sqrt[n]{\prod_{j=1}^{n} \frac{q_{ij}^\gamma}{q_{ji}^\beta}} = \sqrt[n]{\prod_{j=1}^{n} \frac{q_{ij}^\gamma}{1 - q_{ij}^\gamma}}, \tag{4.213}$$

$$u_i^\delta = \sqrt[n]{\prod_{j=1}^{n} \frac{q_{ij}^\delta}{q_{ji}^\alpha}} = \sqrt[n]{\prod_{j=1}^{n} \frac{q_{ij}^\delta}{1 - q_{ij}^\delta}}. \tag{4.214}$$

Note that the formulas (4.211)–(4.214) could be obtained also by simply applying simplified standard fuzzy arithmetic (3.38) to the fuzzy extension of the formula (2.62) since both constrained and standard fuzzy arithmetic give the same results in this particular case. As it will be shown later, such simplification is not possible when extending the formula (2.63) to FAPCMs-M.

Example 66 Let us consider the interval FAPCM-M $\overline{Q}$ given by (4.165). The interval priorities of objects obtained by formulas (4.211)–(4.214) are

$$\overline{u}_1 = [1.1447, 2.0801], \quad \overline{u}_2 = [1, 1.15874], \quad \overline{u}_3 = [0.3467, 0.7631].$$

The same interval priorities are obtained also from the permuted interval FAPCM-M $\overline{Q}^\pi$ given as (4.166), i.e. $\overline{u}_{\pi(1)}^\pi = \overline{u}_1, \overline{u}_{\pi(2)}^\pi = \overline{u}_2, \overline{u}_{\pi(3)}^\pi = \overline{u}_3$. Compare the resulting interval priorities with the interval priorities in Example 65 obtained by applying the method proposed by Xia and Xu (2011). △

Analogously as in the previous sections, formula (2.63) for obtaining normalized priorities from APCMs-M needs to be extended to FAPCMs-M by using the constrained fuzzy arithmetic. Only in this way the preservation of the additive reciprocity of the related PCs and of invariance under permutation of objects can be guaranteed.

By applying constrained fuzzy arithmetic (3.47) to the fuzzy extension of the formula (2.63) for obtaining normalized priorities of objects, the fuzzy priorities $\widetilde{u}_{Ci} = \left(u_{Ci}^\alpha, u_{Ci}^\beta, u_{Ci}^\gamma, u_{Ci}^\delta\right)$, $i = 1, \ldots, n$, (the lower index C stands for the applied concept of constrained fuzzy arithmetic) are obtained from a FAPCM-M $\widetilde{Q} = \{\widetilde{q}_{ij}\}_{i,j=1}^n$, $\widetilde{q}_{ij} = (q_{ij}^\alpha, q_{ij}^\beta, q_{ij}^\gamma, q_{ij}^\delta)$, in this form:

$$u_{Ci}^\alpha = \min \left\{ \frac{\sqrt[n]{\prod_{j=1}^{n} \frac{q_{ij}}{q_{ji}}}}{\sum_{k=1}^{n} \sqrt[n]{\prod_{j=1}^{n} \frac{q_{kj}}{q_{jk}}}} ; \begin{array}{l} q_{rs} \in \left[q_{rs}^\alpha, q_{rs}^\delta\right], \\ q_{sr} = 1 - q_{rs}, \\ r, s = 1, \ldots, n \end{array} \right\}, \tag{4.215}$$

$$u_{Ci}^{\beta} = \min \left\{ \frac{\sqrt[n]{\prod_{j=1}^{n} \frac{q_{ij}}{q_{ji}}}}{\sum_{k=1}^{n} \sqrt[n]{\prod_{j=1}^{n} \frac{q_{kj}}{q_{jk}}}} ; \begin{array}{l} q_{rs} \in [q_{rs}^{\beta}, q_{rs}^{\gamma}], \\ q_{sr} = 1 - q_{rs}, \\ r, s = 1, \dots, n \end{array} \right\}, \tag{4.216}$$

$$u_{Ci}^{\gamma} = \max \left\{ \frac{\sqrt[n]{\prod_{j=1}^{n} \frac{q_{ij}}{q_{ji}}}}{\sum_{k=1}^{n} \sqrt[n]{\prod_{j=1}^{n} \frac{q_{kj}}{q_{jk}}}} ; \begin{array}{l} q_{rs} \in [q_{rs}^{\beta}, q_{rs}^{\gamma}], \\ q_{sr} = 1 - q_{rs}, \\ r, s = 1, \dots, n \end{array} \right\}, \tag{4.217}$$

$$u_{Ci}^{\delta} = \max \left\{ \frac{\sqrt[n]{\prod_{j=1}^{n} \frac{q_{ij}}{q_{ji}}}}{\sum_{k=1}^{n} \sqrt[n]{\prod_{j=1}^{n} \frac{q_{kj}}{q_{jk}}}} ; \begin{array}{l} q_{rs} \in [q_{rs}^{\alpha}, q_{rs}^{\delta}], \\ q_{sr} = 1 - q_{rs}, \\ r, s = 1, \dots, n \end{array} \right\}. \tag{4.218}$$

Theorem 83 *The fuzzy priorities* $\tilde{u}_{Ci} = (u_{Ci}^{\alpha}, u_{Ci}^{\beta}, u_{Ci}^{\gamma}, u_{Ci}^{\delta}), i = 1, \dots, n,$ *obtained from a FAPCM-M* $\tilde{Q} = \{\tilde{q}_{ij}\}_{i,j=1}^{n}$ *by the formulas (4.215)–(4.218) are normalized.*

Proof It is sufficient to prove that the fuzzy priorities $\tilde{u}_{Ci}, i = 1, \dots, n,$ obtained by the formulas (4.215)–(4.218) satisfy the inequalities (3.15). From the formula (4.215) it follows that u_{Ci}^{α} was obtained by applying the formula (2.63) to one particular APCM-M $Q^{\alpha i} = \{q_{rs}\}_{r,s=1}^{n}, q_{rs} \in [q_{rs}^{\alpha}, q_{rs}^{\delta}], r, s = 1, \dots, n.$ Let us denote $u_{k}^{\alpha i}$ the priorities of objects $o_{k}, k \neq i,$ obtainable by the formula (2.63) from the same APCM-M $Q^{\alpha i}$. Obviously, $u_{Ci}^{\alpha} + \sum_{\substack{k=1 \\ k \neq i}}^{n} u_{k}^{\alpha i} = 1,$ and $u_{k}^{\alpha i} \in [u_{Ck}^{\alpha}, u_{Ck}^{\delta}], k \neq i.$ From this it follows that $u_{Ci}^{\alpha} + \sum_{\substack{k=1 \\ k \neq i}}^{n} u_{Ck}^{\delta} \geq 1.$ The remaining inequalities in (3.15) are proved analogously. □

Remark 38 According to Theorem 83, the fuzzy priorities $\tilde{u}_{Ci}, i = 1, \dots, n,$ obtained from a FAPCM-M by the formulas (4.215)–(4.218) are normalized in the sense of Definition 29. Notice that the normality of the fuzzy priorities was again reached naturally by just properly applying constrained fuzzy arithmetic to the fuzzy extension of the formula (2.63) for obtaining normalized priorities from an APCM-M, similarly as in the case of the fuzzy extension of the EVM and the GMM; no forced normalization was needed.

Theorem 84 *The method for obtaining the normalized fuzzy priorities of objects from FAPCMs-M by using the formulas (4.215)–(4.218) is invariant under permutation of objects in FAPCMs-M.*

Proof It is sufficient to show that for a given object $o_{i}, i \in \{1, \dots, n\},$ its priority $\tilde{u}_{Ci}$ obtained by the formulas (4.215)–(4.218) does not change under permutation of objects in a FAPCM-M $\tilde{Q}$.

From the invariance of the formula (2.63) reviewed in Sect. 2.3.3.3 it follows that the priority u_i of object o_i determined by the formula (2.63) from the given APCM-M Q does not change under any permutation $Q^\pi = PQP^T$ of Q, it is just permuted accordingly. This means that the priority u_i obtained from Q is equal to the corresponding priority $u^\pi_{\pi(i)}$ obtained from Q^π.

Therefore, neither the minimum u^α_{Ci} nor the maximum u^δ_{Ci} of the priority u_i of object o_i obtained by (2.63) over all APCMs-M obtainable from the closures of the supports of the trapezoidal fuzzy numbers in the trapezoidal FAPCM-M $\widetilde{Q} = \{\widetilde{q}_{ij}\}^n_{i,j=1}$, $\widetilde{q}_{ij} = (q^\alpha_{ij}, q^\beta_{ij}, q^\gamma_{ij}, q^\delta_{ij})$, change. Analogously, also the minimum u^β_{Ci} and the maximum u^γ_{Ci} of the priority u_i obtained by (2.63) over all APCMs-M obtainable from the cores of the trapezoidal fuzzy numbers in the trapezoidal FAPCM-M $\widetilde{Q} = \{\widetilde{q}_{ij}\}^n_{i,j=1}$ do not change. Therefore, the fuzzy priority $\widetilde{u}_{Ci} = (u^\alpha_{Ci}, u^\beta_{Ci}, u^\gamma_{Ci}, u^\delta_{Ci})$ obtained by the formulas (4.215)–(4.218) does not change under permutation (it is only permuted accordingly), which concludes the proof.   □

The optimization problems solved in (4.215)–(4.218) have $n^2 - n$ variables and $\frac{n^2-n}{2}$ additive-reciprocity constraints (the number of variables and additive-reciprocity constraints gets reduced when crisp numbers are present above and below the main diagonal of the FAPCM-M). Thus, the computational complexity of the optimization problems increases rapidly with an increasing dimension n. However, the following theorem shows that the optimization problems (4.215)–(4.218) can be simplified significantly. First, the additive-reciprocity constraints can be incorporated into the objective functions. Second, when u^α_{Ci} is computed, the variables $q_{ij}, j = 1, \ldots, n$, can be fixed as the lower boundary values of the trapezoidal fuzzy numbers in the i-th row of the FAPCM, i.e. as $q_{ij} := q^\alpha_{ij}$. Analogously also for the representing values $u^\beta_{Ci}, u^\gamma_{Ci}$, and u^δ_{Ci}. In this way, the number of variables is reduced from $n^2 - n$ to $\frac{n^2-n}{2} - (n - 1)$.

Theorem 85 *Let* $\widetilde{Q} = \{\widetilde{q}_{ij}\}^n_{i,j=1}$, $\widetilde{q}_{ij} = \left(q^\alpha_{ij}, q^\beta_{ij}, q^\gamma_{ij}, q^\delta_{ij}\right)$, *be a trapezoidal FAPCM-M. The optimization problems (4.215)–(4.218) can be simplified for $i = 1, \ldots, n$ in the following way:*

$$u^\alpha_{Ci} = \frac{\sqrt[n]{\prod\limits_{j=1}^{n} \frac{q^\alpha_{ij}}{1-q^\alpha_{ij}}}}{\sqrt[n]{\prod\limits_{j=1}^{n} \frac{q^\alpha_{ij}}{1-q^\alpha_{ij}}} + \max\left\{\sum\limits_{\substack{k=1\\k\neq i}}^{n} \sqrt[n]{\frac{1-q^\alpha_{ik}}{q^\alpha_{ik}} \prod\limits_{\substack{l=1\\l\neq i}}^{k-1} \frac{1-q_{lk}}{q_{lk}} \prod\limits_{\substack{l=k+1\\l\neq i}}^{n} \frac{q_{kl}}{1-q_{kl}}} \;;\; \begin{matrix} q_{rs} \in \left[q^\alpha_{rs}, q^\delta_{rs}\right], \\ r = 1, \ldots, n-1, \\ s = r+1, \ldots, n, \\ r, s \neq i \end{matrix}\right\}},$$

(4.219)

$$u_{Ci}^{\beta} = \frac{\sqrt[n]{\prod_{j=1}^{n} \frac{q_{ij}^{\beta}}{1-q_{ij}^{\beta}}}}{\sqrt[n]{\prod_{j=1}^{n} \frac{q_{ij}^{\beta}}{1-q_{ij}^{\beta}}} + \max\left\{ \sum_{\substack{k=1\\k\neq i}}^{n} \sqrt[n]{\frac{1-q_{ik}^{\beta}}{q_{ik}^{\beta}} \prod_{\substack{l=1\\l\neq i}}^{k-1} \frac{1-q_{lk}}{q_{lk}} \prod_{\substack{l=k+1\\l\neq i}}^{n} \frac{q_{kl}}{1-q_{kl}}} \; ; \; \begin{array}{l} q_{rs} \in \left[q_{rs}^{\beta}, q_{rs}^{\gamma}\right], \\ r = 1, \ldots, n-1, \\ s = r+1, \ldots, n, \\ r, s \neq i \end{array} \right\}},$$

(4.220)

$$u_{Ci}^{\gamma} = \frac{\sqrt[n]{\prod_{j=1}^{n} \frac{q_{ij}^{\gamma}}{1-q_{ij}^{\gamma}}}}{\sqrt[n]{\prod_{j=1}^{n} \frac{q_{ij}^{\gamma}}{1-q_{ij}^{\gamma}}} + \min\left\{ \sum_{\substack{k=1\\k\neq i}}^{n} \sqrt[n]{\frac{1-q_{ik}^{\gamma}}{q_{ik}^{\gamma}} \prod_{\substack{l=1\\l\neq i}}^{k-1} \frac{1-q_{lk}}{q_{lk}} \prod_{\substack{l=k+1\\l\neq i}}^{n} \frac{q_{kl}}{1-q_{kl}}} \; ; \; \begin{array}{l} q_{rs} \in \left[q_{rs}^{\beta}, q_{rs}^{\gamma}\right], \\ r = 1, \ldots, n-1, \\ s = r+1, \ldots, n, \\ r, s \neq i \end{array} \right\}},$$

(4.221)

$$u_{Ci}^{\delta} = \frac{\sqrt[n]{\prod_{j=1}^{n} \frac{q_{ij}^{\delta}}{1-q_{ij}^{\delta}}}}{\sqrt[n]{\prod_{j=1}^{n} \frac{q_{ij}^{\delta}}{1-q_{ij}^{\delta}}} + \min\left\{ \sum_{\substack{k=1\\k\neq i}}^{n} \sqrt[n]{\frac{1-q_{ik}^{\delta}}{q_{ik}^{\delta}} \prod_{\substack{l=1\\l\neq i}}^{k-1} \frac{1-q_{lk}}{q_{lk}} \prod_{\substack{l=k+1\\l\neq i}}^{n} \frac{q_{kl}}{1-q_{kl}}} \; ; \; \begin{array}{l} q_{rs} \in \left[q_{rs}^{\alpha}, q_{rs}^{\delta}\right], \\ r = 1, \ldots, n-1, \\ s = r+1, \ldots, n, \\ r, s \neq i \end{array} \right\}}.$$

(4.222)

Proof First, let us show that the formulas (4.215) and (4.219) are identical. For any $i \in \{1, \ldots, n\}$ the formula (4.215) can be written in the following way:

$$u_{Ci}^{\alpha} = \min\left\{ \frac{\sqrt[n]{\prod_{j=1}^{n} \frac{q_{ij}}{1-q_{ij}}}}{\sqrt[n]{\prod_{j=1}^{n} \frac{q_{ij}}{1-q_{ij}}} + \sum_{\substack{k=1\\k\neq i}}^{n} \sqrt[n]{\frac{1-q_{ik}}{q_{ik}} \prod_{\substack{j=1\\j\neq i}}^{n} \frac{q_{kj}}{1-q_{kj}}}} \; ; \; \begin{array}{l} q_{rs} \in \left[q_{rs}^{\alpha}, q_{rs}^{\delta}\right], \\ q_{sr} = 1 - q_{rs}, \\ r, s = 1, \ldots, n \end{array} \right\}.$$

Let us denote $\quad x_i := \sqrt[n]{\prod_{j=1}^{n} \frac{q_{ij}}{1-q_{ij}}} \quad$ and $\quad y_i := \sum_{\substack{k=1\\k\neq i}}^{n} \sqrt[n]{\frac{1-q_{ik}}{q_{ik}} \prod_{\substack{j=1\\j\neq i}}^{n} \frac{q_{kj}}{1-q_{kj}}}.$

Obviously, $x_i > 0$ for $q_{is} \in \left[q_{is}^\alpha, q_{is}^\delta\right]$, $s = 1, \ldots, n$, and $y_i > 0$ for $q_{rs} \in \left[q_{rs}^\alpha, q_{rs}^\delta\right]$, $q_{sr} = 1 - q_{rs}$, $r, s = 1, \ldots, n$. Further, x_i is increasing in all variables $q_{is}, s \neq i$:

$$\frac{\partial x_i}{\partial q_{is}} = \left[\prod_{\substack{j=1 \\ j \neq s}}^{n}\left(\frac{q_{ij}}{1 - q_{ij}}\right)^{\frac{1}{n}}\right]\frac{1}{n}\left(\frac{q_{is}}{1 - q_{is}}\right)^{\frac{1-n}{n}}\frac{1}{(1 - q_{is})^2} > 0, \quad \begin{array}{l} q_{ik} \in [q_{ik}^\alpha, q_{ik}^\delta], \\ k = 1, \ldots, n, \end{array}$$

and y_i is decreasing in variables $q_{is}, s \neq i$:

$$\frac{\partial y_i}{\partial q_{is}} = \left[\prod_{\substack{j=1 \\ j \neq i}}^{n}\left(\frac{q_{sj}}{1 - q_{sj}}\right)^{\frac{1}{n}}\right] \cdot \frac{1}{n} \cdot \left(\frac{1 - q_{is}}{q_{is}}\right)^{\frac{1-n}{n}} \frac{-1}{q_{is}^2} < 0, \quad \begin{array}{l} q_{kj} \in [q_{kj}^\alpha, q_{kj}^\delta], \\ k, j = 1, \ldots, n. \end{array}$$

Further, let us denote $f_i := \frac{x_i}{x_i + y_i}$. Then $\frac{\partial f_i}{\partial x_i} = \frac{y_i}{(x_i + y_i)^2} > 0$ and $\frac{\partial f_i}{\partial y_i} = \frac{-x_i}{(x_i + y_i)^2} < 0$. Hence, f_i is an increasing function of x_i and a decreasing function of y_i. It means that for minimizing the function f_i, we have to minimize x_i and maximize y_i. Since the function x_i is increasing in all the variables, we obtain

$$x_i' := \min\left\{x_i \; ; \; q_{ij} \in \left[q_{ij}^\alpha, q_{ij}^\delta\right], \; j = 1, \ldots, n\right\} = \sqrt[n]{\prod_{j=1}^{n}\frac{q_{ij}^\alpha}{1 - q_{ij}^\alpha}}.$$

Since the function y_i is decreasing in the variables $q_{i1}, \ldots, q_{in}$, we obtain

$$y_i' := \max\left\{y_i \; ; \; \begin{array}{l} q_{rs} \in \left[q_{rs}^\alpha, q_{rs}^\delta\right], q_{sr} = 1 - q_{rs}, \\ r, s = 1, \ldots, n \end{array}\right\} =$$

$$\left\{\sum_{\substack{k=1 \\ k \neq i}}^{n}\sqrt[n]{\frac{1 - q_{ik}^\alpha}{q_{ik}^\alpha}\prod_{\substack{j=1 \\ j \neq i}}^{n}\frac{q_{kj}}{1 - q_{kj}}} \; ; \; \begin{array}{l} q_{rs} \in \left[q_{rs}^\alpha, q_{rs}^\delta\right], q_{sr} = 1 - q_{rs}, \\ r, s = 1, \ldots, n \end{array}\right\}.$$

Finally, thanks to the additive reciprocity of $\widetilde{Q}$, we can also replace all the elements q_{sr}, $r, s = 1, \ldots, n, r < s$, i.e. the elements below the main diagonal, by the reciprocals $1 - q_{rs}$ of the corresponding elements q_{rs} above the main diagonal. By that we obtain formula (4.219).

Analogously, it can be demonstrated that (4.216) is equivalent to (4.220), (4.217) is equivalent to (4.221), and (4.218) is equivalent (4.222).   □

Example 67 Let us consider the trapezoidal FAPCM-M

$$
\widetilde{Q} = \begin{pmatrix}
\dfrac{1}{2} & \left(\dfrac{1}{3},\dfrac{1}{2},\dfrac{1}{2},\dfrac{2}{3}\right) & \left(\dfrac{4}{8},\dfrac{5}{8},\dfrac{6}{8},\dfrac{7}{8}\right) & \left(\dfrac{6}{9},\dfrac{7}{9},\dfrac{7.5}{9},\dfrac{8}{9}\right) \\[2ex]
\left(\dfrac{1}{3},\dfrac{1}{2},\dfrac{1}{2},\dfrac{2}{3}\right) & \dfrac{1}{2} & \left(\dfrac{1}{2},\dfrac{1}{2},\dfrac{3}{5},\dfrac{3}{5}\right) & \left(\dfrac{4.5}{7},\dfrac{5}{7},\dfrac{5.5}{7},\dfrac{6}{7}\right) \\[2ex]
\left(\dfrac{1}{8},\dfrac{2}{8},\dfrac{3}{8},\dfrac{4}{8}\right) & \left(\dfrac{2}{5},\dfrac{2}{5},\dfrac{1}{2},\dfrac{1}{2}\right) & \dfrac{1}{2} & \left(\dfrac{4}{6},\dfrac{4.5}{6},\dfrac{5}{6},\dfrac{5}{6}\right) \\[2ex]
\left(\dfrac{1}{9},\dfrac{1.5}{9},\dfrac{2}{9},\dfrac{3}{9}\right) & \left(\dfrac{1}{7},\dfrac{1.5}{7},\dfrac{2}{7},\dfrac{2.5}{7}\right) & \left(\dfrac{1}{6},\dfrac{1}{6},\dfrac{1.5}{6},\dfrac{2}{6}\right) & 1
\end{pmatrix}.
$$
(4.223)

The fuzzy priorities of objects obtained from this FAPCM-M by the formulas (4.219)–(4.222) are given as

$$
\begin{aligned}
\widetilde{u}_1 &= (0.2096, 0.3358, 0.4251, 0.6036), \\
\widetilde{u}_2 &= (0.1774, 0.2666, 0.3378, 0.4493), \\
\widetilde{u}_3 &= (0.1161, 0.1894, 0.2914, 0.3455), \\
\widetilde{u}_4 &= (0.0438, 0.0666, 0.1001, 0.1472).
\end{aligned}
$$
(4.224)

Let us now examine in detail how the upper boundary value $u_2^\delta = 0.4493$ of the fuzzy priority $\widetilde{u}_2$ was obtained. By applying the formula (4.222) for the fixed $i = 2$, the optimum 0.4493 was obtained from an additively reciprocal matrix, in particular from the APCM-M

$$
Q^* = \begin{pmatrix}
\dfrac{1}{2} & \dfrac{1}{3} & 0.5359 & \dfrac{6}{9} \\[2ex]
\dfrac{2}{3} & \dfrac{1}{2} & \dfrac{3}{5} & \dfrac{6}{7} \\[2ex]
0.4641 & \dfrac{2}{5} & \dfrac{1}{2} & \dfrac{4}{6} \\[2ex]
\dfrac{3}{9} & \dfrac{1}{7} & \dfrac{2}{6} & \dfrac{1}{2}
\end{pmatrix}.
$$
(4.225)

The elements of this APCM-M clearly belong to the closures of the supports of the respective trapezoidal fuzzy numbers in the trapezoidal FAPCM-M (4.223). In the same way, it could be shown that all representing values of all four fuzzy priorities were obtained from APCMs-M by the formulas (4.215)–(4.218). △

4.4 Transformations Between FMPCMs and FAPCMs

In Sect. 2.4, transformations between MPCMs, APCMs-A, and APCMs-M, and between the related consistency conditions and the priority vectors obtainable from these PCMs were reviewed. In this section it will be proved that also FMPCMs,

FAPCMs-A, and FAPCMs-M, and the related methods proposed in Sects. 4.2 and 4.3 are equivalent.

4.4.1   Transformations Between FMPCMs and FAPCMs-A

In this section, transformation between FMPCMs and FAPCMs-A and between the related methods proposed in Sects. 4.2 and 4.3.2 are examined.

Theorem 86 *A trapezoidal FMPCM* $\tilde{M} = \{\tilde{m}_{ij}\}_{i,j=1}^{n}$, $\tilde{m}_{ij} = (m_{ij}^{\alpha}, m_{ij}^{\beta}, m_{ij}^{\gamma}, m_{ij}^{\delta})$, *can be transformed into a trapezoidal FAPCM-A* $\tilde{R} = \{\tilde{r}_{ij}\}_{i,j=1}^{n}$, $\tilde{r}_{ij} = (r_{ij}^{\alpha}, r_{ij}^{\beta}, r_{ij}^{\gamma}, r_{ij}^{\delta})$, *by transformation formulas*

$$
\begin{aligned}
r_{ij}^{\alpha} &= \frac{1}{2}(1 + \log_9 m_{ij}^{\alpha}), & r_{ij}^{\beta} &= \frac{1}{2}(1 + \log_9 m_{ij}^{\beta}), \\
r_{ij}^{\gamma} &= \frac{1}{2}(1 + \log_9 m_{ij}^{\gamma}), & r_{ij}^{\delta} &= \frac{1}{2}(1 + \log_9 m_{ij}^{\delta}).
\end{aligned}
\tag{4.226}
$$

Proof From the transformation formula (2.63) for transforming a MPCM into an APCM-A it is obvious that $\tilde{r}_{ij} \in [0, 1]$ and $\tilde{r}_{ii} = 0.5$. It remains to show that $\tilde{R}$ is additively reciprocal, i.e. $r_{ij}^{\alpha} = 1 - r_{ji}^{\delta}, r_{ij}^{\beta} = 1 - r_{ji}^{\gamma}, i, j = 1, \ldots, n$. Clearly

$$
r_{ij}^{\alpha} = \frac{1}{2}(1 + \log_9 m_{ij}^{\alpha}) = \frac{1}{2}(1 + \log_9 \frac{1}{m_{ji}^{\delta}}) = 1 - \frac{1}{2}(1 + \log_9 m_{ji}^{\delta}) = 1 - r_{ji}^{\delta}.
$$

Analogously, the validity of $r_{ij}^{\beta} = 1 - r_{ji}^{\gamma}$ is proved.   □

Corollary 7 *A trapezoidal FAPCM-A* $\tilde{R} = \{\tilde{r}_{ij}\}_{i,j=1}^{n}$, $\tilde{r}_{ij} = (r_{ij}^{\alpha}, r_{ij}^{\beta}, r_{ij}^{\gamma}, r_{ij}^{\delta})$, *can be transformed into a trapezoidal FMPCM* $\tilde{M} = \{\tilde{m}_{ij}\}_{i,j=1}^{n}$, $\tilde{m}_{ij} = (m_{ij}^{\alpha}, m_{ij}^{\beta}, m_{ij}^{\gamma}, m_{ij}^{\delta})$, *by transformation formulas*

$$
\begin{aligned}
m_{ij}^{\alpha} &= 9^{2r_{ij}^{\alpha}-1}, & m_{ij}^{\beta} &= 9^{2r_{ij}^{\beta}-1}, \\
m_{ij}^{\gamma} &= 9^{2r_{ij}^{\gamma}-1}, & m_{ij}^{\delta} &= 9^{2r_{ij}^{\delta}-1}.
\end{aligned}
\tag{4.227}
$$

Remark 39 The validity of Corollary 7 follows immediately from Theorem 86 by utilizing properties of an inverse function. Note that this form of representing the results is used in the whole section. This means that the transformation of a particular property is formulated in a theorem and proved only in one direction. Afterwards, each such theorem is followed by a corollary showing the transformation of the property in the opposite direction without providing the proof.

In the following, it is proved that the transformation formulas (4.226) and (4.227) transform the multiplicative weak consistency into the additive weak consistency and vice versa.

Theorem 87 *Let* $\tilde{M} = \{\tilde{m}_{ij}\}_{i,j=1}^{n}$, $\tilde{m}_{ij} = (m_{ij}^{\alpha}, m_{ij}^{\beta}, m_{ij}^{\gamma}, m_{ij}^{\delta})$, *be a trapezoidal FMPCM multiplicatively weakly consistent according to Definition 50. Then the trapezoidal FAPCM-A* $\tilde{R} = \{\tilde{r}_{ij}\}_{i,j=1}^{n}$, $\tilde{r}_{ij} = (r_{ij}^{\alpha}, r_{ij}^{\beta}, r_{ij}^{\gamma}, r_{ij}^{\delta})$, *obtained from* $\tilde{M}$ *by the transformations (4.226) is additively weakly consistent according to Definition 59.*

Proof It is sufficient to show that when the inequalities (4.16) are valid for a FMPCM M, then the inequalities (4.125) are valid for the FAPCM-A $\tilde{R}$ obtained from $\tilde{M}$ by the transformations (4.226):

$$\max_{k=1,\dots,n}\left\{r_{ik}^{\alpha}+r_{kj}^{\alpha}-0.5\right\} = \max_{k=1,\dots,n}\left\{\frac{1}{2}\left(1+\log_9 m_{ik}^{\alpha}\right)+\frac{1}{2}\left(1+\log_9 m_{kj}^{\alpha}\right)-0.5\right\} =$$

$$\max_{k=1,\dots,n}\left\{\frac{1}{2}\log_9 m_{ik}^{\alpha}+\frac{1}{2}\log_9 m_{kj}^{\alpha}+\frac{1}{2}\right\} = \max_{k=1,\dots,n}\left\{\frac{1}{2}\log_9\left(m_{ik}^{\alpha}m_{kj}^{\alpha}\right)+\frac{1}{2}\right\} \overset{(4.16)}{\leq}$$

$$\min_{k=1,\dots,n}\left\{\frac{1}{2}\log_9\left(m_{ik}^{\delta}m_{kj}^{\delta}\right)+\frac{1}{2}\right\} = \min_{k=1,\dots,n}\left\{\frac{1}{2}\log_9 m_{ik}^{\delta}+\frac{1}{2}\log_9 m_{kj}^{\delta}+\frac{1}{2}\right\} =$$

$$\min_{k=1,\dots,n}\left\{\frac{1}{2}\left(1+\log_9 m_{ik}^{\delta}\right)+\frac{1}{2}\left(1+\log_9 m_{kj}^{\delta}\right)-0.5\right\} = \min_{k=1,\dots,n}\left\{r_{ik}^{\delta}+r_{kj}^{\delta}-0.5\right\}.$$

□

Corollary 8 *Let* $\tilde{R} = \{\tilde{r}_{ij}\}_{i,j=1}^{n}$, $\tilde{r}_{ij} = (r_{ij}^{\alpha}, r_{ij}^{\beta}, r_{ij}^{\gamma}, r_{ij}^{\delta})$, *be a trapezoidal FAPCM-A additively weakly consistent according to Definition 59. Then the trapezoidal FMPCM* $\tilde{M} = \{\tilde{m}_{ij}\}_{i,j=1}^{n}$, $\tilde{m}_{ij} = (m_{ij}^{\alpha}, m_{ij}^{\beta}, m_{ij}^{\gamma}, m_{ij}^{\delta})$, *obtained from* $\tilde{R}$ *by the transformations (4.227) is multiplicatively weakly consistent according to Definition 50.*

Similarly, also multiplicative consistency is transformed into additive consistency and vice versa by the transformation formulas (4.226) and (4.227), respectively.

Theorem 88 *Let* $\tilde{M} = \{\tilde{m}_{ij}\}_{i,j=1}^{n}$, $\tilde{m}_{ij} = (m_{ij}^{\alpha}, m_{ij}^{\beta}, m_{ij}^{\gamma}, m_{ij}^{\delta})$, *be a trapezoidal FMPCM multiplicatively consistent according to Definition 51. Then the trapezoidal FAPCM-A* $\tilde{R} = \{\tilde{r}_{ij}\}_{i,j=1}^{n}$, $\tilde{r}_{ij} = (r_{ij}^{\alpha}, r_{ij}^{\beta}, r_{ij}^{\gamma}, r_{ij}^{\delta})$, *obtained from* $\tilde{M}$ *by the transformations (4.226) is additively consistent according to Definition 60.*

Proof It is sufficient to show that when the inequalities (4.25) and (4.26) are valid for a FMPCM $\tilde{M}$, then the inequalities (4.133) and (4.134) are valid for the FAPCM-A $\tilde{R}$ obtained from $\tilde{M}$ by the transformations (4.226).

$$r_{ik}^{\alpha} + r_{kj}^{\alpha} - 0.5 = \frac{1}{2}\left(1 + \log_9 m_{ik}^{\alpha}\right) + \frac{1}{2}\left(1 + \log_9 m_{kj}^{\alpha}\right) - 0.5 =$$

$$\frac{1}{2}\left(1 + \log_9\left(m_{ik}^{\alpha} m_{kj}^{\alpha}\right)\right) \overset{(4.25)}{\leq} \frac{1}{2}\left(1 + \log_9 m_{ij}^{\alpha}\right) = r_{ij}^{\alpha}.$$

The remaining inequalities are proved in the same way.   □

Corollary 9 Let $\widetilde{R} = \{\widetilde{r}_{ij}\}_{i,j=1}^{n}$, $\widetilde{r}_{ij} = (r_{ij}^{\alpha}, r_{ij}^{\beta}, r_{ij}^{\gamma}, r_{ij}^{\delta})$, be a trapezoidal FAPCM-A additively consistent according to Definition 60. Then the trapezoidal FMPCM $\widetilde{M} = \{\widetilde{m}_{ij}\}_{i,j=1}^{n}$, $\widetilde{m}_{ij} = (m_{ij}^{\alpha}, m_{ij}^{\beta}, m_{ij}^{\gamma}, m_{ij}^{\delta})$, obtained from $\widetilde{R}$ by the transformations (4.227) is multiplicatively consistent according to Definition 51.

In the following, a relation between fuzzy priorities obtained from FMPCMs and from FAPCMs-A is shown.

Theorem 89 Let $\widetilde{M} = \{\widetilde{m}_{ij}\}_{i,j=1}^{n}$, $\widetilde{m}_{ij} = (m_{ij}^{\alpha}, m_{ij}^{\beta}, m_{ij}^{\gamma}, m_{ij}^{\delta})$, be a trapezoidal FMPCM and let $\underline{\widetilde{w}} = (\widetilde{w}_1, \ldots, \widetilde{w}_n)^T$, $\widetilde{w}_i = (w_i^{\alpha}, w_i^{\beta}, w_i^{\gamma}, w_i^{\delta})$, $i = 1, \ldots, n$, be the fuzzy priority vector obtained from $\widetilde{M}$ by the formulas (4.85). The fuzzy priority vector $\underline{\widetilde{w}} = (\widetilde{w}_1, \ldots, \widetilde{w}_n)^T$ can be transformed into a fuzzy priority vector $\underline{\widetilde{v}} = (\widetilde{v}_1, \ldots, \widetilde{v}_n)^T$, $\widetilde{v}_i = (v_i^{\alpha}, v_i^{\beta}, v_i^{\gamma}, v_i^{\delta})$, $i = 1, \ldots, n$, obtainable by formulas (4.147) from the corresponding FAPCM-A $\widetilde{R} = \{\widetilde{r}_{ij}\}_{i,j=1}^{n}$, $\widetilde{r}_{ij} = (r_{ij}^{\alpha}, r_{ij}^{\beta}, r_{ij}^{\gamma}, r_{ij}^{\delta})$, by using the transformation formulas

$$\begin{array}{ll} v_i^{\alpha} = 1 + \log_9 w_i^{\alpha}, & v_i^{\beta} = 1 + \log_9 w_i^{\beta}, \\ v_i^{\gamma} = 1 + \log_9 w_i^{\gamma}, & v_i^{\delta} = 1 + \log_9 w_i^{\delta}. \end{array} \quad (4.228)$$

Proof The validity of the transformation formulas follows immediately from the transformation formula (2.66) for the crisp case.   □

Corollary 10 Let $\widetilde{R} = \{\widetilde{r}_{ij}\}_{i,j=1}^{n}$, $\widetilde{r}_{ij} = (r_{ij}^{\alpha}, r_{ij}^{\beta}, r_{ij}^{\gamma}, r_{ij}^{\delta})$, be a trapezoidal FAPCM-A and let $\underline{\widetilde{v}} = (\widetilde{v}_1, \ldots, \widetilde{v}_n)^T$, $\widetilde{v}_i = (v_i^{\alpha}, v_i^{\beta}, v_i^{\gamma}, v_i^{\delta})$, $i = 1, \ldots, n$, be the fuzzy priority vector obtained from $\widetilde{R}$ by the formulas (4.147). The fuzzy priority vector $\underline{\widetilde{v}} = (\widetilde{v}_1, \ldots, \widetilde{v}_n)^T$ can be transformed into a fuzzy priority vector $\underline{\widetilde{w}} = (\widetilde{w}_1, \ldots, \widetilde{w}_n)^T$, $\widetilde{w}_i = (w_i^{\alpha}, w_i^{\beta}, w_i^{\gamma}, w_i^{\delta})$, $i = 1, \ldots, n$, obtainable by formulas (4.85) from the corresponding FMPCM $\widetilde{M} = \{\widetilde{m}_{ij}\}_{i,j=1}^{n}$, $\widetilde{m}_{ij} = (m_{ij}^{\alpha}, m_{ij}^{\beta}, m_{ij}^{\gamma}, m_{ij}^{\delta})$, by using the transformation formulas

$$\begin{array}{ll} w_i^{\alpha} = 9^{v_i^{\alpha} - 1}, & w_i^{\beta} = 9^{v_i^{\beta} - 1}, \\ w_i^{\gamma} = 9^{v_i^{\gamma} - 1}, & w_i^{\delta} = 9^{v_i^{\delta} - 1}. \end{array} \quad (4.229)$$

Similarly to the crisp case (see the discussion on p. 49), it is not possible to derive transformation formulas for transforming normalized fuzzy priorities (4.91)–(4.94) obtained from a FMPCM into the normalized fuzzy priorities (4.158) obtained from the corresponding FAPCM-A and vice versa.

4.4.2 Transformations Between FMPCMs and FAPCMs-M

In this section, transformation between FMPCMs and FAPCMs-M and between the related methods proposed in Sects. 4.2 and 4.3.3 are examined.

Theorem 90 A trapezoidal FMPCM $\tilde{M} = \{\tilde{m}_{ij}\}_{i,j=1}^{n}$, $\tilde{m}_{ij} = (m_{ij}^{\alpha}, m_{ij}^{\beta}, m_{ij}^{\gamma}, m_{ij}^{\delta})$, can be transformed into a trapezoidal FAPCM-M $\tilde{Q} = \{\tilde{q}_{ij}\}_{i,j=1}^{n}$, $\tilde{q}_{ij} = (q_{ij}^{\alpha}, q_{ij}^{\beta}, q_{ij}^{\gamma}, q_{ij}^{\delta})$, by transformation formulas

$$q_{ij}^{\alpha} = \frac{m_{ij}^{\alpha}}{1 + m_{ij}^{\alpha}}, \qquad q_{ij}^{\beta} = \frac{m_{ij}^{\beta}}{1 + m_{ij}^{\beta}},$$

$$q_{ij}^{\gamma} = \frac{m_{ij}^{\gamma}}{1 + m_{ij}^{\gamma}}, \qquad q_{ij}^{\delta} = \frac{m_{ij}^{\delta}}{1 + m_{ij}^{\delta}}. \tag{4.230}$$

Proof From the transformation formula (2.75) for transforming a MPCM into an APCM-M it is obvious that $\tilde{q}_{ij} \in]0, 1[$ and $\tilde{q}_{ii} = 0.5$. It remains to show that $\tilde{Q}$ is additively reciprocal, i.e. $q_{ij}^{\alpha} = 1 - q_{ji}^{\delta}$, $q_{ij}^{\beta} = 1 - q_{ji}^{\gamma}$, $i, j = 1, \ldots, n$. Clearly

$$q_{ij}^{\alpha} = \frac{m_{ij}^{\alpha}}{1 + m_{ij}^{\alpha}} = \frac{\frac{1}{m_{ji}^{\delta}}}{1 + \frac{1}{m_{ji}^{\delta}}} = \frac{1}{1 + m_{ji}^{\delta}} = 1 - \frac{m_{ji}^{\delta}}{1 + m_{ji}^{\delta}} = 1 - q_{ji}^{\delta}.$$

Analogously, the validity of $q_{ij}^{\beta} = 1 - q_{ji}^{\gamma}$ is proved. □

Corollary 11 A trapezoidal FAPCM-M $\tilde{Q} = \{\tilde{q}_{ij}\}_{i,j=1}^{n}$, $\tilde{q}_{ij} = (q_{ij}^{\alpha}, q_{ij}^{\beta}, q_{ij}^{\gamma}, q_{ij}^{\delta})$, can be transformed into a trapezoidal FMPCM $\tilde{M} = \{\tilde{m}_{ij}\}_{i,j=1}^{n}$, $\tilde{m}_{ij} = (m_{ij}^{\alpha}, m_{ij}^{\beta}, m_{ij}^{\gamma}, m_{ij}^{\delta})$, by transformation formulas

$$m_{ij}^{\alpha} = \frac{q_{ij}^{\alpha}}{q_{ji}^{\delta}}, \qquad m_{ij}^{\beta} = \frac{q_{ij}^{\beta}}{q_{ji}^{\gamma}},$$

$$m_{ij}^{\gamma} = \frac{q_{ij}^{\gamma}}{q_{ji}^{\beta}}, \qquad m_{ij}^{\delta} = \frac{q_{ij}^{\delta}}{q_{ji}^{\alpha}}. \tag{4.231}$$

In the following, it is proved that the transformation formulas (4.230) and (4.231) transform the multiplicative weak consistency for FMPCMs into the multiplicative weak consistency for FAPCMs-M and vice versa.

Theorem 91 *Let* $\tilde{M} = \{\tilde{m}_{ij}\}_{i,j=1}^{n}$, $\tilde{m}_{ij} = (m_{ij}^{\alpha}, m_{ij}^{\beta}, m_{ij}^{\gamma}, m_{ij}^{\delta})$, *be a trapezoidal FMPCM multiplicatively weakly consistent according to Definition 50. Then the trapezoidal FAPCM-M* $\tilde{Q} = \{\tilde{q}_{ij}\}_{i,j=1}^{n}$, $\tilde{q}_{ij} = (q_{ij}^{\alpha}, q_{ij}^{\beta}, q_{ij}^{\gamma}, q_{ij}^{\delta})$, *obtained from* $\tilde{M}$ *by the transformations (4.230) is multiplicatively weakly consistent according to Definition 66.*

Proof It is sufficient to show that when the inequalities (4.16) are valid for a FMPCM $\tilde{M}$, then the inequalities (4.177) are valid for the FAPCM-M $\tilde{Q}$ obtained from $\tilde{M}$ by the transformations (4.230):

$$\max_{k=1,\ldots,n}\left\{\frac{q_{ik}^{\alpha}q_{kj}^{\alpha}}{q_{ik}^{\alpha}q_{kj}^{\alpha}+(1-q_{ik}^{\alpha})(1-q_{kj}^{\alpha})}\right\} = \max_{k=1,\ldots,n}\left\{\frac{\frac{m_{ik}^{\alpha}}{1+m_{ik}^{\alpha}}\frac{m_{kj}^{\alpha}}{1+m_{kj}^{\alpha}}}{\frac{m_{ik}^{\alpha}}{1+m_{ik}^{\alpha}}\frac{m_{kj}^{\alpha}}{1+m_{kj}^{\alpha}}+\frac{1}{1+m_{ik}^{\alpha}}\frac{1}{1+m_{kj}^{\alpha}}}\right\} =$$

$$\max_{k=1,\ldots,n}\left\{\frac{m_{ik}^{\alpha}m_{kj}^{\alpha}}{1+m_{ik}^{\alpha}m_{kj}^{\alpha}}\right\} = \frac{1}{\min\limits_{k=1,\ldots,n}\left\{\frac{1+m_{ik}^{\alpha}m_{kj}^{\alpha}}{m_{ik}^{\alpha}m_{kj}^{\alpha}}\right\}} = \frac{1}{1+\frac{1}{\max\limits_{k=1,\ldots,n}\left\{m_{ik}^{\alpha}m_{kj}^{\alpha}\right\}}} \overset{(4.16)}{\leq}$$

$$\frac{1}{1+\frac{1}{\min\limits_{k=1,\ldots,n}\left\{m_{ik}^{\delta}m_{kj}^{\delta}\right\}}} = \frac{1}{\max\limits_{k=1,\ldots,n}\left\{\frac{1+m_{ik}^{\delta}m_{kj}^{\delta}}{m_{ik}^{\delta}m_{kj}^{\delta}}\right\}} = \min_{k=1,\ldots,n}\left\{\frac{m_{ik}^{\delta}m_{kj}^{\delta}}{1+m_{ik}^{\delta}m_{kj}^{\delta}}\right\} =$$

$$\min_{k=1,\ldots,n}\left\{\frac{\frac{q_{ik}^{\delta}}{q_{ki}^{\alpha}}\frac{q_{kj}^{\delta}}{q_{jk}^{\alpha}}}{1+\frac{q_{ik}^{\delta}}{q_{ki}^{\alpha}}\frac{q_{kj}^{\delta}}{q_{jk}^{\alpha}}}\right\} = \min_{k=1,\ldots,n}\left\{\frac{q_{ik}^{\delta}q_{kj}^{\delta}}{q_{ik}^{\delta}q_{kj}^{\delta}+(1-q_{ik}^{\delta})(1-q_{kj}^{\delta})}\right\}.$$

$\square$

Corollary 12 *Let* $\tilde{Q} = \{\tilde{q}_{ij}\}_{i,j=1}^{n}$, $\tilde{q}_{ij} = (q_{ij}^{\alpha}, q_{ij}^{\beta}, q_{ij}^{\gamma}, q_{ij}^{\delta})$, *be a trapezoidal FAPCM-M multiplicatively weakly consistent according to Definition 66. Then the trapezoidal FMPCM* $\tilde{M} = \{\tilde{m}_{ij}\}_{i,j=1}^{n}$, $\tilde{m}_{ij} = (m_{ij}^{\alpha}, m_{ij}^{\beta}, m_{ij}^{\gamma}, m_{ij}^{\delta})$, *obtained from* $\tilde{Q}$ *by the transformations (4.231) is multiplicatively weakly consistent according to Definition 50.*

Similarly, also the multiplicative consistency of FMPCMs is transformed into the multiplicative consistency of FAPCMs-M and vice versa by the transformation formulas (4.230) and (4.231), respectively.

Theorem 92 *Let* $\tilde{M} = \{\tilde{m}_{ij}\}_{i,j=1}^{n}$, $\tilde{m}_{ij} = (m_{ij}^{\alpha}, m_{ij}^{\beta}, m_{ij}^{\gamma}, m_{ij}^{\delta})$, *be a trapezoidal FMPCM multiplicatively consistent according to Definition 51. Then the trapezoidal*

FAPCM-M $\widetilde{Q} = \{\widetilde{q}_{ij}\}_{i,j=1}^{n}$, $\widetilde{q}_{ij} = (q_{ij}^{\alpha}, q_{ij}^{\beta}, q_{ij}^{\gamma}, q_{ij}^{\delta})$, *obtained from $\widetilde{M}$ by the transformations (4.230) is multiplicatively consistent according to Definition 67.*

Proof It is sufficient to show that when the inequalities (4.25) and (4.26) are valid for a FMPCM M, then the inequalities (4.190) and (4.191) are valid for the FAPCM-M $\widetilde{Q}$ obtained from $\widetilde{M}$ by the transformations (4.230).

$$\frac{q_{ik}^{\alpha} q_{kj}^{\alpha}}{q_{ik}^{\alpha} q_{kj}^{\alpha} + (1 - q_{ik}^{\alpha})(1 - q_{kj}^{\alpha})} = \frac{m_{ik}^{\alpha} m_{kj}^{\alpha}}{1 + m_{ik}^{\alpha} m_{kj}^{\alpha}} \overset{(4.25)}{\leq} \frac{m_{ij}^{\alpha}}{1 + m_{ij}^{\alpha}} = \frac{\frac{q_{ij}^{\alpha}}{q_{ji}^{\delta}}}{1 + \frac{q_{ij}^{\alpha}}{q_{ji}^{\delta}}} = \frac{q_{ij}^{\alpha}}{q_{ij}^{\alpha} + q_{ji}^{\delta}} = q_{ij}^{\alpha}.$$

The remaining inequalities are proved in the same way. $\qquad\square$

Corollary 13 *Let* $\widetilde{Q} = \{\widetilde{q}_{ij}\}_{i,j=1}^{n}$, $\widetilde{q}_{ij} = (q_{ij}^{\alpha}, q_{ij}^{\beta}, q_{ij}^{\gamma}, q_{ij}^{\delta})$, *be a trapezoidal FAPCM-M multiplicatively consistent according to Definition 67. Then the trapezoidal FMPCM* $\widetilde{M} = \{\widetilde{m}_{ij}\}_{i,j=1}^{n}$, $\widetilde{m}_{ij} = (m_{ij}^{\alpha}, m_{ij}^{\beta}, m_{ij}^{\gamma}, m_{ij}^{\delta})$, *obtained from $\widetilde{Q}$ by the transformations (4.231) is multiplicatively consistent according to Definition 51.*

In the following, a relation between fuzzy priorities obtained from FMPCMs and from FAPCMs-M is shown.

Theorem 93 *Let* $\widetilde{M} = \{\widetilde{m}_{ij}\}_{i,j=1}^{n}$, $\widetilde{m}_{ij} = (m_{ij}^{\alpha}, m_{ij}^{\beta}, m_{ij}^{\gamma}, m_{ij}^{\delta})$, *be a trapezoidal FMPCM and let* $\widetilde{\underline{w}} = (\widetilde{w}_1, \ldots, \widetilde{w}_n)^T$, $\widetilde{w}_i = (w_i^{\alpha}, w_i^{\beta}, w_i^{\gamma}, w_i^{\delta})$, $i = 1, \ldots, n$, *be the fuzzy priority vector obtained from $\widetilde{M}$ by the formulas (4.85). The fuzzy priority vector* $\widetilde{\underline{w}} = (\widetilde{w}_1, \ldots, \widetilde{w}_n)^T$ *is identical to the fuzzy priority vector* $\widetilde{\underline{u}} = (\widetilde{u}_1, \ldots, \widetilde{u}_n)^T$, $\widetilde{u}_i = (u_i^{\alpha}, u_i^{\beta}, u_i^{\gamma}, u_i^{\delta})$, $i = 1, \ldots, n$, *obtainable by the formulas (4.211)–(4.214) from the corresponding trapezoidal FAPCM-M* $\widetilde{Q} = \{\widetilde{q}_{ij}\}_{i,j=1}^{n}$, $\widetilde{q}_{ij} = (q_{ij}^{\alpha}, q_{ij}^{\beta}, q_{ij}^{\gamma}, q_{ij}^{\delta})$, *i.e.*

$$\widetilde{w} = \widetilde{u}. \tag{4.232}$$

Proof The validity of the transformation formulas follows immediately from the transformation formula (2.77) for the crisp case. $\qquad\square$

Moreover, similarly to the crisp case, also the normalized fuzzy priorities obtained from a FMPCM and from the corresponding FAPCM-M are identical.

Theorem 94 *Let* $\widetilde{M} = \{\widetilde{m}_{ij}\}_{i,j=1}^{n}$, $\widetilde{m}_{ij} = (m_{ij}^{\alpha}, m_{ij}^{\beta}, m_{ij}^{\gamma}, m_{ij}^{\delta})$, *be a trapezoidal FMPCM and let* $\widetilde{\underline{w}}_C = (\widetilde{w}_{C1}, \ldots, \widetilde{w}_{Cn})^T$, $\widetilde{w}_{Ci} = (w_{Ci}^{\alpha}, w_{Ci}^{\beta}, w_{Ci}^{\gamma}, w_{Ci}^{\delta})$, $i = 1, \ldots, n$, *be the normalized fuzzy priority vector obtained from R by the formulas (4.91)–(4.94). The normalized fuzzy priority vector* $\widetilde{\underline{w}}_C = (\widetilde{w}_{C1}, \ldots, \widetilde{w}_{Cn})^T$ *is identical to the normalized fuzzy priority vector* $\widetilde{\underline{u}}_C = (\widetilde{u}_{C1}, \ldots, \widetilde{u}_{Cn})^T$, $\widetilde{u}_{Ci} = (u_{Ci}^{\alpha}, u_{Ci}^{\beta}, u_{Ci}^{\gamma}, u_{Ci}^{\delta})$, $i = 1, \ldots, n$, *obtainable by the formulas (4.215)–(4.218) from the corresponding trapezoidal FAPCM-M* $\widetilde{Q} = \{\widetilde{q}_{ij}\}_{i,j=1}^{n}$, $\widetilde{q}_{ij} = (q_{ij}^{\alpha}, q_{ij}^{\beta}, q_{ij}^{\gamma}, q_{ij}^{\delta})$, *i.e.*

$$\widetilde{w}_C = \widetilde{u}_C. \tag{4.233}$$

Proof Let us demonstrate the equality of the lower boundary values of the fuzzy priorities $\widetilde{w}_{Ci}$ and $\widetilde{u}_{Ci}$ obtained by (4.91) and (4.215), respectively. The equality of the remaining representing values would be demonstrated in the same way.

By applying (4.231) and $m_{rs} = \frac{q_{rs}}{q_{sr}}$, $r, s = 1, \ldots, n$, to the optimization problem (4.91), we obtain

$$
w_{Ci}^{\alpha} = \min \left\{ \frac{\sqrt[n]{\prod_{j=1}^{n} m_{ij}}}{\sum_{k=1}^{n} \sqrt[n]{\prod_{j=1}^{n} m_{kj}}} \; ; \; \begin{array}{l} m_{rs} \in \left[m_{rs}^{\alpha}, m_{rs}^{\delta}\right], \\ m_{sr} = \frac{1}{m_{rs}}, \\ r, s = 1, \ldots, n \end{array} \right\} =
$$

$$
\min \left\{ \frac{\sqrt[n]{\prod_{j=1}^{n} \frac{q_{ij}}{q_{ji}}}}{\sum_{k=1}^{n} \sqrt[n]{\prod_{j=1}^{n} \frac{q_{kj}}{q_{jk}}}} \; ; \; \begin{array}{l} \frac{q_{rs}}{1-q_{rs}} \in \left[\frac{q_{rs}^{\alpha}}{q_{sr}^{\delta}}, \frac{q_{rs}^{\delta}}{q_{sr}^{\alpha}}\right], \\ q_{sr} = 1 - q_{rs}, \\ r, s = 1, \ldots, n \end{array} \right\} =
$$

$$
\min \left\{ \frac{\sqrt[n]{\prod_{j=1}^{n} \frac{q_{ij}}{q_{ji}}}}{\sum_{k=1}^{n} \sqrt[n]{\prod_{j=1}^{n} \frac{q_{kj}}{q_{jk}}}} \; ; \; \begin{array}{l} q_{rs} \in \left[q_{rs}^{\alpha}, q_{rs}^{\delta}\right], \\ q_{sr} = 1 - q_{rs}, \\ r, s = 1, \ldots, n \end{array} \right\} = u_{Ci}^{\alpha}.
$$

$\square$

4.4.3 Transformations Between FAPCMs-A and FAPCMs-M

In this section, transformations between FAPCMs-A and FAPCMs-M and between the related methods proposed in Sects. 4.3.2 and 4.3.3 are examined. Analogously as for the transformations between APCMs-A and APCMs-M, the transformation formulas can be derived directly by composing the corresponding formulas from the previous two sections, as specified in the following theorems.

Theorem 95 *A trapezoidal FAPCM-A $\widetilde{R} = \{\widetilde{r}_{ij}\}_{i,j=1}^{n}$, $\widetilde{r}_{ij} = (r_{ij}^{\alpha}, r_{ij}^{\beta}, r_{ij}^{\gamma}, r_{ij}^{\delta})$, can be transformed into a trapezoidal FAPCM-M $\widetilde{Q} = \{\widetilde{q}_{ij}\}_{i,j=1}^{n}$, $\widetilde{q}_{ij} = (q_{ij}^{\alpha}, q_{ij}^{\beta}, q_{ij}^{\gamma}, q_{ij}^{\delta})$, by transformation formulas*

$$
q_{ij}^{\alpha} = \frac{9^{2r_{ij}^{\alpha}-1}}{1 + 9^{2r_{ij}^{\alpha}-1}}, \qquad q_{ij}^{\beta} = \frac{9^{2r_{ij}^{\beta}-1}}{1 + 9^{2r_{ij}^{\beta}-1}},
$$
$$
q_{ij}^{\gamma} = \frac{9^{2r_{ij}^{\gamma}-1}}{1 + 9^{2r_{ij}^{\gamma}-1}}, \qquad q_{ij}^{\delta} = \frac{9^{2r_{ij}^{\delta}-1}}{1 + 9^{2r_{ij}^{\delta}-1}}.
$$

(4.234)

Proof Because the transformation formulas (4.227) transform a FAPCM-A into a FMPCM, and the transformation formulas (4.230) transform a FMPCM into a FAPCM-M, then the composition of these formulas transforms a FAPCM-A into a FAPCM-M. By composing (4.227) and (4.230) we immediately obtain (4.234). $\square$

Corollary 14 *A trapezoidal FAPCM-M* $\widetilde{Q} = \{\widetilde{q}_{ij}\}_{i,j=1}^{n}$, $\widetilde{q}_{ij} = (q_{ij}^{\alpha}, q_{ij}^{\beta}, q_{ij}^{\gamma}, q_{ij}^{\delta})$, *can be transformed into a trapezoidal FAPCM-A* $\widetilde{R} = \{\widetilde{r}_{ij}\}_{i,j=1}^{n}$, $\widetilde{r}_{ij} = (r_{ij}^{\alpha}, r_{ij}^{\beta}, r_{ij}^{\gamma}, r_{ij}^{\delta})$, *by transformation formulas*

$$
r_{ij}^{\alpha} = \frac{1}{2}\left(1 + \log_9 \frac{q_{ij}^{\alpha}}{q_{ji}^{\delta}}\right), \qquad r_{ij}^{\beta} = \frac{1}{2}\left(1 + \log_9 \frac{q_{ij}^{\beta}}{q_{ji}^{\gamma}}\right),
$$
$$
r_{ij}^{\gamma} = \frac{1}{2}\left(1 + \log_9 \frac{q_{ij}^{\gamma}}{q_{ji}^{\beta}}\right), \qquad r_{ij}^{\delta} = \frac{1}{2}\left(1 + \log_9 \frac{q_{ij}^{\delta}}{q_{ji}^{\alpha}}\right).
$$

(4.235)

In the following, it is proved that additive weak consistency is transformed into multiplicative weak consistency and vice versa by the transformation formulas (4.234) and (4.235), respectively.

Theorem 96 *Let* $\widetilde{R} = \{\widetilde{r}_{ij}\}_{i,j=1}^{n}$, $\widetilde{r}_{ij} = (r_{ij}^{\alpha}, r_{ij}^{\beta}, r_{ij}^{\gamma}, r_{ij}^{\delta})$, *be a trapezoidal FAPCM-A additively weakly consistent according to Definition 59. Then the trapezoidal FAPCM-M* $\widetilde{Q} = \{\widetilde{q}_{ij}\}_{i,j=1}^{n}$, $\widetilde{q}_{ij} = (q_{ij}^{\alpha}, q_{ij}^{\beta}, q_{ij}^{\gamma}, q_{ij}^{\delta})$, *obtained from* $\widetilde{R}$ *by the transformations (4.234) is multiplicatively weakly consistent according to Definition 66.*

Proof Because the transformation formulas (4.227) transform the additive weak consistency (4.123) of a FAPCM-A into the multiplicative weak consistency (4.14) of the corresponding FMPCM, and the transformation formulas (4.230) transform multiplicative weak consistency (4.14) of a FMPCM into multiplicative weak consistency (4.175) of the corresponding FAPCM-M, then the composition of these formulas transforms additive weak consistency of a FAPCM-A into multiplicative weak consistency of the corresponding FAPCM-M. By composing (4.227) and (4.230) we immediately obtain (4.234). ∎

Corollary 15 *Let* $\widetilde{Q} = \{\widetilde{q}_{ij}\}_{i,j=1}^{n}$, $\widetilde{q}_{ij} = (q_{ij}^{\alpha}, q_{ij}^{\beta}, q_{ij}^{\gamma}, q_{ij}^{\delta})$, *be a trapezoidal FAPCM-M multiplicatively weakly consistent according to Definition 66. Then the trapezoidal FAPCM-A* $\widetilde{R} = \{\widetilde{r}_{ij}\}_{i,j=1}^{n}$, $\widetilde{r}_{ij} = (r_{ij}^{\alpha}, r_{ij}^{\beta}, r_{ij}^{\gamma}, r_{ij}^{\delta})$, *obtained from* $\widetilde{Q}$ *by the transformations (4.235) is additively weakly consistent according to Definition 59.*

Similarly, also the additive consistency is transformed into the multiplicative consistency and vice versa by the transformation formulas (4.234) and (4.235), respectively.

Theorem 97 *Let* $\widetilde{R} = \{\widetilde{r}_{ij}\}_{i,j=1}^{n}$, $\widetilde{r}_{ij} = (r_{ij}^{\alpha}, r_{ij}^{\beta}, r_{ij}^{\gamma}, r_{ij}^{\delta})$, *be a trapezoidal FAPCM-A additively consistent according to Definition 60. Then the trapezoidal FAPCM-M* $\widetilde{Q} = \{\widetilde{q}_{ij}\}_{i,j=1}^{n}$, $\widetilde{q}_{ij} = (q_{ij}^{\alpha}, q_{ij}^{\beta}, q_{ij}^{\gamma}, q_{ij}^{\delta})$, *obtained from* $\widetilde{R}$ *by the transformations (4.234) is multiplicatively consistent according to Definition 67.*

Proof Because the transformation formulas (4.227) transform the additive consistency (4.127)–(4.128) of a FAPCM-A into the multiplicative consistency (4.18)–(4.19) of the corresponding FMPCM, and the transformation formulas (4.230)

transform multiplicative consistency (4.18)–(4.19) of a FMPCM into multiplicative
consistency (4.179)–(4.180) of the corresponding FAPCM-M, then the composition
of these formulas transforms additive consistency of a FAPCM-A into multiplicative
consistency of the corresponding FAPCM-M. By composing (4.227) and (4.230) we
immediately obtain (4.234).   □

Corollary 16 *Let $\widetilde{Q} = \left\{\widetilde{q}_{ij}\right\}_{i,j=1}^{n}$, $\widetilde{q}_{ij} = (q_{ij}^{\alpha}, q_{ij}^{\beta}, q_{ij}^{\gamma}, q_{ij}^{\delta})$, be a trapezoidal FAPCM-
M multiplicatively consistent according to Definition 67. Then the trapezoidal
FAPCM-A $\widetilde{R} = \left\{\widetilde{r}_{ij}\right\}_{i,j=1}^{n}$, $\widetilde{r}_{ij} = (r_{ij}^{\alpha}, r_{ij}^{\beta}, r_{ij}^{\gamma}, r_{ij}^{\delta})$, obtained from $\widetilde{Q}$ by the transfor-
mations (4.235) is additively consistent according to Definition 60.*

In the following, a relation between fuzzy priorities obtained from FAPCMs-A
and from FAPCMs-M is shown.

Theorem 98 *Let $\widetilde{R} = \left\{\widetilde{r}_{ij}\right\}_{i,j=1}^{n}$, $\widetilde{r}_{ij} = (r_{ij}^{\alpha}, r_{ij}^{\beta}, r_{ij}^{\gamma}, r_{ij}^{\delta})$, be a trapezoidal FAPCM-A
and let $\underline{\widetilde{v}} = (\widetilde{v}_1, \ldots, \widetilde{v}_n)^T$, $\widetilde{v}_i = (v_i^{\alpha}, v_i^{\beta}, v_i^{\gamma}, v_i^{\delta})$, $i = 1, \ldots, n$, be the fuzzy priority vector obtained from $\widetilde{R}$ by the formulas (4.147). The fuzzy priority vector $\underline{\widetilde{v}} =
(\widetilde{v}_1, \ldots, \widetilde{v}_n)^T$ can be transformed into the fuzzy priority vector $\underline{\widetilde{u}} = (\widetilde{u}_1, \ldots, \widetilde{u}_n)^T$,
$\widetilde{u}_i = (u_i^{\alpha}, u_i^{\beta}, u_i^{\gamma}, u_i^{\delta})$, $i = 1, \ldots, n$, obtainable by formulas (4.211)–(4.214) from the
corresponding FAPCM-M by using the transformation formulas*

$$u_i^{\alpha} = 9^{v_i^{\alpha}-1}, \qquad u_i^{\beta} = 9^{v_i^{\beta}-1}, \qquad (4.236)$$
$$u_i^{\gamma} = 9^{v_i^{\gamma}-1}, \qquad u_i^{\delta} = 9^{v_i^{\delta}-1}.$$

Proof Because the transformation formulas (4.229) transform the fuzzy priority vector (4.147) of a FAPCM-A into the fuzzy priority vector (4.85) of the corresponding FMPCM, and the transformation formula (4.232) transforms the fuzzy priority
vector (4.85) of a FMPCM into the fuzzy priority vector (4.211)–(4.214) of the
corresponding FAPCM-M, then the composition of these transformation formulas
transforms the fuzzy priority vector (4.147) of a FAPCM-A into the fuzzy priority
vector (4.211)–(4.214) of the corresponding FAPCM-M. By composing (4.229) and
(4.232) we immediately obtain (4.236).   □

Corollary 17 *Let $\widetilde{Q} = \left\{\widetilde{q}_{ij}\right\}_{i,j=1}^{n}$, $\widetilde{q}_{ij} = (q_{ij}^{\alpha}, q_{ij}^{\beta}, q_{ij}^{\gamma}, q_{ij}^{\delta})$, be a trapezoidal
FAPCM-M and let $\underline{\widetilde{u}} = (\widetilde{u}_1, \ldots, \widetilde{u}_n)^T$, $\widetilde{u}_i = (u_i^{\alpha}, u_i^{\beta}, u_i^{\gamma}, u_i^{\delta})$, $i = 1, \ldots, n$, be the
fuzzy priority vector obtainable from $\widetilde{Q}$ by the formulas (4.211)–(4.214). The fuzzy
priority vector $\underline{\widetilde{u}} = (\widetilde{u}_1, \ldots, \widetilde{u}_n)^T$ can be transformed into the fuzzy priority vec-
tor $\underline{\widetilde{v}} = (\widetilde{v}_1, \ldots, \widetilde{v}_n)^T$, $\widetilde{v}_i = (v_i^{\alpha}, v_i^{\beta}, v_i^{\gamma}, v_i^{\delta})$, $i = 1, \ldots, n$, obtainable by formulas
(4.147) from the corresponding FAPCM-A $\widetilde{R} = \left\{\widetilde{r}_{ij}\right\}_{i,j=1}^{n}$, $\widetilde{r}_{ij} = (r_{ij}^{\alpha}, r_{ij}^{\beta}, r_{ij}^{\gamma}, r_{ij}^{\delta})$, by
using the transformation formulas*

$$v_i^{\alpha} = 1 + \log_9 u_i^{\alpha}, \qquad v_i^{\beta} = 1 + \log_9 u_i^{\beta}, \qquad (4.237)$$
$$v_i^{\gamma} = 1 + \log_9 u_i^{\gamma}, \qquad v_i^{\delta} = 1 + \log_9 u_i^{\delta}.$$

Similarly to the crisp case, it is not possible to derive transformation formulas for transforming normalized fuzzy priorities (4.158) obtained from a FAPCM-A into normalized fuzzy priorities (4.215)–(4.218) obtained from the corresponding FAPCM-M and vice versa; see the discussion on p. 49 and p. 56.

4.5 Conclusion

In this chapter, the first research question– *"Based on a FPCM of objects, how should fuzzy priorities of these objects be determined so that they reflect properly all preference information available in the FPCM?"*– was answered. Three types of FPCMs were examined in this chapter—FMPCMs, FAPCMs-A, and FAPCMs-M. Construction of FPCMs, defining and verifying their consistency, and deriving fuzzy priorities of objects from them have been studied in detail for each of the three types of FPCMs.

First, the relevant methods proposed in the literature based on the fuzzy extension of methods originally proposed for PCMs were reviewed and their major drawbacks were identified (step (1.b) formulated in Sect. 1.3). In particular, it was found out that "equal preference" of two compared objects is very often modeled inappropriately in FPCMs, which results in misinterpretation of the preference information provided by the DM and thus leads to false results. Further, it was found out that most of the definitions of consistency reviewed in this chapter violate the reciprocity of the related PCs in FPCMs or are not invariant under permutation of objects. These are severe drawbacks that lead to false conclusions about consistency/inconsistency of the FPCMs. Similarly, also the reviewed approaches for obtaining fuzzy maximal eigenvalues of FMPCMs violate the reciprocity of the related PCs or the invariance under permutation of objects. This again leads to the distortion of the preference information contained in FMPCMs. In particular, the fuzzy maximal eigenvalue obtained by these approaches is not necessarily greater or equal to the dimension of the FMPCM, which is an inherent property of the maximal eigenvalues of MPCMs. Analogously, also the reviewed methods for deriving fuzzy priorities of objects from FPCMs violate the reciprocity of the related PCs or the invariance under permutation of objects. The consequences in this case are even more critical. The fuzzy priorities of objects obtained by applying these defective methods do not reflect properly the preference information contained in the FPCM. Such fuzzy priorities are not only excessively uncertain, but often completely distorted, which may also lead to a completely different ranking of the compared objects and thus to a decision that is not optimal.

Second, it was shown that in order to reflect appropriately the preference information contained in a FPCM, in particular the reciprocity of the related PCs, it is necessary to apply constrained fuzzy arithmetic to the fuzzy extension of the PC methods instead of standard fuzzy arithmetic (step (1.c) formulated in Sect. 1.3). From the multiplicative reciprocity $m_{ji} = \frac{1}{m_{ij}}$, $i, j = 1, \ldots, n$, for MPCMs, the equality $m_{ij} m_{ji} = 1$, $i, j = 1, \ldots, n$, automatically follows. Similarly, from the additive reciprocity $a_{ji} = 1 - a_{ij}$, $i, j = 1, \ldots, n$, for APCMs, the equality $a_{ij} + a_{ji} = 1$, $i, j = 1, \ldots, n$, follows. These properties are inherent to every MPCM and every APCM, respectively. However, these properties are not preserved for FMPCMs and FAPCMs when standard fuzzy arithmetic is applied to the computations. In particular, the multiplicative reciprocity $\widetilde{m}_{ji} = \frac{1}{\widetilde{m}_{ij}}$, $i, j = 1, \ldots, n$, holds for FMPCMs but $\widetilde{m}_{ij}\widetilde{m}_{ji} \neq 1$, $i, j = 1, \ldots, n$, $i \neq j$. Similarly, the additive reciprocity $\widetilde{a}_{ji} = 1 - \widetilde{a}_{ij}$, $i, j = 1, \ldots, n$, holds for FAPCMs but $\widetilde{a}_{ij} + \widetilde{a}_{ji} \neq 1$, $i, j = 1, \ldots, n$, $i \neq j$. This is a serious drawback. It was shown that validity of the equalities $\widetilde{m}_{ij}\widetilde{m}_{ji} = 1$ and $\widetilde{a}_{ij} + \widetilde{a}_{ji} = 1$, $i, j = 1, \ldots, n$, can be guaranteed by appropriately applying constrained fuzzy arithmetic.

Third, a complete approach based on constrained fuzzy arithmetic was proposed to deal with all three types of FPCMs (step (1.d) formulated in Sect. 1.3). Namely, it was shown how to appropriately model the meaning of the linguistic term "equal preference" used for PCs in FPCMs. Further, definitions of consistency for FMPCMs, FAPCMs-A, and FAPCMs-M were proposed in such a way that they are invariant under permutation of objects and do not violate the reciprocity of the related PCs. Two definitions of consistency, weak version and strong version, were proposed for each type of FPCMs, and useful tools for verifying the consistency according to each definition were provided. Moreover, by using constrained fuzzy arithmetic, it was also possible to extend to FPCMs the properties equivalent to the corresponding consistency conditions for PCMs in such a way that they are still equivalent. This was not possible with standard fuzzy arithmetic. Further, a method for obtaining the fuzzy maximal eigenvalue of a FMPCM was proposed in such a way that it is invariant under permutation of objects and does not violate the reciprocity of the related PCs in a FMPCM. By preserving the reciprocity of the related PCs, the fuzzy maximal eigenvalue is always greater than (or equal to) the dimension of the FMPCM, which is an inherent property of the maximal eigenvalues of MPCMs. Afterwards, methods for obtaining fuzzy priorities of objects from FMPCMs, FAPCMs-A, and FAPCMs-M were proposed based on constrained fuzzy arithmetic so that they preserve the reciprocity of the related PCs and are invariant under permutation of objects. The fuzzy priorities obtained by these methods thus properly represent the preference information contained in the FPCMs. Moreover, applying constrained fuzzy arithmetic to the fuzzy extension of the formulas for obtaining normalized priorities preserves the normality property, i.e., the fuzzy priorities obtained by these methods form a normalized fuzzy vector. This was again not possible by using standard fuzzy arithmetic. The approaches defined for each type of the FPCMs are mutually equivalent; each type of the FPCMs can be transformed into another together with the respective consistency properties. In the same way, fuzzy priorities obtained from each type of the FPCMs can be transformed one into another.

Chapter 5
Incomplete Large-Dimensional Pairwise Comparison Matrices

Abstract This chapter is concerned with large-dimensional pairwise comparison problems and answers the following research question: "How can the amount of preference information required from the decision maker in a large-dimensional pairwise comparison matrix be reduced while still obtaining comparable priorities of objects?" The chapter reviews a real-life case-study motivating the need for methods dealing with large-dimensional pairwise comparison problems and discusses desirable properties of methods for constructing an incomplete pairwise comparison matrix and deriving priorities of objects. A two-phase method ensuring the desirable properties is introduced in detail, and its application to a real-life case study is described. The excellent performance of the method in terms of saving a large number of pairwise comparisons required from the decision maker and obtaining comparable priorities of objects is supported by simulations. In addition, the method is critically compared with another well-known method for constructing incomplete pairwise comparison matrices.

5.1 Introduction to Large-Dimensional Pairwise Comparison Problems

In Chap. 2, PC methods were reviewed. These methods are based on constructing PCMs where every two objects from a given set of n objects are compared pairwisely. As we know from Chap. 2, $n(n-1)/2$ PCs are required to compare n objects pairwisely. Thus, to compare pairwisely 5 objects, for example, 10 PCs need to be provided by the DM. Now imagine that 20 objects need to be compared. In this case, 190 PCs are required from the DM. This number is clearly very high. Such a high number of PCs is not easy to obtain in sufficient quality. In fact, the more PCs need to be made, the less reliable the information expressed by the DMs might be (due to fatigue, time constraints, and similar factors).

In order to avoid a large number of PCs, Saaty (1977, 2008) suggested to split a large-dimensional problem into several subproblems of smaller dimensions. This implies creating supercategories of objects. The objects are then compared pairwisely only within the defined supercategories. Additionally, PCs of the supercategories have to be also provided. This results in the reduction of the complexity of the problem and in making the preference information requirements (the number of PCs needed) feasible. However, this procedure also results in a slight loss of information; the objects from different supercategories are not directly compared pairwisely. This loss of information can be to some extent compensated for by introducing a strong enough consistency condition on the preferences expressed by the experts, which provides means of calculating the missing values in the PCM (Jandová et al. 2017).

For some real-life problems, the above described approach works well. There are, however, situations when splitting the problem into several smaller ones renders parts of the problem too abstract and hence intractable for the experts providing the information on the preferences among objects. Stoklasa et al. (2013) provided a real-life example of such a problem in the area of arts evaluation. Stoklasa et al. (2013) refers to the development of the evaluation model for the Registry of Artistic Performances that has been used in the Czech Republic within the principles and rules of financing public universities; see Ministry of Education of the Czech Republic (2011). This model has been used since 2012 to provide a basis for the distribution of a part of the subsidy from the state budget among public universities in the Czech Republic. The mathematical model presented by Stoklasa et al. (2013) is designed to compute evaluations (priorities) for different categories of works of art (currently 27 categories) based on combination of expert assessment of the significance of the respective work of art and two more objective criteria (extent and institutional reception). The authors dealt with a 27×27 PCM representing a problem that could not be split into several smaller ones due to partial dependencies among the evaluation criteria and due to the necessity of providing real-life examples to all the compared categories for the experts to be able to express their intensities of preference.

In large-dimensional PC problems which cannot be split into smaller subproblems the required priorities of objects may be obtained from incomplete PCMs. In that case, focusing on an appropriate reduction of the number of PCs which have to be provided by the DM and obtaining enough preference information in the incomplete PCM to be able to compute the priorities of objects are of paramount importance. When using incomplete PCMs, it is necessary to deal adequately with two key steps: (i) finding a method for efficiently selecting the subset of the $n(n-1)/2$ PCs that should be provided by the DM, and (ii) finding an appropriate method for deriving the priority vector from the incomplete PCM.

Harker (1987a, b, c) and later Harker and Millet (1990) were the first to deal with the problem of reducing the number of PCs. They proposed to perform only a part of the $n(n-1)/2$ PCs by means of an algorithm which iteratively selects the PCs to be submitted to the DM. This selection is made according to the largest modification in the priority vector. The process of inputting PCs is then stopped when the provided PC changes the priority vector by less than a fixed threshold. Wedley et al. (1993) focused on the choice of only $n-1$ PCs, which is the minimum number required for comparing n objects, and they compared and discussed several methods of entering them. Sanchez and Soyer (1998) proposed to use entropy-based measures of the information content to evaluate judgment accuracy and to state a stopping rule of the process of inputting PCs. Ra (1999) worked with n PCs which form a closed chain.

Fedrizzi and Giove (2013) proposed a method for selecting PCs in an incomplete PCM which takes into account both the robustness of the collected data and the consistency of the expressed preferences.

For what concerns the methods for deriving the priority vector from incomplete PCMs, several different approaches have been proposed; see, e.g., Alonso et al. (2008); Chen and Triantaphillou (2001); Fedrizzi and Giove (2007, 2013); Harker (1987a, b); Harker and Millet (1990); Kwiesielewicz (1996); Kwiesielewicz and van Uden (2003); Ramík (2016); Shiraishi et al. (1998); Xu (2004, 2005). Some of these methods are aimed at automatically determining the missing PCs in order to complete the incomplete PCM. Once the PCM is completed, one of the known methods for deriving priorities from a complete PCM can be used. Conversely, other methods compute the priorities from the incomplete PCM directly. Clearly, having first computed the priorities, every missing PCs in the incomplete PCM can then be determined accordingly, thus completing the PCM.

In this chapter, the method for deriving priorities of objects from large-dimensional PCMs developed by Jandová et al. (2017) is reviewed. For the original description of the method and the related study, the readers are asked to refer to Jandová et al. (2017). In this method, the consistency preservation plays a crucial role. In particular, the weak consistency defined in Sect. 2.2.2.2 for MPCMs and in Sect. 2.3.3.2 for APCMs is employed in the method as a minimum requirement of consistency that has to be satisfied. The method differs from the other PC methods mentioned above since the weak consistency of the incomplete PCM is preserved in every step of the method; a similar property is not required in any other known PC method.

The objective of the PC method introduced by Jandová et al. (2017) is not simply reducing the number of PCs required from the DM. It is known that this number could be radically reduced to $n-1$, as proposed by Wedley et al. (1993); Herrera-Viedma et al. (2004) and others. Such choice completely fulfills the requirement of maximally reducing the number of PCs required from the DM. However, it gives up the fundamental characterizing property of the PC methods–the ability to use the redundancy of information contained in a PCM in order to suitably manage the unavoidable inconsistency of human judgments. The numerical example in Sect. 5.3.2 demonstrates that the methods requiring only $n-1$ PCs do not always result in reliable outcomes.

The main objective of the PC method proposed by Jandová et al. (2017) is to find an ideal compromise between requiring as little preference information from the DM as possible and still obtaining enough information to calculate priorities of objects that are close to the hypothetical full-information case, i.e., the case when the DM provides all PCs in the PCM. Moreover, the final priorities of objects provided by the novel PC method are computed in such a way that they contain information concerning the uncertainty stemming from the fact that some PCs are not provided by the DM, nor are they entered automatically by the proposed algorithm. The priorities of objects are computed as intervals in order to reflect the missing information in the incomplete PCM and to provide ranges for the values of the crisp priorities of objects obtainable from any weakly consistent completion of the incomplete PCM. The range of the interval priorities depends on the amount of preference information that is missing in the incomplete PCM. One of the formulas for calculating fuzzy priorities introduced in Sects. 4.3, 4.4.2, and 4.4.3 is applied to the incomplete PCM, depending on the type of the PC used for expressing DM's preferences.

In the following section, preliminaries indispensable for describing the novel method for large-dimensional PC problems developed by Jandová et al. (2017) and for demonstrating its performance are given.

5.2 Background

In this section the large-dimensional evaluation model for the registry of artistic performances proposed by Stoklasa et al. (2013) is described in more detail as its results are used for confrontation with the results obtained by the PC method proposed by Jandová et al. (2017). Further, an overview of the algorithm for the optimal choice of PCs in incomplete large-dimensional PCMs proposed by Fedrizzi and Giove (2013) is given here as this algorithm is partially utilized in the method of Jandová et al. (2017).

5.2.1 Case Study: Evaluation Model for the Registry of Artistic Performances

The evaluation model for the Registry of Artistic Performances, as mentioned in Sect. 5.1, was a motivation for Jandová et al. (2017) to develop the novel method for large-dimensional PC problems. This evaluation model was also used by Jandová et al. (2017) for validating the performance of the novel method, as it will be shown in Sect. 5.3. Thus, a short description of the model is provided in this section. For a more detailed description, the readers can refer to Stoklasa et al. (2013).

The outputs of artistic performance are currently evaluated in the Czech Republic based on the following three criteria, for each of which there are three levels distinguished:

Criterion 1 – Relevance or significance of the piece of art

    A – a new piece of art or a performance of crucial significance

    B – a new piece of art or a performance containing numerous important innovations

    C – a new piece of art or a performance pushing forward modern trends

Criterion 2 – Extent of the piece of art

    K – a piece of art or a performance of large extent

    L – a piece of art or a performance of medium extent

    M – a piece of art or a performance of limited extent

Criterion 3 – Institutional and media reception/impact of the piece of art
  X – international reception/impact
  Y – national reception/impact
  Z – regional reception/impact

Criterion 1 is an expertly assessed criterion that brings a peer-review element into the evaluation. Each segment of art provided a general linguistic specification for each level of this criterion to be made available for the expert evaluators, real-life (historical) examples for levels A, B, and C are also available. Also the levels of Criterion 2 are specified linguistically. This criterion was, however, intended to be measurable for each segment on such a level of accuracy that most of the ambiguity in categorizing works of art according to this criterion is removed. For Criterion 3, lists of institutions corresponding to level X, Y, and Z are provided. Hence, there is no subjectivity in evaluation against this criterion in the process.

By combining the various levels of the three criteria, 27 categories of works of art can be defined. These categories are represented in the model by triplets of the capital letters identifying the levels (e.g. AKY, BLZ, or CMZ). Each of these 27 categories needs to be assigned a score (priority). The original idea was to obtain all PCs of the 27 well-represented (that is represented by real-life examples) categories of works of art (351 PCs in total) using Saaty's scale given in Table 2.1, and afterwards, to compute the score for each category using the GMM (2.24).

Because the MPCMs and Saaty's scale were not intended for large-dimensional problems, Saaty (1977) proposed to approach these problems by splitting them into subproblems of lower dimensions. However, this approach was not applicable to the problem in question for the following reasons: (a) there are some dependencies among the criteria which are not easy to describe or capture, (b) to compare various levels of one criterion (e.g. big, medium, and small) without any real-life representatives (good representatives for such broad categories proved to be difficult to find) is not easy for the experts, (c) the experts were not able to express their preferences between the criteria (these too proved to be too abstract to provide enough representation for the experts to be able to express their preferences). For these reasons, all 27 categories were compared pairwisely. Since the multiplicative-consistency condition (2.4) is almost impossible to achieve for large-dimensional MPCMs, the much more relaxed weak-consistency condition (2.11) was used to control the consistency of the PCs of the categories provided by the experts. The weak consistency was considered as a minimum requirement on the consistency of the MPCM.

As the weak-consistency condition (2.11) is easy to check during the process of inputting preferences, and even more so when the rows and columns of the MPCM are ordered in accordance with the preference ordering of the categories (from the most preferred to the least preferred one), the 27 categories were first ordered according to their preference; see Stoklasa et al. (2013) for more details. Afterwards, the experts provided 351 PCs of the categories using the elements from Saaty's scale given in Table 2.1, and the normalized priorities of the categories were obtained from the complete MPCM by using the GMM (2.24). The MPCM and the derived priorities of all 27 categories are shown later in Sect. 5.3.3, where these results are compared with the results obtained by Jandová et al. (2017) by applying the novel method.

After two years of using the described model and the computed evaluations, minor adjustments to the evaluation methodology proved to be necessary. Adding one more level to one of the criteria and changing the initial preference ordering of the categories were considered; see Stoklasa et al. (2016). Changing the preference ordering of the categories would result in the need of inputting the large MPCM again. In the case of adding one level of one of the criteria, the dimension of the MPCM would increase, thus dramatically increasing the number of PCs required. Generally, it has to be expected that these changes might occur in the model in the future in order to meet new requirements. If such an adaptation results in the need of providing all the PCs again (or even in providing more of them), an algorithm capable of reducing the number of PCs that need to be provided, such as the algorithm proposed by Jandová et al. (2017) and reviewed in this chapter, would be most needed in order to reduce the strain and the time consumption for the experts without substantial loss of information.

5.2.2   Overview of the Algorithm of Fedrizzi and Giove (2013) for Optimal Sequencing in Incomplete Large-Dimensional PCMs

In this section, the algorithm for optimal sequencing (i.e. the optimal choice of PCs) in incomplete large-dimensional PCMs proposed by Fedrizzi and Giove (2013) is briefly summarized as its part is utilized in the method for dealing with incomplete large-dimensional PCMs developed by Jandová et al. (2017) and described in this chapter. For a more detailed description of the algorithm, interested readers can refer to Fedrizzi and Giove (2013).

Fedrizzi and Giove (2013) proposed an algorithm for iteratively selecting PCs that should be provided by the DM in an incomplete PCM. The algorithm was presented in the form for APCMs-A. Nevertheless, the authors themselves emphasized that the approaches based on APCMs and MPCMs are equivalent (this was also shown in detail in Sect. 2.5).

The algorithm uses a selection rule based on two criteria. The first criterion, quantified by y_{ij}, is used to achieve enough indirect PCs (r_{ik}, r_{kj}) for every missing PC r_{ij} of the APCM-A $R = \{r_{ij}\}_{i,j=1}^{n}$. The second criterion, quantified by z_{ij}, is used to reduce possible inconsistency of judgments. A scoring function F is defined to determine the usefulness of selecting a particular pair of not yet mutually compared objects o_i and o_j, $i, j \in \{1, \ldots, n\}$. A high value of the scoring function indicates high necessity to compare o_i with o_j, $i, j \in \{1, \ldots, n\}$. Thus, at each step of the algorithm, the pair of objects with the maximal value of F is selected. The scoring function is defined as

$$F(y_{ij}, z_{ij}) = \lambda y_{ij} + (1 - \lambda)z_{ij}, \tag{5.1}$$

where $\lambda \in [0, 1]$ is the parameter quantifying the importance of the criterion y_{ij} over the criterion z_{ij}. Using the simplified notation $f(o_i, o_j) := F(y_{ij}, z_{ij})$ to refer directly to the pair of objects, the selection rule is defined as

$$(o_i, o_j) = \arg \max_{(o_k, o_l) \in \Omega \backslash Q} f(o_k, o_l), \tag{5.2}$$

where Q is the set of PCs that were already provided by the DM during the questioning process, and $\Omega = \{(o_i, o_j); i, j = 1, \ldots, n, i < j\}$ is the set of all PCs between the n objects. The criteria used in the scoring function (5.1) are defined by the following formulas:

$$y_{ij} = 1 - \frac{|s_i| + |s_j|}{2(n - 2)}, \tag{5.3}$$

$$z_{ij} = \frac{\varphi_{ij}}{|s_i \cap s_j| + 1} \frac{1}{3} = \frac{3}{|s_i \cap s_j| + 1} \varphi_{ij}. \tag{5.4}$$

First, let us analyze the expression (5.3), where $s_i = \{k; (o_i, o_k) \in Q \vee (o_k, o_i) \in Q\}$ and $|s_i|$ is the cardinality[1] of the set s_i. Then $|s_i| + |s_j|$ is the number of PCs involving object o_i or object o_j. The maximum value of $|s_i|$ is $n - 2$ since (o_i, o_j) was not yet provided and (o_i, o_i) is excluded. Thus, the maximum value of $|s_i| + |s_j|$ is $2(n - 2)$, and $\frac{|s_i| + |s_j|}{2(n-2)}$ represents the normalized number of PCs involving objects o_i or o_j. Criterion y_{ij} is defined by (5.3) in order to have the scoring function F increasing in both variables. Criterion y_{ij} determines the lack of PCs suffered by objects o_i and o_j.

Now, let us analyze the expression (5.4), where φ_{ij} is the mean inconsistency of the indirect PCs of objects o_i and o_j. First, let us define the variable μ_{ij} representing the mean value of all indirect PCs of o_i and o_j based on the additive-consistency condition (2.28):

$$\mu_{ij} = \begin{cases} 0 & \text{if } s_i \cap s_j = \emptyset, \\ \sum_{k \in s_i \cap s_j} \frac{r_{ik} + r_{kj} - 0.5}{|s_i \cap s_j|} & \text{if } s_i \cap s_j \neq \emptyset. \end{cases} \tag{5.5}$$

Because indirect PCs of objects o_i and o_j are usually not completely consistent, the mean inconsistency φ_{ij} of indirect PCs of o_i and o_j is defined as

$$\varphi_{ij} = \begin{cases} 0 & \text{if } s_i \cap s_j = \emptyset, \\ \sum_{k \in s_i \cap s_j} \frac{(r_{ik} + r_{kj} - 0.5 - \mu_{ij})^2}{|s_i \cap s_j|} & \text{if } s_i \cap s_j \neq \emptyset. \end{cases} \tag{5.6}$$

---

[1]Cardinality $|s|$ of the set s is the number of its elements; e.g. $|\{2, 4, 5\}| = 3$.

Note that for $s_i \cap s_j \neq \emptyset$, φ_{ij} is the variance of $(r_{ih} + r_{hj} - 0.5)$, and it holds that $\varphi_{ij} = 0$ if and only if all the indirect PCs of o_i and o_j are additively consistent according to Definition 9.

The maximum achievable reduction $\Delta\varphi_{ij}$ of φ_{ij} is obtained if the direct PC is $r_{ij} = \mu_{ij}$ and, in such a case, $\Delta\varphi_{ij} = \frac{\varphi_{ij}}{|s_i \cap s_j|+1}$. In the formula (5.4), $\Delta\varphi_{ij}$ is normalized, i.e., it is divided by $\frac{1}{3}$ as it is the maximum achievable value of $\Delta\varphi_{ij}$; see Fedrizzi and Giove (2013). The criterion z_{ij} expresses the normalized maximum achievable reduction of the inconsistency φ_{ij} that can be reached by means of the direct PCs of o_i and o_j.

The algorithm for selecting the PCs that should be provided by the DM in an incomplete APCM-A given by Fedrizzi and Giove (2013) consists of the following steps:

1. At the beginning, no PCs are performed and $Q = \emptyset$. Thus, $y_{ij} = 1$, $z_{ij} = 0$, and $f(o_i, o_j) = \lambda$ for all $i, j = 1, \ldots, n$. Instead of a random selection, recommended initial PCs are $\{(o_{2i-1}, o_{2i}); i = 1, \ldots, \frac{n}{2}\}$ if n is even and $\{(o_{2i-1}, o_{2i}); i = 1, \ldots, \frac{n-1}{2}\}$ if n is odd.
2. In each step of the selection process, the value of the scoring function f is quantified for each missing PC (o_i, o_j) by using the formula (5.1). According to (5.2), the suitable PC (o_i, o_j) is selected. In the case of equal values of $f(o_i, o_j)$, the pair of objects o_{i^*}, o_{j^*} such that $i^* + j^*$ minimizes $i + j$ is selected. In the case of equal values of $i + j$, the pair containing the minimum index is selected.
3. The selection is stopped when the value of the scoring function becomes lower than the threshold $\delta \in [0, 1]$ that is subjectively defined by the DM, i.e.

$$\max_{(o_i, o_j) \in \Omega \setminus Q} f(o_i, o_j) \leq \delta. \tag{5.7}$$

5.3 Novel Method for Incomplete Large-Dimensional PCMs

In this section the novel PC method for large-dimensional PCMs as introduced by Jandová et al. (2017) is described. In particular, Sect. 5.3.1 provides a detailed description of the PC method. In Sect. 5.3.2, the application of the method is demonstrated on an illustrative example and compared with another well-known method for large-dimensional PCMs. In Sect. 5.3.3, the method is applied to the evaluation model for the Registry of Artistic Performances, and the results are confronted with the results obtained by the original model proposed by Stoklasa et al. (2013). In Sect. 5.3.4, the results of numerical simulations are provided in order to analyze the performance of the novel PC method.

5.3.1 Description of the Method

In this section, the novel method developed by Jandová et al. (2017) is described in detail. The method combines the concept of weak consistency with the PC-selection process proposed by Fedrizzi and Giove (2013). The interactive algorithm guides the DM through the PC-input phase by identifying the pair of objects that should be compared next. This way, the increase of preference information in the incomplete PCM is maximized, and the compliance with the weak-consistency condition is ensured in each step of the algorithm. This results in a weakly consistent incomplete PCM after each input. Moreover, information on all feasible preference intensities of each missing PC of an incomplete PCM (that is such values that would not violate the weak consistency when put in the PCM) is available in each step of the algorithm. Values that are unambiguous are input automatically into the PCM, and the DM is not bothered to provide these. This way, the amount of information contained in the incomplete PCM can increase after each step without the effort of the DM. When enough information is provided by the DM, the algorithm stops asking the DM for inputs and determines the preference ordering of the objects and their priorities, which are in this case in the form of intervals.

Let us consider objects $o_1, o_2, \ldots, o_n$ to which priorities need to be assigned. The PC of a pair of objects o_i and o_j will be denoted as (o_i, o_j). Considering that the approaches based on MPCMs, APCMs-A, and APCMs-M are equivalent (transformation of one representation into the other can be done using the formulas reviewed in Sect. 2.5), the DM can express the preference intensities in any of these forms. For the sake of the algorithm presentation and without any loss of generality by presenting the algorithm only for one particular type of a PCM, the MPCMs are chosen in this section to describe the algorithm. This approach is used also because the practical application of large-dimensional PCMs described in Sect. 5.2.1 was actually done using a MPCM. In this way, it is possible to confront the outputs of the algorithm described in this section with the results of the practical application directly, as it was done by Jandová et al. (2017).

Saaty's scale given in Table 2.1 is used here for expressing the PCs in the MPCM $M = \{m_{ij}\}_{i,j=1}^{n}$, i.e. $m_{ij} \in \{\frac{1}{9}, \frac{1}{8}, \ldots, \frac{1}{2}, 1, 2, \ldots, 8, 9\}$, $i, j = 1, \ldots, n$, with the meanings described in Table 2.1. Since MPCM M is multiplicatively reciprocal, it is sufficient to enter only the PCs above the main diagonal of M or alternatively only the PCs below the main diagonal of M. In this algorithm, without any loss of generality, the PCs above the main diagonal of M are required from the DM. Hence, the set Ω of all PCs required to complete the PCM is $\Omega = \{(o_i, o_j); i, j = 1, \ldots, n, \ i < j\}$, the cardinality of Ω being $|\Omega| = n(n-1)/2$. The objective of this algorithm is twofold: (i) finding such a set $\bar{\Omega} \subset \Omega$ that its cardinality (i.e., the number of the PCs required from the DM) allows for the computation of all the priorities of objects, and (ii) proposing a way of generating the elements of this set in such an order that minimizes the cardinality of $\bar{\Omega}$.

The set of all PCs already performed will be denoted by Q, and the set of PCs not yet entered into the MPCM will be denoted $\Omega \setminus Q$. For each $(o_i, o_j) \in \Omega \setminus Q$, the

set $FV_{ij} \subseteq \{\frac{1}{9}, \frac{1}{8}, \ldots, \frac{1}{2}, 1, 2, \ldots, 8, 9\}$ of all feasible values that are in compliance
with the weak-consistency condition (2.11) will be always given. For simplicity and
in the figures, the notation $[\min FV_{ij}, \max FV_{ij}]$ is used where there is no risk of
ambiguity. The notation $[\min FV_{ij}, \max FV_{ij}]$ represents a range of the values from
Saaty's scale from $\min FV_{ij}$ to $\max FV_{ij}$ for a given $(o_i, o_j) \in \Omega \setminus Q$. For example,
the set $\{6, 7, 8, 9\}$ will be denoted as $[6, 9]$ and interpreted as a range of values of
Saaty's scale from 6 to 9. An incomplete MPCM will be denoted $\widehat{M} = \{\widehat{m}_{ij}\}_{i,j=1}^n$,
where

$$\widehat{m}_{ij} = \begin{cases} [m_{ij}^L, m_{ij}^U] & \text{for } (o_i, o_j) \in \Omega \setminus Q, \\ m_{ij} & \text{for } (o_i, o_j) \in Q. \end{cases}$$

It is obvious that $[m_{ij}^L, m_{ij}^U] = [\min FV_{ij}, \max FV_{ij}]$ for each $(o_i, o_j) \in \Omega \setminus Q$.

The process of guided input of the preference information and computation of the
priorities of n compared objects can be summarized based on Jandová et al. (2017)
in the following steps:

1. The DM chooses which PCM will be used to express the preference intensities
   (MPCM is considered for the purpose of the description of the algorithm). The
   diagonal elements (o_i, o_i) of MPCM $\widehat{M} = \{\widehat{m}_{ij}\}_{i,j=1}^n$ are set, i.e. $\widehat{m}_{ii} = 1$ for
   all $i = 1, \ldots, n$. The sets of feasible values (FV sets) FV_{ij} are established for
   $(o_i, o_j) \in \Omega \setminus Q$. At the beginning, $FV_{ij} = [\frac{1}{9}, 9]$ for $(o_i, o_j) \in \Omega$.
2. The DM provides initial PCs. In this algorithm, the setting proposed by Fedrizzi
   and Giove (2013) is used, i.e., the set of initial PCs $\{(o_{2i-1}, o_{2i}), i = 1, \ldots, \lfloor \frac{n}{2} \rfloor\}$,
   where $\lfloor \frac{n}{2} \rfloor$ is the floor[2] of $\frac{n}{2}$, is required from the DM. However, also a different
   set of initial PCs can be selected. The only restriction is that these initial PCs do
   not violate the weak-consistency condition (2.11).

The following Steps 3–5 are repeated until the stopping criterion is met:

3. Based on the algorithm of Fedrizzi and Giove (2013), it is determined iteratively
   which PC $(o_i, o_j) \in \Omega \setminus Q$ is to be provided next by the DM. The PC (o_i, o_j)
   that maximizes the scoring function (5.1) is selected, and the DM is asked to
   provide the corresponding preference intensity into the incomplete PCM $\widehat{M}$. The
   DM selects the value of the PC (o_i, o_j) from its FV set FV_{ij}.
4. Based on the weak-consistency requirement, the FV set FV_{ij} is recalculated for
   each missing PC $(o_i, o_j) \in \Omega \setminus Q$. The weak-consistency rules (2.11)–(2.14) for
   MPCMs are used in this step of the algorithm to determine $[\min FV_{ij}, \max FV_{ij}]$.
   Obviously, the FV set is restricted only when an indirect PC exists. That is when
   for a PC (o_i, o_j) not yet entered into the MPCM there exists at least one object
   with index k, $k \neq i, j$, such that the PCs (o_i, o_k) and (o_k, o_j) are already entered
   into the incomplete MPCM $\widehat{M}$ or restricted FV sets are determined for them.
5. Missing PCs $(o_i, o_j) \in \Omega \setminus Q$, for which FV_{ij} contain just a single element, are
   entered into the incomplete MPCM $\widehat{M}$ automatically. Obviously, the occurrence
   of such single-element FV_{ij} sets is far more frequent when a discrete scale is used

---

[2]Floor $\lfloor x \rfloor$ of $x \in \mathbb{R}$ is the largest integer lower or equal to x; e.g., $\lfloor 5.7 \rfloor = 5$.

for making PCs of objects. In real-life applications, the requirement of a discrete scale rather than a continuous scale is not a limitation to the decision-making problem. That is because in real-life applications discrete scales of numbers (either crisp of fuzzy) with assigned linguistic terms expressing the intensities of preference are used far more frequently than continuous scales. Discrete scales are more natural for DMs as they provide the required simplifying granularity for continuous universes similar to the common language. The algorithm, however, remains valid also for continuous scales. Here, discrete Saaty's scale as given in Table 2.1 is assumed for the description of the algorithm. The sets FV_{ij} are recalculated (Step 5 is performed) after each such input and Step 4 is performed again.

Steps 4 and 5 are repeated until there are no elements of the incomplete MPCM $\widehat{M}$ that could be entered automatically this way.

6. Stopping criterion: For every missing PC in the incomplete MPCM $\widehat{M}$, there exists at least one indirect PC.

   Note that this stopping criterion varies from the stopping criterion proposed by Fedrizzi and Giove (2013) in their algorithm. Since the scope of the novel method is computing the interval priorities of objects, the stopping criterion is designed in order to be able to determine for each missing PC $\widehat{m}_{ij}$ a (restricted) set FV_{ij} of feasible intensities of preference that can be entered in order to preserve the weak-consistency condition (2.11).

7. The so-called reciprocal FV sets are identified, i.e., such FV_{ij}, $(o_i, o_j) \in V \subseteq \Omega \setminus Q$, that contain at least one of the values of the respective scale along with its reciprocal value. As an example, a set containing the two numbers 3 and $\frac{1}{3}$ is a reciprocal FV set. From a reciprocal FV set FV_{ij}, it is not possible to derive which object from the pair (o_i, o_j) is preferred to the other. This ambiguity is not desired. Thus, all reciprocal FV sets need to be replaced by a specific value provided by the DM or by a non-reciprocal FV set (as a consequence of filling in a value from another reciprocal FV set), so that $V = \emptyset$. The DM is asked to provide a PC $(o_k, o_l) \in V$ such that $(o_k, o_l) = \arg \max_{(o_i, o_j) \in V} |FV_{ij}|$. In the case that there are more pairs of objects with the same maximal cardinality of their reciprocal FV sets, one of them is chosen randomly. Alternatively, to make the algorithm more user friendly, the DM can be asked to provide the PC of one pair of objects of his/her choice. After such PC is provided, i.e., after the DM chooses one value from the given reciprocal FV set, FV_{ij}, $(o_i, o_j) \in \Omega \setminus Q$, are recalculated using Steps 4 and 5. This step is repeated until there are no reciprocal FV sets left.

   Described technique enables us to reduce the amount of information required from the DM as much as possible since providing the PC of the pair of objects with the maximal cardinality of the problematic set adds the most information to the MPCM.

8. The preference ordering of objects is derived from the incomplete MPCM $\widehat{M}$. For each object (represented by the corresponding row of the MPCM), we determine the number of elements in the given row of the MPCM that are greater than or

equal to the indifference value or for which the elements of the FV set are all greater than or equal to the indifference value, which is 1 for MPCMs. Based on this information, the objects $o_1, o_2, \ldots, o_n$ can be reordered from the most preferred one to the least preferred one, i.e. $o_{(1)} \succeq o_{(2)} \succeq \cdots \succeq o_{(n)}$. The respectively permuted MPCM with rows and columns ordered from the most preferred object to the least preferred one will be denoted $\widehat{M}^o$.

9. In order to obtain the priorities of objects from the incomplete MPCM $\widehat{M}$, the sets FV_{ij} of feasible intensities of preference for all missing PCs are considered to be intervals given by the minimal and the maximal value in the set (for example, the set $\{3, 4, 5\}$ is now considered to be the interval $[3, 5]$). This allows us to obtain the priorities of objects in the form of intervals. The interval priorities can be obtained either from the preference-ordered MPCM $\widehat{M}^o$ or from the non-preference-ordered MPCM $\widehat{M}$. It is obvious that in both cases the same interval priorities would be obtained as the matrices are the same up to a permutation. To obtain the interval priorities $\overline{w}_1, \ldots, \overline{w}_n$ of objects, the fuzzy extension of the GMM proposed in Sect. 4.2.3.2 is used here. Specifically, either the formulas (4.91)–(4.94) or the formulas (4.97)–(4.100) are applied to the incomplete MPCM $\widehat{M} = \{\widehat{m}_{ij}\}_{i,j=1}^n$. Alternatively, the fuzzy extension of the EVM proposed in Sect. 4.3.3.1 can be applied. Note that the incomplete MPCM $\widehat{M}$ is in fact an interval FMPCM; the filled-in PCs are crisp numbers, which are a special case of intervals, and there are intervals of feasible values for all missing PCs. All the formulas for obtaining fuzzy priorities provided in Chap. 4 are explicitly written for trapezoidal FPCMs. Nevertheless, recall that interval FPCMs are a particular case of trapezoidal FPCMs. Thus, keeping this in mind, the formulas (4.91)–(4.94) or (4.97)–(4.100) can be easily applied to the incomplete MPCM $\widehat{M} = \{\widehat{m}_{ij}\}_{i,j=1}^n$.

Note that in the case when the DM provides preference information utilizing APCMs-A or APCMs-M, the formulas (4.158) and the formulas (4.219)–(4.222), respectively, are used for deriving the interval priorities.

From the formulas (4.91)–(4.94) and from the argumentation preceding their construction (see p. 144) it is obvious that the resulting interval priorities contain all the priorities that would be computed for any particular selection of real values from the sets FV_{ij} corresponding to the missing PCs in $\widehat{M}$ (that is if $\widehat{M}$ was completed) preserving the weak-consistency condition. This means that if the DM provided all the missing PCs preserving the weak consistency, the crisp priorities computed from such a MPCM would lie within the computed interval priorities.

Furthermore, because the interval priorities $\overline{w}_i = [w_i^L, w_i^U], i = 1, \ldots, n$, obtained by the formulas (4.97)–(4.100) from an incomplete weakly consistent MPCM are normalized according to Definition 28, i.e., $\overline{w}_i \subseteq [0, 1]$ and

$$w_i^L + \sum_{\substack{j=1 \\ j \neq i}}^n w_j^U \geq 1, \qquad w_i^U + \sum_{\substack{j=1 \\ j \neq i}}^n w_j^L, \leq 1 \qquad i = 1, \ldots, n,$$

the interval priorities get very narrow with an increasing dimension n of a MPCM. Further, the interval priorities obtained from the preference-ordered MPCM $\widehat{M}^o$ by the formulas (4.97)–(4.100) have the following property. From the weak consistency and particularly from the property of a non-decreasing sequence of elements in each row and a non-increasing sequence of elements in each column of an ordered weakly consistent MPCM it follows that any two interval priorities $\overline{w}_i, \overline{w}_j, i, j \in \{1, \ldots, n\}$, obtained by formulas (4.97)–(4.100) can be ordered according to the standard partial order $\leq$ on intervals; $[a, b] \leq [c, d]$ if $a \leq c, b \leq d$. Therefore, $\leq$ is a total order on the set of all interval priorities $\overline{w}_i, i = 1, \ldots, n$. Recall that according to Step 8 the preference ordering of objects is derived immediately from the preference information in $\widehat{M}^o$ without the need of computing the interval priorities. Moreover, for any two objects o_i, o_j such that $o_i \succ o_j$, it holds that $\overline{w}_i > \overline{w}_j$; for the case when $o_i \succeq o_j$ and $o_j \succeq o_i$ it holds that $\overline{w}_i = \overline{w}_j$.

5.3.2 Illustrative Example and Comparison Study

For better understanding of how the novel method proposed by Jandová et al. (2017) works, a simple illustrative example with a weakly consistent MPCM of seven objects provided by Jandová et al. (2017) is presented here. In addition, the comparison of the performance of the novel method with the well-known method for incomplete MPCMs proposed by Herrera-Viedma et al. (2004) as done by Jandová et al. (2017) is described here.

Obviously, applying the method to a PCM of just several (in this case seven) objects has only limited significance in practice as such a PCM does not require many PCs from the DM in the first place. However, for better visual illustration of each step of the novel method, an example with just several objects is more convenient.

Let $o_1, o_2, \ldots, o_7$ be objects that need to be compared and whose priorities need to be determined by the DM. In order to compare seven objects pairwisely, the DM would have to provide 21 PCs in the full-information case. By applying the new algorithm, only a part of these 21 PCs will be required from the DM. In order to evaluate the performance of the novel algorithm of Jandová et al. (2017), it is necessary to confront its results with the results obtainable in the hypothetical full-information case. Thus, the MPCM given in Table 5.1 is considered as the full-information MPCM M that would be obtained if the DM provided all 21 PCs. For better illustration, easier understanding, and an easy check of the compliance with the weak-consistency condition (2.11), the objects in the MPCM M given in Table 5.1 are ordered from the most preferred one to the least preferred one. For the sake of simplicity, only the PCs above the main diagonal are given since the PCs below the main diagonal are the reciprocals of the corresponding PCs above the main diagonal. The priorities $w_1, \ldots, w_7$ of objects $o_1, \ldots, o_7$ obtainable from the full-information MPCM M by the GMM (2.24) are given in the second column of Table 5.9.

Table 5.1 MPCM with all
PCs provided by the DM

	o_1	o_2	o_3	o_4	o_5	o_6	o_7
o_1	1	9	9	9	9	9	9
o_2		1	2	2	9	9	9
o_3			1	1	5	8	8
o_4				1	5	8	8
o_5					1	7	8
o_6						1	7
o_7							1

Table 5.2 Starting empty
MPCM $\widehat{M}$

	o_3	o_4	o_6	o_1	o_2	o_5	o_7
o_3	1						
o_4		1					
o_6			1				
o_1				1			
o_2					1		
o_5						1	
o_7							1

The novel method introduced by Jandová et al. (2017) is designed to be applicable to general PC problems with no information about the preference ordering of the objects that are to be compared pairwisely. This means that the method can be applied to any random initial ordering of objects in the MPCM. Let us therefore assume that the preference ordering of objects is not known in advance, and instead, the objects are ordered randomly. Let us assume the random initial order of the objects as given in Table 5.2. The empty MPCM $\widehat{M} = \{\widehat{m}_{ij}\}_{i,j=1}^{n}$ in Table 5.2 is the starting matrix where the PCs identified by the algorithm are going to be provided by the DM or entered automatically based on the weak-consistency condition.

Notice that the labeling of objects in the incomplete MPCM $\widehat{M} = \{\widehat{m}_{ij}\}_{i,j=1}^{n}$ in Table 5.2 does not correspond to the numbering of rows and columns of the MPCM anymore. For example, object o_1 is not in the first row of the MPCM, but instead, it is in the fourth row now. Therefore, it is important to realize that from now on when we refer to a PC (o_i, o_j), this does not necessarily correspond to the PC $\widehat{m}_{ij}$ in the i-th row and the j-th column of $\widehat{M}$.

At the beginning of the algorithm, the diagonal elements are set to the value 1 and the DM is asked to provide initial PCs (o_3, o_4), (o_6, o_1), and (o_2, o_5) as it is required in Step 2 of the algorithm. Any value from Saaty's scale can be chosen in this step. This is because the FV sets for all missing PCs are $[\frac{1}{9}, 9]$. For easier orientation in the tables the initial FV sets $[\frac{1}{9}, 9]$ are replaced by empty fields. Only the FV sets calculated from indirect PCs in the following steps of the algorithm will be entered into the incomplete MPCM $\widehat{M}$.

Steps 3 to 5 are repeated until the stopping criterion is met. In Step 3, we apply the algorithm based on searching for a missing PC (o_i, o_j) with the maximum value of the scoring function (5.1), i.e., the missing PC that should be provided by the DM. In this illustrative example, both criteria of the scoring function (5.1) are considered to have the same importance, therefore the parameter $\lambda = 0.5$ is set.

As already mentioned in the previous section, in contrast to the method proposed by Fedrizzi and Giove (2013), the incomplete MPCM $\widehat{M}$ is required to be weakly consistent in the method proposed by Jandová et al. (2017); it has to satisfy the properties (2.11)–(2.14). According to this requirement, in Step 4, it is possible to restrict the sets FV_{ij} of feasible intensities of preference for some missing PCs. If any set FV_{ij} contains only one value, this value is entered automatically into the incomplete MPCM $\widehat{M}$ as suggested in Step 5.

Table 5.3 demonstrates the incomplete MPCM $\widehat{M}$ after the initial PCs $(o_3, o_4) = 1$, $(o_6, o_1) = \frac{1}{9}$, and $(o_1, o_5) = 9$, and after the first iteration of the algorithm. The first PC chosen in the first iteration and provided by the DM is $(o_3, o_7) = 8$. As it can be seen from the incomplete MPCM $\widehat{M}$, the PC $(o_4, o_7) = 8$ was filled in automatically according to the weak consistency since $(o_3, o_4) = 1$ and $(o_3, o_7) = 8$. The legend explaining the notation used in the tables in this section is provided in Table 5.4.

Table 5.5 shows the incomplete MPCM $\widehat{M}$ after the second iteration of the algorithm. The PC $(o_4, o_6) = 8$ was provided by the DM, and according to the weak consistency, one missing PC and ranges for other three missing PCs were added automatically. For example, the range $[1/9, 1/2]$ for the missing PC (o_3, o_1) was

Table 5.3 Incomplete MPCM $\widehat{M}$ after the first iteration

	o_3	o_4	o_6	o_1	o_2	o_5	o_7
o_3	1	1					8
o_4		1					8
o_6			1	1/9			
o_1				1			
o_2					1	9	
o_5						1	
o_7							1

Table 5.4 Legend

1/9	initial PC provided by the DM
8	PC provided by the DM during the algorithm
8	PC filled in automatically according to the weak consistency
[1/9,1/2]	FV set
[1/9.9]	reciprocal FV set
1/9	value from the reciprocal FV set provided by the DM

Table 5.5 Incomplete MPCM $\widehat{M}$ after the second iteration

	o_3	o_4	o_6	o_1	o_2	o_5	o_7
o_3	1	1	8	[1/9,1/2]			8
o_4		1	8	[1/9,1/2]			8
o_6			1	1/9			[1/8,8]
o_1				1			
o_2					1	9	
o_5						1	
o_7							1

Table 5.6 Incomplete MPCM $\widehat{M}$ after the stopping criterion is met

	o_3	o_4	o_6	o_1	o_2	o_5	o_7
o_3	1	1	8	1/9	[1/9,1/2]	5	8
o_4		1	8	1/9	[1/9,1/2]	5	8
o_6			1	1/9	[1/9,9]	[1/8,1/2]	[1/8,8]
o_1				1	9	9	9
o_2					1	9	9
o_5						1	[2,8]
o_7							1

derived from the PCs $(o_3, o_6) = 8$ and $(o_6, o_1) = 1/9$ according to the first rule of the weak-consistency property (2.13). The DM continues providing the missing PCs until the stopping criterion is met. The incomplete MPCM obtained at the moment of meeting the stopping criterion is given in Table 5.6.

Two reciprocal FV sets are present in the incomplete MPCM $\widehat{M}$ in Table 5.6; see the PCs $(o_6, o_2) = [\frac{1}{9}, 9]$ and $(o_6, o_7) = [\frac{1}{8}, 8]$. This means that it is not possible to conclude which object is preferred to the other one for these pairs of objects; the information obtained from the indirect PCs is too vague. Therefore, according to Step 7, the DM needs to determine the intensities of preference for these pairs of objects.

First the DM is asked to provide the PC (o_6, o_2) as its reciprocal FV set has the biggest cardinality $(|FV_{62}| = |\{\frac{1}{9}, \frac{1}{8}, \ldots, \frac{1}{2}, 1, 2, \ldots, 9\}| = 17)$. In this particular case no restriction of the other FV sets occurs. Afterwards, the DM provides the PC (o_6, o_7), and as a consequence, the FV set of (o_5, o_7) is reduced from [2, 8] to [7, 8]. Table 5.7 shows the incomplete MPCM $\widehat{M}$ after Step 7.

Once the reciprocal FV sets are removed, it is possible to order the compared objects from the most preferred one to the least preferred one according to Step 8 and to reorder the whole incomplete MPCM $\widehat{M}$ accordingly. Table 5.8 demonstrates the preference-ordered incomplete MPCM $\widehat{M}^o$ with FV sets for all missing PCs. The reader can verify that by choosing any value from any of the FV sets the weak consistency of the incomplete MPCM $\widehat{M}^o$ is not violated.

According to Step 9 of the algorithm, the interval priorities of objects are obtained from the incomplete MPCM $\widehat{M}^o$. The interval priorities are summarized in Table 5.9

Table 5.7 Incomplete MPCM $\widehat{M}$ after removing the reciprocal FV sets

	o_3	o_4	o_6	o_1	o_2	o_5	o_7
o_3	1	1	8	1/9	[1/9,1/2]	5	8
o_4		1	8	1/9	[1/9,1/2]	5	8
o_6			1	1/9	1/9	[1/8,1/2]	7
o_1				1	9	9	9
o_2					1	9	9
o_5						1	[7,8]
o_7							1

Table 5.8 Final preference-ordered incomplete MPCM $\widehat{M}^o$

	o_1	o_2	o_3	o_4	o_5	o_6	o_7
o_1	1	9	9	9	9	9	9
o_2		1	[2,9]	[2,9]	9	9	9
o_3			1	1	5	8	8
o_4				1	5	8	8
o_5					1	[2,8]	[7,8]
o_6						1	7
o_7							1

Table 5.9 Priorities of objects

Objects	Crisp priorities obtained by the GMM	Interval priorities according to Jandová et al. (2017)	Priorities obtained according to Herrera-Viedma et al. (2004)
o_1	0.5083	[0.4840, 0.5102]	0.2377
o_2	0.1765	[0.1765, 0.2594]	0.1128
o_3	0.1166	[0.0896, 0.1170]	0.2284
o_4	0.1166	[0.0896, 0.1170]	0.2284
o_5	0.0463	[0.0363, 0.0472]	0.0535
o_6	0.0228	[0.0213, 0.0273]	0.1128
o_7	0.0128	[0.0122, 0.0131]	0.0264

along with the crisp priorities computed from the full-information MPCM M given in Table 5.1.

Let us summarize the results of this illustrative example. In order to have complete preference information and to compute crisp priorities of objects, the DM would have to provide 21 PCs. Using the algorithm for incomplete PCMs proposed by Jandová et al. (2017), the DM had to provide only 10 PCs (approx. 48%). Other 7 PCs (approx. 33%) were added automatically based on the weak-consistency condition, and 4 PCs (approx. 19%) were missing but with FV sets whose elements do not violate the weak consistency of the incomplete MPCM $\widehat{M}^o$. The novel algorithm did not only spare the DM more than half of the PCs, but it also provided very good results. The calculated interval priorities are quite narrow and contain the original priorities; see Table 5.9. Recall that it was mentioned in Step 9 of the algorithm that this is a general property that always holds.

To emphasize the advantage and the significant contribution of the novel method to the decision-making theory, Jandová et al. (2017) compared it with another well-known method for incomplete MPCMs. Particularly, the method proposed by Herrera-Viedma et al. (2004) was applied to the illustrative example for the comparison. Notice that the paper of Herrera-Viedma et al. (2004) has been cited over 740-times, which suggests wide recognition of their method. In the method proposed by Herrera-Viedma et al. (2004), only $n-1$ PCs above the main diagonal, i.e., $\{(o_i, o_{i+1}); i = 1, \ldots, n - 1\}$, are required from the DM. The remaining PCs are completed automatically so that the resulting MPCM $M = \{m_{ij}\}_{i,j=1}^{n}$ is multiplicatively consistent according to (2.14). Clearly, in most of the cases, the missing PCs completed by this automatic procedure exceed Saaty's scale $\left[\frac{1}{9}, 9\right]$. That is why Herrera-Viedma et al. (2004) suggested to transform the obtained MPCM M given on scale $\left[\frac{1}{c}, c\right], c > 9$, into the MPCM $M' = \{m'_{ij}\}_{i,j=1}^{n}$ given on scale $\left[\frac{1}{9}, 9\right]$ by using transformation formula

$$m'_{ij} = m_{ij}^{1/\log_9 c}, \qquad i, j = 1, \ldots, n. \tag{5.8}$$

In Table 5.10, the completed, transformed, and ordered MPCM M' after providing the 6 initial PCs above the main diagonal is given. The 6 PCs provided by the DM are highlighted in bold. Obviously, unlike the incomplete MPCM $\widehat{M}^o$ in Table 5.8, the MPCM M' in Table 5.10 differs substantially from the original MPCM in Table 5.1. Thus, also the priorities obtained from this MPCM given in the last column of Table 5.9 vary essentially from the original priorities given in the second column. Even the ranking of the objects based on these priorities varies from the ranking obtained in the full-information case.

In order too demonstrate how far the MPCM M' in Table 5.10 obtained by the method proposed by Herrera-Viedma et al. (2004) is from the original MPCM M in comparison to the incomplete MPCM $\widehat{M}^o$ obtained by the novel method described

Table 5.10 MPCM M' obtained by the approach of Herrera-Viedma et al. (2004)

	o_1	o_2	o_3	o_4	o_5	o_6	o_7
o_1	1	**2.108**	1.041	1.041	4.444	**2.108**	9
o_2		1	0.494	0.494	**2.108**	1	4.270
o_3			1	1	4.270	2.025	8.647
o_4				1	4.270	**2.025**	8.647
o_5					1	0.474	**2.025**
o_6						1	4.270
o_7							1

in Sect. 5.3, Jandová et al. (2017) measured their distance. In particular, the distance for MPCMs defined by Cook and Kress (1988) was utilized for this scope. Since the incomplete MPCM $\widehat{M}^o$ in Table 5.8 contains intervals, it was necessary to generalize the distance of Cook and Kress (1988) to interval FMPCMs first. For two interval FMPCMs $\overline{M}^1 = \{\overline{m}^1_{ij}\}^n_{i,j=1}$, $\overline{m}^1 = \left[m^{1L}_{ij}, m^{1U}_{ij}\right]$, $\overline{M}^2 = \{\overline{m}^2_{ij}\}^n_{i,j=1}$, $\overline{m}^2 = \left[m^{2L}_{ij}, m^{2U}_{ij}\right]$, the interval distance based on the distance defined by Cook and Kress (1988) is given as $\overline{D}(\overline{M}^1, \overline{M}^2) = \left[d^L, d^U\right]$:

$$d^L = \min_{\substack{m^1_{ij}\in\left[m^{1L}_{ij},m^{1U}_{ij}\right]\\ m^2_{ij}\in\left[m^{2L}_{ij},m^{2U}_{ij}\right]}} \sum_{i=1}^{n-1}\sum_{j=i+1}^{n} |\ln(m^1_{ij}/m^2_{ij})|$$

$$= \sum_{i=1}^{n-1}\sum_{j=i+1}^{n} \min_{\substack{m^1_{ij}\in\left[m^{1L}_{ij},m^{1U}_{ij}\right]\\ m^2_{ij}\in\left[m^{2L}_{ij},m^{2U}_{ij}\right]}} |\ln(m^1_{ij}/m^2_{ij})|, \tag{5.9}$$

$$d^U = \max_{\substack{m^1_{ij}\in\left[m^{1L}_{ij},m^{1U}_{ij}\right]\\ m^2_{ij}\in\left[m^{2L}_{ij},m^{2U}_{ij}\right]}} \sum_{i=1}^{n-1}\sum_{j=i+1}^{n} |\ln(m^1_{ij}/m^2_{ij})|$$

$$= \sum_{i=1}^{n-1}\sum_{j=i+1}^{n} \max_{\substack{m^1_{ij}\in\left[m^{1L}_{ij},m^{1U}_{ij}\right]\\ m^2_{ij}\in\left[m^{2L}_{ij},m^{2U}_{ij}\right]}} |\ln(m^1_{ij}/m^2_{ij})|. \tag{5.10}$$

Note that for crisp MPCMs, the interval distance given by (5.9) and (5.10) is identical to the distance originally defined by Cook and Kress (1988). By applying the formulas (5.9) and (5.10), the distance of the MPCM obtained by the method proposed by Herrera-Viedma et al. (2004) given in Table 5.10 and the original MPCM given in Table 5.1 is $D = 22.8941$. The interval distance of the interval FMPCM in Table 5.8 from the original MPCM in Table 5.1 is $\overline{D} = [0, 4.3934]$. Clearly, $\overline{D} = [0, 4.3934]$ is significantly smaller than $D = 22.8941$, which demonstrates better performance of the novel method proposed by Jandová et al. (2017).

Notice that the lower boundary value d^L of the distance of any incomplete MPCM with intervals of feasible values for all missing PCs obtained by the new method from the hypothetical full-information MPCM is always 0. This follows from the fact that the incomplete MPCM always contains the hypothetical full-information MPCM obtainable if all PCs were provided by the DM, which is the substance and the main advantage of the proposed method.

5.3.3 Application of the Method to the Evaluation Model for the Registry of Artistic Performances

This section describes the application of the novel method to the large-dimensional problem of evaluating outcomes of artistic performance in the Czech Republic as done by Jandová et al. (2017). The outcome of the novel method is compared with the outcome given by Stoklasa et al. (2013). It is drawn from the knowledge of the complete MPCM and a numerical experiment is conducted. In particular, it is started with an empty MPCM of randomly ordered 27 categories of works of art, the novel method described in Sect. 5.3.1 is utilized, and, whenever a PC is required from the DM, the appropriate value is found in the complete MPCM given in Table 5.11.

The randomly generated initial order of the categories (i.e., the categories are not ordered according to their preference but randomly) given in the matrix of Table 5.12 is assumed. For better orientation and simpler notation, the number of the corresponding row was assigned to each category. First, the DM was asked to provide 13 initial PCs $\{(2i - 1, 2i); \ i = 1, \ldots, 13\}$. Subsequently, the algorithm for selecting the missing PCs $(i, j), i, j \in \{1, 2, \ldots, 27\}, i < j$, that should be provided by the DM was applied. The parameter $\lambda = 0.5$ was used in the scoring function (5.1) as both its criteria were considered to have the same importance. The algorithm was stopped after just 109 PCs provided by the DM.

Because missing PCs with reciprocal FV sets were present in the incomplete MPCM at that stage, it was not possible to order the categories from the most preferred one to the least preferred one immediately. First, it was necessary to remove all reciprocal FV sets FV_{ij}. This was done iteratively and after the replacement of every single reciprocal FV set FV_{ij} either by a PC provided by the DM or by a non-reciprocal FV set, all the remaining missing elements were recalculated. In order to eliminate all the reciprocal FV sets, 23 PCs were required from the DM overall. The incomplete MPCM obtained after this step is shown in Table 5.12. Finally, the categories compared in the incomplete MPCM were ordered from the most preferred one to the least preferred one. The preference-ordered weakly consistent incomplete MPCM is given in Table 5.13.

In the original method proposed by Stoklasa et al. (2013), the experts had to provide all 351 PCs. When the novel method was applied to the problem, only 145 PCs (approx. 41%) were required. Other 153 PCs (approx. 44%) were added automatically according to the weak consistency and, for the remaining 53 PCs (approx. 15%), sets of feasible intensities of preference were derived from the weak-consistency properties. These FV sets are relatively narrow containing at most 4 values. Furthermore, the incomplete MPCM contains the original complete MPCM, i.e., all the filled-in PCs are the same and the FV set provided for each missing PC in the incomplete MPCM always contains the preference intensity of the corresponding PC in the complete MPCM; compare Tables 5.11 and 5.13.

Interval priorities of the categories were obtained from the incomplete MPCM in Table 5.13 by using the formulas (4.97)–(4.100). The interval priorities together with the crisp priorities obtained from the complete MPCM in Table 5.11 by the GMM

Table 5.11 Complete weakly consistent MPCM obtained by Stoklasa et al. (2013)

		1	2	3	4	5	6	7	8	9	10	11	12	13	14	15	16	17	18	19	20	21	22	23	24	25	26	27
		AKX	AKY	AKZ	ALX	AMX	ALY	ALZ	BKX	AMY	AMZ	BKY	BKZ	BLX	BMX	BLY	BLZ	BMY	BMZ	CKX	CLX	CKY	CKZ	CMX	CLY	CLZ	CMY	CMZ
1	AKX	1	5	5	5	5	5	5	5	5	5	5	5	5	5	5	7	7	9	9	9	9	9	9	9	9	9	9
2	AKY		1	5	5	5	5	5	5	5	5	5	5	5	5	5	5	5	7	7	7	7	9	9	9	9	9	9
3	AKZ			1	3	3	5	5	5	5	5	5	5	5	5	5	5	5	7	7	7	7	9	9	9	9	9	9
4	ALX				1	3	5	5	5	5	5	5	5	5	5	5	5	5	5	5	7	7	7	9	9	9	9	9
5	AMX					1	5	5	5	5	5	5	5	5	5	5	5	5	5	5	5	5	7	7	7	9	9	9
6	ALY						1	3	3	5	5	5	5	5	5	5	5	5	5	5	5	5	5	7	7	7	9	9
7	ALZ							1	3	5	5	5	5	5	5	5	5	5	5	5	5	5	5	5	7	7	7	9
8	BKX								1	5	5	5	5	5	5	5	5	5	5	5	5	5	5	5	5	7	7	7
9	AMY									1	3	5	5	5	5	5	5	5	5	5	5	5	5	5	5	7	7	7
10	AMZ										1	5	5	5	5	5	5	5	5	5	5	5	5	5	5	7	7	7
11	BKY											1	5	5	5	5	5	5	5	5	5	5	5	5	5	7	7	7
12	BKZ												1	3	5	5	5	5	5	5	5	5	5	5	5	7	7	7
13	BLX													1	5	5	5	5	5	5	5	5	5	5	5	7	7	7
14	BMX														1	3	5	5	5	5	5	5	5	5	5	7	7	7
15	BLY															1	3	3	5	5	5	5	5	5	5	7	7	7
16	BLZ																1	3	5	5	5	5	5	5	5	5	7	7
17	BMY																	1	5	5	5	5	5	5	5	5	7	7
18	BMZ																		1	5	5	5	5	5	5	5	5	5
19	CKX																			1	3	5	5	5	5	5	5	5
20	CLX																				1	5	5	5	5	5	5	5
21	CKY																					1	3	5	5	5	5	5
22	CKZ																						1	3	6	5	5	5
23	CMX																							1	5	5	5	5
24	CLY																								1	3	5	5
25	CLZ																									1	3	3
26	CMY																										1	3
27	CMZ																											1

Table 5.12 Incomplete weakly consistent MPCM after filling in the PCs by the DM

		1 ALZ	2 ALX	3 AMX	4 CLX	5 AMZ	6 CLY	7 BLX	8 CLZ	9 CMX	10 CKY	11 CKX	12 AKZ	13 AKY	14 BLZ	15 AMY	16 BMY	17 BLY	18 ALY	19 AKX	20 BKZ	21 BMZ	22 CMZ	23 BKY	24 BKX	25 CKZ	26 CMY	27 BMX
1	ALZ	1	1/5	3	5	5	[5,9]	5	[5,9]	[7,9]	5	5	1/5	1/5	5	[3,5]	5	5	[1/5, 1/2]	1/5	5	5	9	5	5	5	9	5
2	ALX		1	5	7	5	[7,9]	5	[7,9]	[7,9]	[5,7]	[5,7]	1/3	1/5	5	5	5	5	[2,3]	1/5	5	[5,7]	9	5	5	7	9	5
3	AMX			1	5	5	[5,7]	5	[5,7]	7	5	5	1/5	1/5	5	3	5	5	[1/5, 1/2]	1/5	5	5	[7,9]	5	5	5	7	5
4	CLX				1	1/5	3	1/5	5	5	1/5	1/5	[1/9, 1/7]	[1/9, 1/7]	1/5	1/5	1/5	1/5	[1/9, 1/5]	1/7	1/5	1/5	5	1/5	1/5	1/3	5	1/5
5	AMZ					1	5	5	[5,7]	5	5	5	1/5	1/5	5	[1/5, 1/2]	5	[2,5]	[1/5, 1/2]	1/5	5	5	[7,9]	5	3	5	7	5
6	CLY						1	1/5	5	5	5	5	[1/9, 1/7]	[1/9, 1/7]	5	1/5	1/5	[2,5]	[1/9, 1/5]	1/7	1/3	1/5	5	1/5	1/5	[1/5, 1/3]	5	1/5
7	BLX							1	5	7	5	5	1/5	1/5	5	[1/5, 1/2]	5	[2,5]	1/5	[1/7, 1/5]	[1/7, 1/5]	5	7	1/5	1/5	5	5	5
8	CLZ								1	3	1/5	1/5	[1/9, 1/7]	1/5	1/5	[1/5, 1/2]	1/5	1/5	[1/9, 1/5]	1/7	[1/7, 1/5]	1/5	[3,5]	[1/7, 1/5]	[1/7, 1/5]	1/5	[3,5]	1/5
9	CMX									1	1/5	1/5	[1/9, 1/7]	[1/9, 1/7]	1/7	1/7	1/5	1/7	[1/9, 1/7]	1/7	1/5	1/5	3	1/7	1/7	1/5	3	[1/7, 1/5]
10	CKY										1	1/3	1/7	[1/9, 1/7]	1/5	1/5	1/5	1/5	[1/7, 1/5]	1/7	1/5	1/5	5	1/5	1/5	5	5	1/5
11	CKX											1	1/7	1/5	[1/5, 1/2]	[1/5, 1/2]	[1/5, 1/2]	[1/5, 1/2]	[1/7, 1/5]	1/5	1/5	1/5	5	1/5	1/5	[2,5]	[5,7]	[1/5, 1/2]
12	AKZ												1	1/5	5	5	5	5	3	1/5	5	[5,7]	9	5	5	[7,9]	9	5
13	AKY													1	5	5	5	5	5	1/5	5	[7,9]	9	5	5	[7,9]	9	5
14	BLZ														1	5	3	1/3	1/5	[1/7, 1/5]	1/5	5	7	1/5	1/5	5	7	3
15	AMY															1	5	5	1/5	1/5	[2,5]	[2,5]	[5,7]	[2,5]	[2,5]	[2,5]	[5,7]	5
16	BMY																1	1/3	1/5	[1/9, 1/7]	1/5	5	[5,7]	1/5	1/5	5	7	5
17	BLY																	1	1/5	[1/9, 1/7]	[1/5, 1/2]	5	[7,9]	[1/5, 1/2]	[1/5, 1/2]	5	7	3
18	ALY																		1	1/5	5	9	9	5	5	5	[7,9]	5
19	AKX																			1	5	9	9	5	5	9	9	7

(continued)

Table 5.12 (continued)

		1	2	3	4	5	6	7	8	9	10	11	12	13	14	15	16	17	18	19	20	21	22	23	24	25	26	27
		ALZ	ALX	AMX	CLX	AMZ	CLY	BLX	CLZ	CMX	CKY	CKX	AKZ	AKY	BLZ	AMY	BMY	BLY	ALY	AKX	BKZ	BMZ	CMZ	BKY	BKX	CKZ	CMY	BMX
20	BKZ																				1	5	7	1/5	1/5	5	7	5
21	BMZ																					1	5	1/5	1/5	[2,5]	5	1/5
22	CMZ																						1	1/7	1/7	[1/7, 1/5]	1/3	1/7
23	BKY																							1	1/5	5	7	5
24	BKX																								1	5	7	5
25	CKZ																									1	5	1/5
26	CMY																										1	1/7
27	BMX																											1

Table 5.13 Preference-ordered incomplete weakly consistent MPCM

	1	2	3	4	5	6	7	8	9	10	11	12	13	14	15	16	17	18	19	20	21	22	23	24	25	26	27
	AKX	AKY	AKZ	ALX	AMX	ALY	ALZ	BKX	AMY	AMZ	BKY	BKZ	BLX	BMX	BLY	BLZ	BMY	BMZ	CKX	CLX	CKY	CKZ	CMX	CLY	CLZ	CMY	CMZ
1 AKX	1	5	5	5	5	5	5	5	5	5	5	5	5	5	[5,7]	7	[7,9]	9	9	9	9	9	9	9	9	9	9
2 AKY		1	5	5	5	5	5	5	5	5	5	5	5	5	5	5	5	[5,7]	7	[7,9]	[7,9]	[7,9]	[7,9]	[7,9]	[7,9]	9	9
3 AKZ			1	3	3	5	5	5	5	5	5	5	5	5	5	5	5	[5,7]	7	7	[7,9]	[7,9]	[7,9]	[7,9]	[7,9]	9	9
4 ALX				1	[2,3]	[2,5]	5	[3,5]	5	5	5	5	5	5	5	5	5	[5,7]	[5,7]	[5,7]	7	7	[7,9]	[5,9]	[7,9]	9	9
5 AMX					1		[3,5]	[3,5]	5	5	5	5	5	5	5	5	5	5	5	5	5	[5,7]	[5,9]	[5,9]	[7,9]	9	9
6 ALY						1	3	3	5	5	5	5	5	5	5	5	5	5	5	5	5	5	[5,9]	[5,9]	[7,9]	9	9
7 ALZ							1	1	5	5	5	5	5	5	5	5	5	5	5	5	5	5	[5,7]	[5,7]	7	7	[7,9]
8 BKX								1	[2,5]	[3,5]	5	5	5	5	5	5	5	5	5	5	5	5	5	5	5	7	[7,9]
9 AMY									1	3	5	5	5	5	5	5	5	5	5	5	5	5	5	5	5	7	[7,9]
10 AMZ										1	1	5	5	5	5	5	5	5	5	5	5	5	5	5	5	7	7
11 BKY											1	1	5	5	5	5	5	5	5	5	5	5	5	5	5	7	7
12 BKZ												1	3	[3,5]	5	5	5	5	5	5	5	5	5	5	5	7	7
13 BLX													1	[2,5]	5	5	5	5	5	5	5	5	5	5	5	7	7
14 BMX														1	3	5	5	5	5	5	5	5	5	5	5	7	7
15 BLY															1	3	5	5	5	5	5	5	5	5	5	7	7
16 BLZ																1	3	5	5	5	5	5	5	5	5	7	7
17 BMY																	1	[2,5]	[2,5]	5	5	5	5	5	5	[5,7]	7
18 BMZ																		1	[2,5]	5	5	5	5	5	5	5	7
19 CKX																			1	3	[3,5]	5	5	5	5	5	[5,7]
20 CLX																				1	[2,5]	5	3	5	5	5	5
21 CKY																					1	3	5	5	5	5	5
22 CKZ																						1	[3,5]	5	5	5	5
23 CMX																							1	3	5	5	5
24 CLY																								1	5	5	5
25 CLZ																									1	3	[3,5]
26 CMY																										1	3
27 CMZ																											1

(2.24) are given in Table 5.14. Obviously, the crisp priorities of the categories lie within the intervals delimited by the interval priorities. This result is natural since the FV sets for missing PCs in the incomplete PCM obtain all feasible intensities of preference that preserve the weak consistency. Therefore, the complete MPCM given in Table 5.11 provided by Stoklasa et al. (2013) can be obtained from the weakly consistent incomplete MPCM in Table 5.13 by a particular combination of values from the FV sets.

Using the novel method, it was possible to obtain interval priorities of the categories that represent very well the actual priorities obtained from the complete

Table 5.14 Interval and crisp priorities of the categories

	Categories	Crisp priorities	Interval priorities
1	AKX	0.1357	[0.1314, 0.1370]
2	AKY	0.1132	[0.1126, 0.1166]
3	AKZ	0.0967	[0.0917, 0.0995]
4	ALX	0.0862	[0.0829, 0.0895]
5	ALY	0.0761	[0.0687, 0.0799]
6	ALZ	0.0612	[0.0593, 0.0660]
7	AMX	0.0552	[0.0542, 0.0573]
8	AMY	0.0498	[0.0495, 0.0509]
9	AMZ	0.0418	[0.0415, 0.0423]
10	BKX	0.0385	[0.0382, 0.0390]
11	BKY	0.0335	[0.0333, 0.0340]
12	BKZ	0.0292	[0.0280, 0.0296]
13	BLX	0.0269	[0.0258, 0.0273]
14	BLY	0.0222	[0.0211, 0.0249]
15	BLZ	0.0204	[0.0194, 0.0215]
16	BMX	0.0184	[0.0176, 0.0192]
17	BMY	0.0167	[0.0160, 0.0175]
18	BMZ	0.0134	[0.0133, 0.0140]
19	CKX	0.0117	[0.0114, 0.0125]
20	CKY	0.0106	[0.0102, 0.0112]
21	CKZ	0.0088	[0.0088, 0.0092]
22	CLX	0.0080	[0.0077, 0.0082]
23	CLY	0.0072	[0.0067, 0.0074]
24	CLZ	0.0057	[0.0053, 0.0066]
25	CMX	0.0047	[0.0045, 0.0051]
26	CMY	0.0042	[0.0040, 0.0045]
27	CMZ	0.0038	[0.0035, 0.0040]

MPCM (compare the results in Table 5.14). In contrast to the original method, how-ever, only 145 PCs were required from the DMs instead of 351. This means that the amount of the information required from the DM was reduced to only 41% of the information required by Stoklasa et al. (2013). This is a very significant reduction of the information required from the DM, that reduces considerably the strain and time demands and raises the quality of the information provided.

5.3.4   Simulations and Numerical Results

Jandová et al. (2017) performed simulations to evaluate the benefit of the proposed method from the point of view of sparing a part of PCs required from the DM. MPCMs of six different dimensions–$n = 5, 10, \ldots, 30$–were considered. The pro-posed method was applied to 600 randomly generated weakly consistent MPCMs, 100 of each dimension. For each such a MPCM, an empty initial MPCM of the given dimension n was considered, and the novel method was applied to the empty MPCM in order to identify iteratively the missing PCs that should be provided by the DM. Whenever such a missing PC was identified, the value of the PC was taken from the complete MPCM and entered into the incomplete MPCM. The number x of PCs required in the iterative algorithm from the DM was computed. Consequently, also the number of spared PCs was computed as $n(n - 1)/2 - x$. After applying the method to all 100 MPCMs of the given dimension n, an average number of spared PCs was computed as well as an average % of spared PCs. The numerical results are shown in Table 5.15.

According to the results in Table 5.15, the average percentage of the spared PCs increases with the increasing dimension of the MPCM. For MPCMs of dimension 15 and greater, more than 60% of the PCs are spared on average. However, it is necessary to point out that, despite this huge reduction in the number of PCs required from the DM, the resulting interval priorities obtained by the formulas (4.97)–(4.100) from the final incomplete weakly consistent MPCM always contain the crisp priorities obtainable from the original randomly generated weakly consistent MPCM. More-over, as discussed on p. 147, the interval priorities get very narrow with the increasing dimension of the MPCM.

Table 5.15 Average number of spared PCs required from the DM

Dimension of PCMs	$n = 5$	$n = 10$	$n = 15$	$n = 20$	$n = 25$	$n = 30$
Number of PCs required in the full-information case	10	45	105	190	300	435
Average number of spared PCs	4	24	64	123	207	312
Average % of spared PCs	42%	53%	61%	65%	69%	72%

5.4 Conclusion

An answer to the second research question, *"How can the amount of preference information required from the DM in a large-dimensional PCM be reduced while still obtaining comparable priorities of objects?"*, was provided in this chapter. In particular, it was argued that the novel method for dealing with incomplete large-dimensional PCMs proposed by Jandová et al. (2017) has all desirable properties and thus provides an answer to the research question. The method is applicable to all three types of PCMs examined in this book, i.e., to MPCMs, APCMs-A, as well as APCMs-M.

The novel method developed by Jandová et al. (2017) strives to identify the trade-off between decreasing the number of PCs required from the DM and obtaining a sufficient amount of information to compute relevant priorities of objects. The method is suggested as a possible solution to large-dimensional problems where the complete information (i.e. providing all PCs) is either costly, too time consuming, or infeasible to obtain, or where the preference intensities in the large-dimensional PCMs require (frequent) revisions.

In the first part of the method, an iterative algorithm for optimal choice of PCs that should be provided by the DM in an incomplete large-dimensional PCM is applied. The algorithm is based on the combination of the weak-consistency condition introduced by Jandová and Talašová (2013) for MPCMs and by Jandová et al. (2017) for APCMs, with the modified version of the optimal PC-selection algorithm proposed by Fedrizzi and Giove (2013). The weak consistency is imposed as a minimum consistency requirement on the incomplete PCM, and it is required during the whole process of entering PCs into the incomplete PCM. As a consequence, some PCs and intervals of feasible values for some missing PCs are entered automatically without violating the weak consistency, based on the PCs previously provided by the DM. The algorithm is stopped when there exists an interval of feasible values for every missing PC in the incomplete PCM. Afterwards, interval priorities of objects are derived from such an incomplete PCM by means of formulas proposed in Chap. 4.

The whole method is designed in such a way that the interval priorities derived from the incomplete large-dimensional PCM include the priorities of objects obtainable from any weakly consistent completion of the incomplete PCM. This means that the interval priorities contain the crisp priorities that would be obtained from the hypothetical complete PCM obtainable if the DM provided all PCs in the PCM.

The numerical results of the performed simulations demonstrate that the application of the novel method can significantly reduce the number of PCs required from the DM and thus results in significant resource savings. At the same time, a high accuracy of the output is guaranteed by the algorithm as the resulting interval priorities contain the priorities that would be obtained from the hypothetical complete PCM. For randomly generated MPCMs of dimension $n \geq 15$, even more than 60% of PCs required (with respect to the full-information case) were spared on average. The numerical example of a small 7×7 PCM exhibited a reduction of ca. 50% in the number of PCs required (from 21 to 10). In the real-life case study of the works-of-art evaluation model utilizing a 27×27 PCM, the number of PCs required from the experts was reduced by ca. 60% (from the overall 351 PCs only 145 were required). The obtained interval priorities of the 27 categories of works of art contain the crisp priorities obtained from the original complete PCM, and they are very narrow.

Chapter 6
Discussion and Future Research

Abstract This chapter answers the research questions posed at the beginning of the book, summarizes the research results obtained in the book, and provides directions for future research in the field of fuzzy multi-criteria decision making methods based on pairwise comparisons.

6.1 Discussion

"Traditional" PC methods were not designed to cope with MCDM problems under uncertainty. However, uncertainty is integral to human mind, and thus it is necessarily closely related to decision making. In order to properly handle uncertainty, PC methods were extended to fuzzy numbers that allow for better modeling of uncertain PCs of objects. When extending PC methods to fuzzy PC methods it is of paramount importance to extend appropriately the key properties of PCMs and of the related PC methods in order to reflect properly the preference information contained in FPCMs.

Beside the inability to capture uncertainty, the "traditional" PC methods are also unable to deal with incompleteness of preference information. The problem of incomplete preference information concerns especially large-dimensional PCMs where it is not possible or reasonable to obtain complete preference information from the DM, e.g., due to time or cost limitations. When dealing with incomplete large-dimensional PCMs, compromise between maximally reducing the number of PCs required from the DM and obtaining reasonable priorities of objects from the incomplete PCM is of crucial importance.

Motivated by the above mentioned issues, the book was aiming at answering two research questions:

(1) *Based on a FPCM of objects, how should fuzzy priorities of these objects be determined so that they reflect properly all preference information available in the FPCM?*
(2) *How can the amount of preference information required from the DM in a large-dimensional PCM be reduced while still obtaining comparable priorities of objects?*

© Springer International Publishing AG, part of Springer Nature 2018   261
J. Krejčí, *Pairwise Comparison Matrices and their Fuzzy Extension*, Studies
in Fuzziness and Soft Computing 366, https://doi.org/10.1007/978-3-319-77715-3_6

The first research question was answered by pursuing four steps identified in Sect. 1.3. Each step and the related findings are summarized as follows:

(1.a) Well-known and in practice most often applied PC methods were critically reviewed in Chap. 2. In particular, three types of PCMs were examined—MPCMs, APCMs-A, and APCMs-M.

Two key properties of PCMs and of the related PC methods were identified—reciprocity of the related PCs and invariance of PC methods under permutation of objects. Reciprocity of the related PCs is an inherent property of every PCM that results from the meaning of PCs in the PCM (multiplicative reciprocity for MPCMs and additive reciprocity for APCMs-A and APCMs-M). Invariance under permutation of objects had been introduced as a property that every "good" method should satisfy. Therefore, it is necessary to extended properly both these properties also to FPCMs.

(1.b) In Chap. 4, critical review of the approaches to the fuzzy extension of the PC methods reviewed in Chap. 2 within step (1.a) was done, and two main drawbacks were identified.

The reviewed approaches are mostly based on applying standard fuzzy arithmetic to the fuzzy extension of the PC methods, and they violate the reciprocity of the related PCs or the invariance of PC methods under permutation of objects. This leads to false results (resulting fuzzy priorities of objects in particular) that distort the preference information contained in the FPCM.

(1.c) Necessity of applying constrained fuzzy arithmetic to the fuzzy extension of PC methods in order to reflect properly the preference information contained in FPCMs was demonstrated in Chap. 4.

Constrained fuzzy arithmetic allows for imposing constraints on operands of arithmetic operations with fuzzy numbers. Thus, reciprocity of the related PCs, which is an inherent property of PCMs, is introduced as a constraint in the computations with PCs in a FPCM. Applying constrained fuzzy arithmetic with reciprocity constraints to the fuzzy extension of the PC methods reviewed in Chap. 2 also automatically ensures invariance of the PC methods under permutation of objects.

(1.d) The fuzzy extension of the PC methods critically reviewed within step (1.a) was proposed in Chap. 4 in such a way that it reflects properly all preference information contained in the FPCM.

Specifically, a whole set of PC methods based on constrained fuzzy arithmetic was proposed in the book to deal with three types of FPCMs—FMPCMs, FAPCMs-A, and FAPCMs-M. FPCMs were defined properly, and two definitions of consistency were given for each type of FPCMs. Formulas for obtaining the fuzzy maximal eigenvalue of a FMPCM were proposed, and properties of the fuzzy maximal eigenvalues were identified. The fuzzy maximal eigenvalue is indispensable in order to define fuzzy extension of Consistency Index and Consistency Ratio for verifying acceptable multiplicative consistency of FMPCMs and to define a fuzzy extension of the EVM. Finally, methods for deriving fuzzy priorities of objects from FPCMs were proposed.

The methods proposed for each type of FPCMs are mutually equivalent. FMPCMs, FAPCMs-A, and FAPCMs-M can be transformed one into another together with the respective consistency properties. Similarly, fuzzy priorities obtained from FMPCMs, FAPCMs-A, and FAPCMs-M can be transformed one into another. The proposed PC methods were compared with the PC methods critically reviewed within step (1.c). Further, it was proved that all new PC methods based on constrained fuzzy arithmetic preserve the reciprocity of the related PCs and are invariant under permutation of objects. By preserving these two key properties, the fuzzy priorities obtained by the new PC methods reflect better the preference information contained in FPCMs in comparison to the fuzzy priorities obtained by the PC methods reviewed within step (1.c).

The PC methods based on constrained fuzzy arithmetic introduced within the answer to the research question (1) require the same amount of preference information from the DM as the reviewed PC methods based on standard fuzzy arithmetic. However, unlike them, they are invariant under permutation of objects and they preserve the reciprocity of the related PCs. This means that, based on the same amount of preference information from the DM, the new PC methods provide results that better reflect the preference information contained in the FPCMs, which leads to a better quality of decisions.

The second research question was answered by pursuing two steps identified in Sect. 1.3. The steps were

(2.a) proposing an efficient method for partially filling an incomplete large-dimensional PCM that minimizes the number of PCs required from the DM but provides a sufficient amount of preference information;

(2.b) proposing a suitable method for deriving priorities from an incomplete large-dimensional PCM that reflect the incompleteness of preference information and that are "close" to the priorities obtainable from the hypothetical complete PCM.

The steps resulted to be highly interconnected. In particular, development of the method in step (2.a) was substantially influenced by the requirement to obtain priorities that are "close" to the priorities obtainable from the hypothetical complete PCM. Thus, it is difficult to draw a clear line between the two steps, and consequently, it is not possible to represent the findings separately for each step.

Steps (2.a) and (2.b) were carried out by proposing an iterative algorithm for optimal choice of PCs that should be provided by the DM in an incomplete large-dimensional PCM. The algorithm is based on the concept of weak consistency. The weak-consistency condition is a minimum requirement of consistency that has to be satisfied in each step of the algorithm. Based on the weak-consistency condition, some missing PCs are entered into the PCM automatically, and for some, intervals of feasible values are provided. The whole process is based on searching for a compromise between minimizing the number of PCs provided by the DM and maximizing the amount of preference information contained in the incomplete PCM. At the end of the process, interval priorities of objects are computed using the formulas proposed in Chap. 4. The interval priorities include the priorities of objects obtainable from any weakly consistent completion of the incomplete PCM. This means that the interval priorities contain the priorities that would be obtained from the hypothetical complete PCM obtainable if the DM provided all PCs in the PCM. The average percentage of spared PCs in an incomplete PCM increases with the increasing dimension of the PCM; for PCMs of 15 or more objects, more than 60% of PCs are spared on average. Despite this great reduction of the number of PCs required from the DM, the resulting interval priories are very narrow for large-dimensional PCMs.

The novel PC method is particularly useful for real-life decision-making problems where providing all PCs is either costly, too time consuming, or infeasible to obtain. It is also very effective in dealing with large-dimensional problems where PCs provided by the DM require frequent revisions. By applying the novel method, the preference information required from the DM is significantly reduced, which leads to cost reduction and time saving. Despite this reduction, the method provides results (resulting interval priorities) that are very close to the hypothetical results obtainable from the complete preference information.

Naturally, the PC methods introduced in this book have some limitations. In Chap. 4, new fuzzy PC methods were introduced. The fuzzy PC methods were developed by applying constrained fuzzy arithmetic to the fuzzy extension of well-known and in practice most often applied PC methods (that were critically reviewed in Chap. 2). Unlike the fuzzy PC methods based on standard fuzzy arithmetic, the new fuzzy PC methods preserve both the reciprocity of the related PCs and the invariance under permutation of objects, which are two key properties identified for PCMs and for the related PC methods. Thus, it is justifiable to claim that the fuzzy priorities of objects obtained by the new fuzzy PC methods reflect the preference information contained in FPCMs better in comparison to the fuzzy priorities obtained by the fuzzy PC methods based on standard fuzzy arithmetic. However, the whole idea of "properly reflecting" the preference information contained in FPCMs by means of the fuzzy extension of the PC methods in this book is based on the assumption that the original PC methods are a suitable means of representing the preference information contained in PCMs.

The limitation of the PC method for dealing with large-dimensional PCMs described in Chap. 5 is that the method is based on the assumption that the preference system of the DM is in compliance with the weak-consistency condition. The weak-consistency condition is imposed as a minimal and very natural requirement of consistency in the PC method, and the DM is expected to provide weakly consistent preference information. Nevertheless, it is not guaranteed that every DM is able or willing to keep weak consistency during the process of providing PCs. If the DM refuses weak consistency as not reflecting properly his or her preference system, the PC method described in Chap. 5 cannot be used.

Another limitation might be that some of the formulas proposed in the book are based on highly non-linear optimization problems and should be, therefore, carefully managed by numerical computation. In this book, optimization methods predefined in MATLAB® were used.

6.2 Future Research

Despite the effort, the book could not cover all issues related to the fuzzy extension of PC methods. Therefore, there is still a lot of space for future research. In the following, some ideas are presented.

Calibration: In Sect. 2.2.1, Saaty's scale of linguistic terms with assigned integers for expressing intensities of preference in MPCMs was reviewed, and its fuzzy extension was studied in Sect. 4.2.1. However, as mentioned in Sect. 2.2.1, the linguistic terms do not correspond very well to the respective numerical values that are distributed uniformly in the interval [1, 9]. This problem naturally concerns also the fuzzy extension of Saaty's scale. Thus, as mentioned in 4.2.1, it would be appropriate to customize Saaty's scale for each DM with respect to the given decision problem. The first attempt of customizing the scale by using fuzzy numbers was done by Ishizaka and Nguyen (2013). However, as mentioned in Sect. 4.2.1, this process for customizing the scale is not designed well, which results in an inappropriate calibration. Therefore, this area still needs to be explored more thoroughly in order to design an appropriate calibration process.

Consistency: In Chap. 4, two consistency conditions were proposed for each type of FPCMs, one very weak and easy to reach and one very strong and difficult to reach. For real-life applications, a compromise between these two definitions of consistency might be useful. Therefore, searching for such a "compromise" definition of consistency that is again invariant under permutation and that preserves the reciprocity property of PCs is a subject for future research.

Weak consistency: As discussed in Chap. 2, weak-consistency conditions for MPCMs and APCMs provide an intuitive minimum consistency requirement. These definitions of consistency are less restrictive than traditional definitions of consistency reviewed in Chap. 4 and fuzzified in Chap. 4, and they provide DMs with some space for expressing their preferences. Further, weak consistency is much easier to reach and to control during the process of entering PCs into a PCM. This is especially convenient for real-life applications. Therefore, it would be useful to have such definitions of weak consistency also for FPCMs. The first step towards the fuzzy extension of the weak-consistency condition was done by Krejčí and Stoklasa (2016) who applied a fuzzy extension of the weak-consistency condition for MPCMs in the evaluation of scientific monographs.

Aggregation: Because of the excessive extent of the topic, the fuzzy extension of aggregation methods for obtaining final priorities of alternatives representing their final multi-criteria evaluations was not dealt with in the book. However, it is a very important part of the fuzzy PC methods. Similarly as for the definitions of consistency and the methods for deriving fuzzy priorities of objects from FPCMs, also the aggregation methods have to be extended properly to FPCMs by applying constrained fuzzy arithmetic in order to preserve the reciprocity property of the related PCs. Such fuzzy extension becomes considerably more complex in comparison to the fuzzy extension of the consistency conditions and of the methods for deriving priorities of objects since fuzzy priorities of criteria and alternatives obtained from several FPCMs are involved in the aggregation process. Nevertheless, for the completeness of the fuzzy PC methods proposed in this book, it is necessary to deal also with this issue. The fuzzy extension of the weighted average for aggregating fuzzy priorities obtained from FMPCMs was already introduced by Krejčí et al. (2017). The fuzzy extension of the aggregation methods for FAPCMs-A and FAPCMs-M is still left for future research.

Multiple DMs: Another issue that was not addressed in this book is considering multiple DMs. A large number of PC methods has been proposed in the literature to deal with MCDM problems involving multiple DMs, and some of the methods have been extended also to FPCMs. With the fuzzy extension of these PC methods the very same challenges that have been approached in this book arise. In particular, it is again necessary to preserve the reciprocity of the related PCs in FPCMs as well as the invariance under permutation in order to reflect appropriately the preference information provided by multiple DMs.

Incomplete large-dimensional FPCMs: The last but not least topic for future research is the fuzzy extension of the PC method for dealing with incomplete large-dimensional PCMs described in Chap. 5. The PC method described in Chap. 5 is designed for large-dimensional problems where the DM provides PCs by means of crisp numbers. However, as discussed in the book, crisp numbers cannot model properly uncertainty stemming from subjectivity of human thinking and from vagueness of information about the problem that are very often related to MCDM problems. A fuzzy extension of the PC method described in Chap. 5 is needed in order to handle properly large-dimensional problems with uncertainty as well as with incompleteness of preference information provided by the DM. Thus, the last but not least topic for future research is the fuzzy extension of the PC method for dealing with incomplete large-dimensional PCMs described in Chap. 5. The fuzzy extension of the method requires a fuzzy extension of the weak-consistency condition, on which the method is based. Besides that, a large number of rules derived from the weak-consistency condition has to be fuzzified accordingly and employed in the iterative process of identifying PCs that should be provided by the DM. At the end of the process of entering PCs into the incomplete large-dimensional FPCM, we would obtain a FPCM instead of an interval PCM, which is the output of the current PC method. Afterwards, in order to derive fuzzy priorities from such a FPCM, it would be again sufficient to apply one of the methods proposed in Chap. 4.

References

Alonso, J.A., and M.T. Lamata. 2006. Consistency in the analytic hierarchy process: a new approach. *International Journal of Uncertainty, Fuzziness and Knowledge-Based Systems* 14 (4): 445–459.

Alonso, S., F. Chiclana, F. Herrera, E. Herrera-Viedma, E. Alcalá-Fdez, and C. Porcel. 2008. A consistency-based procedure to estimate missing pairwise preference values. *International Journal of Intelligent Systems* 23: 155–175.

Baas, S.M., and H. Kwakernaak. 1977. Rating and ranking of multiple aspect alternative using fuzzy sets. *Automatica* 13 (1): 47–58.

De Baets, B., H. De Meyer, and B. De Schuymer. 2006. Cyclic evaluation of transitivity of reciprocal relations. *Social Choice and Welfare* 26 (2): 217–238.

Barzilai, J. 1997. Deriving weights from pairwise comparison matrices. *The Journal of the Operational Research Society* 48 (12): 1226–1232.

Basile, L., and L. D'Apuzzo. 2002. Weak consistency and quasi-linear means imply the actual ranking. *International Journal of Uncertainty, Fuzziness and Knowledge-Based Systems* 10 (3): 227–239.

Bellman, R., and L.A. Zadeh. 1970. Decision making in a fuzzy environment. *Management Science* 17B (4): 141–164.

Belton, V., T. Stewart. 2002. *Multiple criteria decision analysis*. Springer. ISBN 978-1-4613-5582-3.

Bezdek, J.C., B. Spillman, and R. Spillman. 1978. A fuzzy relation space for group decision theory. *Fuzzy Sets and Systems* 1 (4): 255–268.

Blaquero, R., E. Carrizosa, and E. Conde. 2006. Inferring efficient weights from pairwise comparison matrices. *Mathematical Methods of Operations Research* 64 (2): 271–284.

Bond, S.D., K.A. Carlson, and R.L. Keeney. 2008. Generating objectives: Can decision makers articulate what they want? *Management Science* 54 (1): 56–70.

Brunelli, M., and M. Fedrizzi. 2015. Axiomatic properties of inconsistency indices for pairwise comparisons. *Journal of Operational Research Society* 66 (1): 1–15.

Buckley, J.J. 1985a. Fuzzy hierarchical analysis. *Fuzzy Sets and Systems* 17 (3): 233–247.

Buckley, J.J. 1985b. Ranking alternatives using fuzzy numbers. *Fuzzy Sets and Systems* 15 (1): 21–31.

Cabrerizo, J., M.R. Ureña, W. Pedrycz, and E. Herrera-Viedma. 2014. Building consensus in group decision making with an allocation of information granularity. *Fuzzy Sets and Systems* 255 (16): 115–127.

Cavallo, B., and L. D'Apuzzo. 2016. Ensuring reliability of the weighting vector: Weak consistent pairwise comparison matrices. *Fuzzy Sets and Systems* 296 (1): 21–34.

Chang, D.Y. 1996. Applications of the extent analysis method on fuzzy AHP. *European Journal of Operational Research* 95 (3): 649–655.

Chang, P.T., and E.S. Lee. 1995. The estimation of normalized fuzzy weights. *Computers and Mathematics with Applications* 29 (5): 21–42.

© Springer International Publishing AG, part of Springer Nature 2018
J. Krejčí, *Pairwise Comparison Matrices and their Fuzzy Extension*, Studies
in Fuzziness and Soft Computing 366, https://doi.org/10.1007/978-3-319-77715-3

Chen, Q., and E. Triantaphillou. 2001. Estimating data for multi-criteria decision making problems: optimization techniques. In *Encyclopedia of optimization*, ed. P.M. Pardalos, and C. Floudas. Boston: Kluwer Academic Publishers.

Chen, S.J., and C.L. Hwang. 1992. *Fuzzy multiple attribute decision making—methods and applications*. Berlin: Springer.

Chiclana, F., F. Herrera, and E. Herrera-Viedma. 1998. Integrating three representation models in fuzzy multipurpose decision making based on fuzzy preference relations. *Fuzzy Sets and Systems* 97 (1): 33–48.

Chiclana, F., E. Herrera-Viedma, S. Alonso, and F. Herrera. 2009. Cardinal consistency of reciprocal preference relations: A characterization of multiplicative transitivity. *IEEE Transactions on Fuzzy Systems* 17 (1): 14–23.

Cook, D., and M. Kress. 1988. Deriving weights from pairwise comparison matrices: An axiomatic approach. *European Journal of Operational Research* 37 (1): 35–362.

Crawford, G., and C. Williams. 1985. A note on the analysis of subjective judgment matrices. *Journal of Mathematical Psychology* 29 (2): 387–405.

Csutora, R., and J.J. Buckley. 2001. Fuzzy hierarchical analysis: The Lambda-Max method. *Fuzzy Sets and Systems* 120 (2): 181–195.

Dijkstra, T.K. 2013. On the extraction of weights from pairwise comparison matrices. *Central European Journal of Operational Research* 21 (1): 103–123.

Dubois, D., and H. Prade. 1986. Recent models of uncertainty and imprecision as a basis for decision theory: Towards less normative frameworks. *Intelligent decision support in process environments*. Berlin: Springer.

Efstathiou, J., and C. Rajkovic. 1979. Multiattribute decisionmaking using a fuzzy heuristic approach. *IEEE Transactions on Systems, Man, Cybernetics* 9 (6): 326–333.

Enea, M., and T. Piazza. 2004. Project selection by constrained fuzzy AHP. *Fuzzy Optimization and Decision Making* 3 (1): 39–62.

Fedrizzi, M. 1990. On a consensus measure in a group MCDM problem. In *Multiperson decision making models using fuzzy sets and possibility theory*, ed. J. Kacprzyk, and M. Fedrizzi, 231–241. Dordrecht: Kluwer Academic Publishers.

Fedrizzi, M., and M. Brunelli. 2009. On the normalisation of a priority vector associated with a reciprocal relation. *International Journal of General Systems* 38 (5): 579–586.

Fedrizzi, M., and M. Brunelli. 2010. On the priority vector associated with a reciprocal relation and a pairwise comparison matrix. *Soft Computing* 14 (6): 639–645.

Fedrizzi, M., and S. Giove. 2007. Incomplete PC and consistency optimization. *European Journal of Operational Research* 183 (1): 303–313.

Fedrizzi, M., and S. Giove. 2013. Optimal sequencing in incomplete pairwise comparisons for large-dimensional problems. *International Journal of General Systems* 42 (4): 366–375.

Fedrizzi, M., and J. Krejčí. 2015. A note on the paper 'Fuzzy analytic hierarchy process: Fallacy of the popular methods'. *International Journal of Uncertainty, Fuzziness and Knowledge-Based Systems* 23 (6): 965–970.

Fichtner, J. 1986. On deriving priority vectors from matrices of pairwise comparisons. *Socio-Economic Planning Sciences* 20 (6): 341–345.

Figueira, J., S. Greco, and M. Ehrogott. 2005. *Multiple criteria decision analysis: State of the art surveys*. New York: Springer.

Gavalec, M., J. Ramík., and K. Zimmermann. 2015. *Decision making and optimization*. Springer. ISBN 978-3-319-08322-3.

Genç, S., F.E. Boran, D. Akay, and Z. Xu. 2010. Interval multiplicative transitivity for consistency, missing values and priority weights of interval fuzzy preference relations. *Information Sciences* 180 (24): 4877–4891.

Harker, P.T. 1987a. Alternative modes of questioning in the Analytic Hierarchy Process. *Mathematical Modelling* 9 (3–5): 353–360.

Harker, P.T. 1987b. Incomplete PCs in the analytic hierarcy process. *Mathematical Modelling* 9 (11): 837–848.

Harker, P.T. 1987c. Derivatives of the Perron root of a positive reciprocal matrix with application to the analytic hierarchy process. *Applied Mathematics and Computation* 22 (1): 217–232.

Harker, P.T., and I. Millet. 1990. Globally effective questioning in the analytic hierarchy process. *European Journal of Operational Research* 48 (1): 88–97.

Herrera-Viedma, E., F. Herrera, F. Chiclana, and M. Luque. 2004. Some issues on consistency of fuzzy preference relations. *European Journal of Operational Research* 154 (1): 98–109.

Hu, M., P. Ren, J. Lan, J. Wang, and W. Zheng. 2014. Note on"Some models for deriving the priority weights from interval fuzzy preference relations". *European Journal of Operational Research* 237 (2): 771–773.

Ishizaka, A. 2014. Comparison of Fuzzy logic, AHP, FAHP and Hybrid Fuzzy AHP for new supplier selection and its performance analysis. *International Journal of Integrated Supply Management* 9 (1–2): 1–22.

Ishizaka, A., and N.H. Nguyen. 2013. Calibrated fuzzy AHP for current bank account selection. *Expert Systems with Applications* 40 (9): 3775–3783.

Jandová, V., and J. Talašová. 2013. Weak consistency: A new approach to consistency in the Saaty's Analytic Hierarchy Process. *Acta Universitatis Palacianae Olomucensis, Facultas Rerum Naturalium, Mathematica* 52 (2): 71–83.

Jandová, V., J. Krejčí., J. Stoklasa., and M. Fedrizzi. 2017. Computing interval weights from incomplete pairwise comparison matrices of large dimension - a weak consistency based approach. *IEEE Transactions on Fuzzy Systems*, 25(6):1714–1728.

Javanbarg, M.B., C. Scawthorn, J. Kiyono, and B. Shahbodaghkhan. 2012. Fuzzy AHP-based multicriteria decision making systems using particle swarm optimization. *Expert Systems with Applications* 39 (3): 6960–9665.

Jiménez, A., S. Ríos-Insua, and A. Mateos. 2003. A decision support systems for multiattribute utility evaluation based on imprecise assignments. *Decision Support Systems* 36 (1): 65–79.

Kacprzyk, J. 1986. Group decision making with a fuzzy linguistic majority. *Fuzzy Sets and Systems* 18 (2): 105–118.

Klir, G.J. 1997. Fuzzy arithmetic with requisite constraints. *Fuzzy Sets and Systems* 91 (2): 165–175.

Klir, G.J., and Y. Pan. 1998. Constrained fuzzy arithmetic: Basic questions and some answers. *Soft Computing* 2 (2): 100–108.

Klir, J., and B. Yuan. 1995. *Fuzzy sets and fuzzy logic*. Prentice Hall PTR. ISBN 0-13-101171-5.

Krejčí, J. 2016. Obtaining fuzzy priorities from additive fuzzy pairwise comparison matrices. *IMA Journal of Management Mathematics*. https://doi.org/10.1093/imaman/dpw006.

Krejčí, J. 2017a. On additive consistency of interval fuzzy preference relations. *Computers and Industrial Engineering* 107 (1): 128–140.

Krejčí, J. 2017b. On multiplicative consistency of interval and fuzzy reciprocal preference relations. *Computers and Industrial Engineering* 111 (1): 67–78.

Krejčí, J. 2017c. Fuzzy eigenvector method for obtaining normalized fuzzy weights from fuzzy pairwise comparison matrices. *Fuzzy Sets and Systems* 315 (1): 26–43.

Krejčí, J. 2017d. On extension of multiplicative consistency to interval fuzzy preference relations. *Operational Research*. https://doi.org/10.1007/s12351-017-0307-8.

Krejčí, J. 2017e. Additively reciprocal fuzzy pairwise comparison matrices and multiplicative fuzzy priorities. *Soft Computing* 21 (12): 3177–3192.

Krejčí, J. 2018. Fuzzy eigenvector method for deriving normalized fuzzy priorities from fuzzy multiplicative pairwise comparison matrices. *Fuzzy Optimization and Decision Making*, resubmitted after minor revisions.

Krejčí, J., and J. Stoklasa. 2016. Fuzzified AHP in the evaluation of scientific monographs. *Central European Journal of Operational Research* 24 (2): 353–370.

Krejčí, J., and J. Talašová. 2013. A proper fuzzification of Saaty's scale and an improved method for computing fuzzy weights in fuzzy AHP. In *Proceedings of the 31th international conference on mathematical methods in economics 2013*, 452–457. Jihlava. ISBN 978-80-87035-76-4.

Krejčí, J., O. Pavlačka, and J. Talašová. 2017. A fuzzy extension of Analytic Hierarchy Process based on the constrained fuzzy arithmetic. *Fuzzy Optimization and Decision Making* 16 (1): 89–110.

Kubler, S., J. Robert, W. Derigent, A. Voisin, and Y.L. Traon. 2016. A state-of the-art survey and testbed of fuzzy AHP (FAHP) applications. *Expert Systems with Applications* 65 (1): 398–422.

Kwakernaak, H. 1979. An algorithm for rating multiple-aspect alternatives using fuzzy sets. *Automatica* 15 (5): 615–616.

Kwiesielewicz, M. 1996. The logarithmic least squares and the generalised pseudoinverse in estimating ratios. *European Journal of Operational Research* 93 (1): 611–619.

Kwiesielewicz, M., and E. van Uden. 2003. Ranking decision variants by subjective paired comparisons in cases with incomplete data. In *Computational science and its applications ICCSA 2003*, ed. V. Kumar, et al., 208–215. Lecture Notes in Computer Science Berlin: Springer.

Van Laarhoven, P.J.M., and W. Pedrycz. 1983. A fuzzy extension of Saaty's priority theory. *Fuzzy Sets and Systems* 11 (1–3): 199–227.

Li, K.W., Z.J. Wag, and X. Tong. 2016. Acceptability analysis and priority weight elicitation for interval multiplicative comparison matrices. *European Journal of Operational Research* 250 (2): 628–638.

Liu, F. 2009. Acceptable consistency analysis of interval reciprocal comparison matrices. *Applied Mathematics and Computation* 160 (18): 2686–2700.

Liu, F., W.G. Zhang, and J.H. Fu. 2012a. A new method of obtaining the priority weights from an interval fuzzy preference relation. *Information Sciences* 185 (1): 32–42.

Liu, F., W.-G. Zhang, and L.-H. Zhang. 2014. Consistency analysis of triangular fuzzy reciprocal preference relations. *European Journal of Operational Research* 235 (3): 718–726.

Liu, X.W., Y.W. Pan, Y.J. Xu, and S. Yu. 2012b. Least square completion and inconsistency repair methods for additively consistent fuzzy preference relations. *Fuzzy Sets and Systems* 198 (1): 1–19.

Lodwick, W.A., and O.A. Jenkins. 2013. Constrained intervals and interval spaces. *Soft Computing* 17 (8): 1393–1402.

Luce, R.D., and P. Suppes. 1965. Preferences, utility and subject probability. In *Handbook of mathematical psychology*, vol. III, ed. R.D. Luce, et al., 249–410. New York: Wiley.

Meng, F., X. Chen, and Y. Zhang. 2016. Consistency-based linear programming models for generating the priority vector from interval fuzzy preference relations. *Applied Soft Computing* 41 (1): 247–264.

Mikhailov, L. 2003. Deriving priorities from fuzzy pairwise comparison judgements. *Fuzzy Sets and Systems* 134 (3): 365–385.

Miller, G.A. 1956. The magical number seven plus or minus two: some limits on our capacity for processing information. *The Psychological Review* 63 (2): 81–97.

Nurmi, N. 1981. Approaches to collective decision making with fuzzy preference relations. *Fuzzy Sets and Systems* 6 (3): 249–259.

Ministry of Education of the Czech Republic. 2011. *Zásady a pravidla financování veřejných vysokých škol pro rok 2012, [Principles and rules of financing public universities for 2012]*. Praha: Czech Republic.

Orlovski, S.A. 1978. Decision-making with a fuzzy preference relation. *Fuzzy Sets and Systems* 1 (3): 155–167.

Pan, N.F. 2008. Fuzzy AHP approach for selecting the suitable bridge construction method. *Automation in Construction* 17 (8): 958–965.

Pavlačka, O. 2014. On various approaches to normalization of interval and fuzzy weights. *Fuzzy Sets and Systems* 243 (1): 110–130.

Qian, W.Y., K.W. Li, and Z.J. Wang. 2014. Approaches to improving consistency of interval fuzzy preference relations. *Journal of Systems Science and Systems Engineering* 23 (4): 460–479.

Ra, J.W. 1999. Chainwise paired comparisons. *Decision Sciences* 30 (2): 581–599.

Ramík, J. 2016. Incomplete preference matrix on alo-group and its application to ranking of alternatives. *International Journal of Mathematics in Operational Research* 9 (4): 412–422.

Ramík, J., and P. Korviny. 2010. Inconsistency of pair-wise comparison matrix with fuzzy elements based on geometric mean. *Fuzzy Sets and Systems* 161 (11): 1604–1613.

Ruspini, E. 1969. A new approach to clustering. *Information Control* 15 (1): 22–32.

Saaty, T.L. 1977. A scaling method for priorities in hierarchical structures. *Journal of Mathematical Psychology* 15 (3): 234–281.

Saaty, T.L. 1980. *The analytic hierarchy process*. New York: McGraw Hill.

Saaty, T.L. 1994. *Fundamentals of decision making and priority theory with the AHP*. Pittsburgh: RWS Publications.

Saaty, T.L. 2006. There is no mathematical validity for using fuzzy number crunching in the analytic hierarchy process. *Journal of Systems Science and Systems Engineering* 15 (4): 457–464.

Saaty, T.L. 2008. Decision making with the analytic hierarchy process. *Journal of Systems Science and Systems Engineering* 1 (1): 83–98.

Saaty, T.L., and G. Hu. 1998. Ranking by eigenvector versus other methods in the analytic hierarchy process. *Applied Mathematics Letters* 11 (4): 121–125.

Saaty, T.L., and L.T. Tran. 2007. On the invalidity of fuzzifying numerical judgments in the analytic hierarchy process. *Mathematical and Computer Modelling* 46 (7–8): 962–975.

Saaty, T.L., and L.G. Vargas. 1984. Comparison of eigenvalue, logarithmic least squares and least squares methods in estimating ratios. *Mathematical Modelling* 5 (5): 309–324.

Sanchez, P.P., and R. Soyer. 1998. Information concepts and pairwise comparison matrices. *Information Processing Letters* 68 (1): 185–188.

Sevastjanov, P., L. Dymova, and P. Bartosiewicz. 2012. A new approach to normalization of interval and fuzzy weights. *Fuzzy Sets and Systems* 198 (1): 34–45.

Shiraishi, S., T. Obata, and M. Daigo. 1998. Properties of a positive reciprocal matrix and their application to AHP. *Journal of the Operations Research Society of Japan* 41 (3): 412–422.

Siebert, J., and R.L. Keeney. 2015. Creating more and better alternatives for decisions using objectives. *Operations Research* 63 (5): 1144–1158.

Stoklasa, J. 2014. *Linguistic models for decision support*. Acta Universitatis Lappeenrantaesis 604. ISBN 978-952-265-687-2. Dissertation.

Stoklasa, J., V. Jandová, and J. Talašová. 2013. Weak consistency in Saaty's AHP - evaluating creative work outcomes of Czech Art Colleges. *Neural Network* 23 (1): 61–77.

Stoklasa, J., T. Talášek, and J. Talašová. 2016. AHP and weak consistency in the evaluation of works of art - a case study of a large problem. *International Journal of Business Innovation and Research* 11 (1): 75–60.

Takagi, T., and M. Sugeno. 1985. Fuzzy identification of systems and its application to modeling and control. *IEEE Transitions on Systems, Man, and Cybernetics* 15 (1): 116–132.

Tanino, T. 1984. Fuzzy preference orderings in group decision making. *Fuzzy Sets and Systems* 12 (2): 117–131.

Tesfamariam, S., and R. Sadiq. 2006. Risk-based environmental decision-making using fuzzy analytic hierarchy process (F-AHP). *Stochastic Environmental Research and Risk Assessment* 21 (1): 35–50.

Thurstone, L.L. 1927. A low of comparative judgment. *Psychological Review* 34 (4): 273–286.

Tsaur, S.-H., T.-Y. Chang, and C.-H. Yen. 2002. The evaluation of airline service quality by fuzzy MCDM. *Tourism Management* 23 (1): 107–115.

Vahidnia, M.H., A.A. Alesheikh, and A. Alimohammadi. 2009. Hospital site selection using fuzzy AHP and its derivatives. *Journal of Environmental Management* 90 (10): 3048–3056.

Wang, J., J. Lan, P. Ren, and Y. Luo. 2012. Some programming models to derive priority weights from additive interval fuzzy preference relations. *Knowledge-Based Systems* 27 (1): 67–77.

Wang, Y.M., and K.S. Chin. 2006. An eigenvector method for generating normalized interval and fuzzy weights. *Applied Mathematics and Computation* 181 (2): 1257–1275.

Wang, Y.M., and T.M.S. Elhag. 2006. On the normalization of interval and fuzzy weights. *Fuzzy Sets and Systems* 157 (18): 2456–2471.

Wang, Y.M., J.B. Yang, and D.L. Xu. 2005a. A two-stage logarithmic goal programming method for generating weights from interval comparison matrices. *Fuzzy Sets and Systems* 152 (3): 475–498.

Wang, Y.M., J.B. Yang, and D.L. Xu. 2005b. Interval weight generation approaches based on consistency test and interval comparison matrices. *Applied Mathematics and Computation* 167 (1): 252–273.

Wang, Y.M., Y. Luo, and Z. Hua. 2008. On the extent analysis method for fuzzy AHP and its applications. *European Journal of Operational Research* 186 (2): 735–747.

Wang, Z.J. 2014. A note on "Incomplete interval fuzzy preference relations and their applications". *Computers and Industrial Engineering* 77 (1): 65–69.

Wang, Z.J. 2015a. A note on "A goal programming model for incomplete interval multiplicative preference relations and its application in group decision-making". *European Journal of Operational Research* 247 (3): 867–871.

Wang, Z.J. 2015b. Consistency analysis and priority derivation of triangular fuzzy preference relations based on modal value and geometric mean. *Information Sciences* 314 (1): 169–183.

Wang, Z.J., and K.W. Li. 2012. Goal programming approaches to deriving interval weights based on interval fuzzy preference relations. *Information Science* 193 (1): 180–198.

Wedley, W.C., B. Schoner, and T.S. Tang. 1993. Starting rules for incomplete comparisons in the Analytic Hierarchy Process. *Mathematical and Computer Modelling* 17 (4–5): 93–100.

Wu, J., and F. Chiclana. 2014a. A social network analysis trust-consensus based approach to group decision-making problems with interval-valued fuzzy reciprocal preference relations. *Knowledge-Based Systems* 59 (1): 97–107.

Wu, J., and F. Chiclana. 2014b. Multiplicative consistency of intuitionistic reciprocal preference relations and its application to missing values estimation and consensus building. *Knowledge-Based Systems* 71 (1): 187–200.

Xia, M., and Z. Xu. 2011. Some issues on multiplicative consistency of interval reciprocal relations. *International Journal of Information Technology and Decision Making* 10 (6): 1043–1065.

Xu, Y., K.W. Li, and H. Wang. 2014a. Incomplete interval fuzzy preference relations and their applications. *Computers and Industrial Engineering* 67 (1): 93–103.

Xu, Y.J., Q.L. Da, and L.H. Liu. 2009. Normalizing rank aggregation method for priority of fuzzy preference relation and its effectiveness. *International Journal of Approximate Reasoning* 50 (8): 1287–1297.

Xu, Y.J., Q.L. Da, and H.M. Wang. 2011. A note on group decision-making procedure based on incomplete reciprocal relations. *Soft Computing* 15 (7): 1289–1300.

Xu, Y.J., K.W. Li, and H. Wang. 2014b. Consistency test and weight generation for additive interval fuzzy preference relations. *Soft Computing* 18 (8): 1499–1513.

Xu, Z. 2004. Goal programming models for obtaining the priority vector of incomplete fuzzy preference relation. *International Journal of Approximate Reasoning* 36 (3): 261–270.

Xu, Z. 2007a. Multiple-attribute group decision making with different formats of preference information on attributes. *IEEE Transactions on Systems, Man and Cybernetics—Part B* 37 (6): 1500–1511.

Xu, Z. 2007b. A survey of preference relations. *International Journal of General Systems* 36 (2): 179–203.

Xu, Z., and J. Chen. 2008a. Some models for deriving the priority weights from interval fuzzy preference relations. *European Journal of Operational Research* 184 (1): 266–280.

Xu, Z., and J. Chen. 2008b. Group decision-making procedure based on incomplete reciprocal relations. *Soft Computing* 12 (6): 515–521.

Xu, Z.S. 2005. A procedure for decision making based on incomplete fuzzy preference relation. *Fuzzy Optimization and Decision Making* 4 (3): 175–189.

Zadeh, L.A. 1965. Fuzzy sets. *Information and Control* 8 (3): 338–353.

Zadeh, L.A. 1975a. Concept of a linguistic variable and its application to approximate reasoning I. *Information Science* 8 (3): 199–249.

Zadeh, L.A. 1975b. Concept of a linguistic variable and its application to approximate reasoning II. *Information Science* 8 (4): 301–357.

Zadeh, L.A. 1975c. Concept of a linguistic variable and its application to approximate reasoning III. *Information Science* 9 (1): 43–80.

Zheng, G., N. Zhu, Z. Tian, Y. Chen, and B. Sun. 2012. Application of a trapezoidal fuzzy AHP method for work safety evaluation and early warning rating of hot and humid environments. *Safety Science* 50 (18): 228–239.

Zhü, K. 2014. Fuzzy analytic hierarchy process: Fallacy of the popular methods. *European Journal of Operational Research* 236 (1): 209–217.

Zimmermann, H.J. 1987. *Fuzzy set, decision making, and expert system.* The Netherlands: Kluwer.

Printed in the United States
By Bookmasters